T0291741

The Allure of Machinic Life

THE ALLURE OF MACHINIC LIFE

Cybernetics, Artificial Life, and the New AI

John Johnston

A Bradford Book

The MIT Press
Cambridge, Massachusetts
London, England

First MIT Press paperback edition, 2010

This book was set in Times New Roman and Syntax on 3B2 by Asco Typesetters, Hong Kong.

Library of Congress Cataloging-in-Publication Data

Johnston, John Harvey, 1947–
The allure of machinic life : cybernetics, artificial life, and the new AI / John Johnston
p. cm.
Includes bibliographical references and index.
ISBN 978-0-262-10126-4 (hardcover : alk. paper), 978-0-262-51502-3 (pb : alk. paper)
1. Cybernetics. 2. Artificial life. 3. Artificial intelligence. I. Title.
Q310.J65 2008
003′.5—dc22 2008005358

For Heidi

Contents

Preface

This book explores a single topic: the creation of new forms of "machinic life" in cybernetics, artificial life (ALife), and artificial intelligence (AI). By *machinic life* I mean the forms of nascent life that have been made to emerge in and through technical interactions in human-constructed environments. Thus the webs of connection that sustain machinic life are material (or virtual) but not directly *of* the natural world. Although automata such as the eighteenth-century clockwork dolls and other figures can be seen as precursors, the first forms of machinic life appeared in the "lifelike" machines of the cyberneticists and in the early programs and robots of AI. Machinic life, unlike earlier mechanical forms, has a capacity to alter itself and to respond dynamically to changing situations.

More sophisticated forms of machinic life appear in the late 1980s and 1990s, with computer simulations of evolving digital organisms and the construction of mobile, autonomous robots. The emergence of ALife as a scientific discipline—which officially dates from the conference on "the synthesis and simulation of living systems" in 1987 organized by Christopher Langton—and the growing body of theoretical writings and new research initiatives devoted to autonomous agents, computer immune systems, artificial protocells, evolutionary robotics, and swarm systems have given the development of machinic life further momentum, solidity, and variety. These developments make it increasingly clear that while machinic life may have begun in the mimicking of the forms and processes of natural organic life, it has achieved a complexity and autonomy worthy of study in its own right. Indeed, this is my chief argument.

While excellent books and articles devoted to these topics abound, there has been no attempt to consider them within a single, overarching theoretical framework. The challenge is to do so while respecting the very significant historical, conceptual, scientific, and technical differences in this material and the diverse perspectives they give rise to. To meet this challenge I have tried to establish an inclusive vantage point that can be shared by specialized and general readers alike. At first view, there are obvious relations of precedence and influence in the distinctive histories of cybernetics, AI, and ALife. Without the groundbreaking discoveries and theoretical orientation of cybernetics, the sciences of AI and ALife would simply not have arisen and developed as they have. In both, moreover, the digital computer was an essential condition of possibility. Yet the development of the stored-program electronic computer was also contemporary with the birth of cybernetics and played multiple roles of instigation, example, and relay for many of its most important conceptualizations. Thus the centrality of the computer results in a complicated nexus of historical and conceptual relationships among these three fields of research.

But while the computer has been essential to the development of all three fields, its role in each has been different. For the cyberneticists the computer was first and foremost a physical device used primarily for calculation and control; yet because it could exist in a nearly infinite number of states, it also exhibited a new kind of complexity. Early AI would demarcate itself from cybernetics precisely in its highly abstract understanding of the computer as a symbol processor, whereas ALife would in turn distinguish itself from AI in the ways in which it would understand the role and function of computation. In contrast to the top-down computational hierarchy posited by AI in its effort to produce an intelligent machine or program, ALife started with a highly distributed population of computational machines, from which complex, lifelike behaviors could emerge.

These different understandings and uses of the computer demand a precise conceptualization. Accordingly, my concept of *computational assemblage* provides a means of pinpointing underlying differences of form and function. In this framework, every computational machine is conceived of as a material assemblage (a physical device) conjoined with a unique discourse that explains and justifies the machine's operation and purpose. More simply, a computational assemblage is comprised of both a machine and its associated discourse, which together determine how and why this machine does what it does. The concept of computational assemblage thus functions as a differentiator within a large set of family resemblances, in contrast to the general term *computer*, which is too vague for my purposes. As with my concept of machinic life, these family resemblances must be spelled out in detail. If computational assemblages comprise a larger unity, or indeed if forms of machinic life can be said to possess a larger unity, then in both cases they are unities-in-difference, which do not derive from any preestablished essence or ideal form. To the contrary, in actualizing new forms of computation and life, the machines and programs I describe constitute novel ramifications of an idea, not further doublings or repetitions of a prior essence.

This book is organized into three parts, which sketch conceptual histories of the three sciences. Since I am primarily concerned with how these sciences are both unified and differentiated in their productions of machinic life, my presentation is not strictly chronological. As I demonstrate, machinic life is fully comprehensible only in relation to new and developing notions of complexity, information processing, and dynamical systems theory, as well as theories of emergence and evolution; it thus necessarily crosses historical and disciplinary borderlines. The introduction traces my larger theoretical trajectory, focusing on key terms and the wider cultural context. Readers of N. Katherine Hayles, Manual DeLanda, Ansel Pearson, Paul Edwards, and Richard Doyle as well as books about Deleuzian philosophy, the posthuman, cyborgs, and cyberculture more generally will find that this trajectory passes over familiar ground. However, my perspective and purpose are distinctly different. For me, what remains uppermost is staying close to the objects at hand—the machines, programs, and processes that constitute machinic life. Before speculating about the cultural implications of these new kinds of life and intelligence, we need to know precisely how they come about and operate as well as how they are already changing.

In part I, I consider the cybernetic movement from three perspectives. Chapter 1 makes a case for the fundamental complexity of cybernetic machines as a new species of automata, existing both "in the metal and in the flesh," to use Norbert Wiener's expression, as built and theorized by Claude Shannon, Ross Ashby, John von Neumann, Grey Walter, Heinz von Foerster, and Valentino Braitenberg. Chapter 2 examines the "cybernetic subject" through the lens of French psychoanalyst Jacques Lacan and his participation (along with others, such as Noam Chomsky) in a new discourse network inaugurated by the confluence of cybernetics, information theory, and automata theory. The chapter concludes with a double view of the chess match between Gary Kasparov and Deep Blue, which suggests both the power and limits of classic AI. Chapter 3 extends the cybernetic perspective to what I call machinic philosophy, evident in Deleuze and Guattari's concept of the assemblage and its intersections with nonlinear dynamical systems (i.e., "chaos") theory. Here I develop more fully the concept of the computational assemblage, specifically in relation to Robert Shaw's "dripping faucet as a model chaotic system" and Jim Crutchfield's ϵ-machine (re)construction.

Part II focuses on the new science of ALife, beginning with John von Neumann's theory of self-reproducing automata and Christopher Langton's self-reproducing digital loops. Langton's theory of ALife as a new science based on computer simulations whose theoretical underpinnings combine information theory with dynamical systems theory is contrasted with Francisco Varela and Humberto Maturana's theory of autopoiesis, which leads to a consideration of both natural and artificial immune systems and computer viruses. Chapter 5 charts the history of ALife after Langton in relation to theories of evolution, emergence, and complex adaptive systems by examining a series of experiments carried out on various software platforms, including Thomas Ray's Tierra, John Holland's Echo, Christoph Adami's Avida, Andrew Pargellis's Amoeba, Tim Taylor's Cosmos, and Larry Yaeger's PolyWorld. The chapter concludes by considering the limits of the first phase of ALife research and the new research initiatives represented by "living computation" and attempts to create an artificial protocell.

Part III takes up the history of AI as a series of unfolding conceptual conflicts rather than a chronological narrative of achievements and failures. I first sketch out AI's familiar three-stage development, from symbolic AI as exemplified in Newel and Simon's physical symbol system hypothesis to the rebirth of the neural net approach in connectionism and parallel distributed processing and to the rejection of both by a "new AI" strongly influenced by ALife but concentrating on building autonomous mobile robots in the noisy physical world. At each of AI's historical stages, I suggest, there is a circling back to reclaim ground or a perspective rejected earlier—the biologically oriented neural net approach at stage two, cybernetics and embodiment at stage three. The decodings and recodings of the first two stages lead inevitably to philosophical clashes over AI's image of thought—symbol manipulation versus a stochastically emergent mentality—and the possibility of robotic consciousness. On the other hand, the behavior-based, subsumption-style approach to robotics that characterizes the new AI eventually has to renege on its earlier rejection of simulation when it commits to artificial evolution as a necessary method of development. Finally, in the concluding chapter, I indicate why further success in the building of intelligent machines will most likely be tied to progress in our understanding of how the human brain actually works, and describe recent examples of robotic self-modeling and communication.

In writing this book I have been stimulated, encouraged, challenged, and aided by many friends, colleagues, and scientists generous enough to share their time with me. Among the latter I would especially like to thank Melanie Mitchell, whose encouragement and help at the project's early stages were essential, Luis Rocha, Jim Crutchfield, Cosma Shalizi, Christoph Adami, David Ackley, Steen Rasmussen, Steve Grand, and Mark Bedau. Among friends and colleagues who made a difference I would like to single out Katherine Hayles, Michael Schippling, Tori Alexander, Lucas Beeler, Gregory Rukavina, Geoff Bennington, Tim Lenoir, Steve Potter, Jeremy Gilbert-Rolfe, and Bob Nelson. This work was facilitated by a one-semester grant from the Emory University Research Committee and a one-semester sabbatical leave. Warm appreciation also goes to Bob Prior at MIT Press for his always helpful and lively commitment to this project.

This book would not have seen the light of day without the always surprising resourcefulness, skills as a reader and critical thinker, and unflagging love and support of my wife, Heidi Nordberg. I dedicate it to her.

Introduction

The electric things have their lives, too.
—Philip K. Dick, *Do Androids Dream of Electric Sheep?*

Liminal Machines

In the early era of cybernetics and information theory following the Second World War, two distinctively new types of machine appeared. The first, the computer, was initially associated with war and death—breaking secret codes and calculating artillery trajectories and the forces required to trigger atomic bombs. But the second type, a new kind of liminal machine, was associated with life, inasmuch as it exhibited many of the behaviors that characterize living entities—homeostasis, self-directed action, adaptability, and reproduction. Neither fully alive nor at all inanimate, these liminal machines exhibited what I call machinic life, mirroring in purposeful action the behavior associated with organic life while also suggesting an altogether different form of "life," an "artificial" alternative, or parallel, not fully answerable to the ontological priority and sovereign prerogatives of the organic, biological realm. First produced under the aegis of cybernetics and proliferating in ALife research and contemporary robotics, the growing list of these machines would include John von Neumann's self-reproducing automata, Claude Shannon's maze-solving mouse, W. Ross Ashby's self-organizing homeostat, W. Grey Walter's artificial tortoises, the digital organisms that spawn and mutate in ALife virtual worlds, smart software agents, and many autonomous mobile robots. In strong theories of ALife these machines are understood not simply to *simulate* life but to *realize* it, by instantiating and actualizing its fundamental principles in another medium or material substrate. Consequently, these machines can be said to inhabit, or "live," in a strange, newly animated realm, where the biosphere and artifacts from the human world touch and pass into each other, in effect constituting a "machinic phylum."[1] The increasing number and variety of forms of machinic life suggest, moreover, that this new realm is steadily expanding and that we are poised on the brink of a new era in which nature and technology will no longer be distinctly opposed.

Conjoining an eerie and sometimes disturbing abstractness with lifelike activity, these liminal machines are intrinsically alluring. Yet they also reveal conceptual ambitions and tensions that drive some of the most innovative sectors of contemporary science. For as we shall see, these forms of machinic life are characterized not by any exact imitation of natural life but by complexity of behavior.[2] Perhaps it is no longer surprising that many human creations—including an increasing numbers of machines and smart systems—exhibit an order of complexity arguably equal to or approaching that of the simplest natural organisms. The conceptual reorientation this requires—that is, thinking in terms of the complexity of automata, whether natural or artificial, rather than in terms of a natural biological hierarchy—is part of the legacy of cybernetics. More specifically, in the progression from the cybernetic machines of von Neumann, Ross Ashby, and Grey Walter to the computer-generated digital organisms in ALife research and the autonomous mobile robots of the 1990s, we witness a developmental trajectory impelled by an interest in how interactions among simple, low-level elements produce the kinds of complex behavior we associate with living systems. As the first theorist of complexity in this sense, von Neumann believed that a self-reproducing automaton capable of evolution would inevitably lead to the breaking of the "complexity barrier." For Ashby, complexity resulted from coupling a simple constructed dynamical system to the environment, thereby creating a larger, more complex system. For Walter, the complex behavior of his mobile electromechanical tortoises followed from a central design decision to make simple elements and networks of connections serve multiple purposes. For Christopher Langton, Thomas Ray, Chris Adami, and many others who have used computers to generate virtual worlds in which digital organisms replicate, mutate, and evolve, complexity emerges from the bottom up, in the form of unpredictable global behaviors resulting from the simultaneous interactions of many highly distributed local agents or "computational primitives."[3] Relayed by the successes of ALife, the "new AI" achieves complexity by embedding the lessons of ALife simulations in autonomous machines that move about and do unexpected things in the noisy material world. More recently, several initiatives in the building of intelligent machines have reoriented their approach to emulate more exactly the complex circuits of information processing in the brain.

For the most part, discussion of these liminal machines has been defined and limited by the specific scientific and technological contexts in which they were constructed. Yet even when discussion expands into the wider orbits of cultural and philosophical analysis, all too often it remains bound by the ligatures of a diffuse and seldom questioned anthropomorphism. In practice this means that questions about the functionality and meaning of these machines are always framed in mimetic, representational terms. In other words, they are usually directed toward "life" as the ultimate reference and final arbiter: how well do these machines model or simulate life and thereby help us to understand its (usually assumed) inimitable singularity? Thus if a mobile robot can move around and avoid obstacles, or a digital organism replicate and evolve, these activities and the value of the machinic life in question are usually gauged in relation to what their natural organic counterparts can do in what phenomenologists refer to as the lifeworld. Yet *life* turns out to be very difficult to define, and rigid oppositions like organic versus nonorganic are noticeably giving way to sliding scales based on complexity of organization and adaptability. While contemporary biologists have reached no consensus on a definition of life, there is wide agreement that two basic processes are involved: some kind of metabolism by which energy is extracted from the environment, and reproduction with a hereditary mechanism that will evolve adaptations for survival.[4] In approaches to the synthesis of life, however, the principal avenues are distinguished by the means employed: hardware (robotics), software (replicating and evolving computer programs), and wetware (replicating and evolving artificial protocells).

By abstracting and reinscribing the logic of life in a medium other than the organic medium of carbon-chain chemistry, the new "sciences of the artificial" have been able to produce, in various ways I explore, a completely new kind of entity.[5] As a consequence these new sciences necessarily find themselves positioned between two perspectives, or semantic zones, of overlapping complexity: the metaphysics of life and the history of technical objects. Paradoxically, the new sciences thus open a new physical and conceptual space between realms usually assumed to be separate but that now appear to reciprocally codetermine each other. Just as it doesn't seem farfetched in an age of cloning and genetic engineering to claim that current definitions of life are determined in large part by the state of contemporary technology, so it would also seem plausible that the very differences that allow and support the opposition between life and technical objects—the organic and inorganic (or fluid and flexible versus rigid and mechanical), reproduction and replication, *phusis* and *technē*—are being redefined and redistributed in a biotechnical matrix out of which machinic life is visibly emerging.[6] This redistribution collapses boundaries and performs a double inversion: nonorganic machines become self-reproducing, and biological organisms are reconceived as autopoietic machines. Yet it is not only a burgeoning fecundity of machinic life that issues from this matrix, but a groundbreaking expansion of the theoretical terrain on which the interactions and relations among computation (or information processing), nonlinear dynamical systems, and evolution can be addressed. Indeed, that artificial life operates as both relay for and privileged instance of new theoretical orientations like complexity theory and complex adaptive systems is precisely what makes it significant in the eyes of many scientists.

As with anything truly new, the advent of machinic life has been accompanied by a slew of narratives and contextualizations that attempt to determine how it is to be received and understood. The simplest narrative, no doubt, amounts to a denial that artificial life can really exist or be anything more than a toy world artifact or peripheral tool in the armoire of theoretical biology, software engineering, or robotics. Proceeding from unquestioned and thoroughly conventionalized assumptions about life, this narrative can only hunker down and reassert age-old boundaries, rebuilding fallen barriers like so many worker ants frenetically shoring up the sides of a crumbling ant hill. The message is always the same: artificial life is not *real* life. All is safe. There is no need to rethink categories and build new conceptual scaffoldings. Yet it was not so long ago that Michel Foucault, writing about the conditions of possibility for the science of biology, reminded us that "life itself did not exist" before the end of the eighteenth century; instead, there were only living beings, understood as such because of "the grid of knowledge constituted by natural history."[7] As Foucault makes clear, life could only emerge as a unifying concept by becoming invisible as a process, a secret force at work within the body's depths. To go very quickly, this notion of life followed from a more precise understanding of death, as revealed by a new mode of clinical perception made possible by anatomical dissection.[8] Indeed, for Xavier Bichat, whose *Treatise on Membranes* (1807) included the first analysis of pathological tissue, life was simply "the sum of the functions that oppose death." One of the first modern cultural narratives about artificial life, Mary Shelley's *Frankenstein* (1819), was deeply influenced by the controversies this new perspective provoked.[9]

At its inception, molecular biology attempted to expunge its remaining ties to a vestigial vitalism—life's secret force invisibly at work—by reducing itself to analysis of genetic programming and the machinery of cell reproduction and growth. But reproduction only perpetuates life in its unity; it does not create it. Molecular biology remains metaphysical, however, insofar as it disavows the conditions of its own possibility, namely, its complete dependence on information technology or bioinformatics.[10] The Human Genome Project emblazons this slide from science to metaphysics in its very name, systematically inscribing "the human" in the space of the genetic code that defines the *anthropos.* In *La technique et le temps*, Bernard Stiegler focuses on this disavowal, drawing attention to a performative dimension of scientific discourse usually rendered invisible by the efficacy of science itself.[11] Stiegler cites a passage from François Jacob's *The Logic of Life: A History of Heredity*, in which Jacob, contrasting the variations of human mental memory with the invariance of genetic memory, emphasizes that the genetic code prevents any changes in its "program" in response to either its own actions or any effects in the environment. Since only random mutation can bring about change, "the programme does not learn from experience" (quoted in Stiegler, 176). Explicitly, for Jacob, it is the autonomy and inflexibility of the DNA code, not the contingencies of cultural memory, that ensure the continued identity of the human. Jacob's position, given considerable weight by the stunning successes of molecular biology—including Jacob's own Nobel Prize–winning research with Jacques Monod and André Lwoff on the genetic mechanisms of *E. coli*—soon became the new orthodoxy. Yet, as Stiegler points out, within eight years of Jacob's 1970 pronouncement the invention of gene-splicing suspended this very axiom. (Jacob's view of the DNA code is axiomatic because it serves as a foundation for molecular biology and generates a specific set of experimental procedures.) Thus since 1978 molecular biology has proceeded with its most fundamental axiom held to be true in theory even while being violated in practice.[12]

A great deal of more recent research, however, has challenged this orthodoxy, both in terms of the "invariance" of the genome and the way in which the genome works as a "program." And in both cases these challenges parallel and resonate with ALife research. In regard to the supposed invariance, Lynn Helena Caporale has presented compelling evidence against the view that the genome is rigidly fixed except for chance mutations. Species' survival, she argues, depends more on *diversity* in the genome than inflexibility. In this sense the genome itself is a complex adaptive system that can anticipate and respond to change. Caporale finds that certain areas of the genome, like those that encode immune response, are in fact "creative sites of focused mutation," whereas other sites, like those where genetic variation is most likely to prove damaging, tend to be much less volatile.[13] With regard to the genetic program, theoretical biologist Stuart Kauffman has suggested that thinking of the development of an organism as a program consisting of serial algorithms is limiting and that a "better image of the genetic program—as a *parallel distributed regulatory network*—leads to a more useful theory."[14] Kauffman's alternative view—that the genetic program works by means of a parallel and highly distributed rather than serial and centrally controlled computational mechanism—echoes the observation made by Christopher Langton that computation in nature is accomplished by large numbers of simple processors that are only locally connected.[15] The neurons in the brain, for example, are natural processors that work concurrently and without any centralized, global control. The immune system similarly operates as a highly evolved complex adaptive system that functions by means of highly distributed computations without any central control structure. Langton saw that this alternative form of computation—later called "emergent computation"—provided the key to understanding how artificial life was possible, and the concept quickly became the basis of ALife's computer simulations.

I stated earlier that artificial life is necessarily positioned in the space it opens between molecular biology—as the most contemporary form of the science of life—and the history of technical objects. And I have begun to suggest that a new, nonstandard theory of computation provides the conceptual bridge that allows us to discuss all three within the same framework. At this point there is no need to return to Stiegler's analysis of Jacob in order to understand that *life* as defined by molecular biology is neither untouched by metaphysics nor monolithic; for the most part, in fact, molecular biology simply leaves detailed definitions of life in abeyance in order to attack specific problems, like protein synthesis and the regulatory role of enzymes. Stiegler's two-volume *La technique et le temps* becomes useful, however, when we consider this other side of artificial life, namely, its place and significance in relation to the history and mode of being of technical objects. Specifically, his discussion of the "dynamic of the technical system" following the rise of industrialism provides valuable historical background for theorizing the advent of machinic self-reproduction and self-organization in cybernetics and artificial life.[16]

Very generally, a technical system forms when a technical evolution stabilizes around a point of equilibrium concretized by a particular technology. Tracking the concept from its origins in the writings of Bertrand Gille and development in those of André Leroi-Gourhan and Gilbert Simondon, Stiegler shows that what is at stake is the extent to which the biological concept of evolution can be applied to the technical system. For example, in *Du mode d'existence des objets techniques* (1958), Simondon argues that with the Industrial Revolution a new kind of technical object, distinguished by a quasi-biological dynamic, is born. Strongly influenced by cybernetics, Simondon understands this "becoming-organic" of the technical object as a tendency among the systems and subsystems that comprise it toward a unity and constant adaptation to itself and to the changing conditions it brings about. Meanwhile, the human role in this process devolves from that of an active subject whose intentionality directs this dynamic to that of an operator who functions as part of a larger system. In this perspective, experiments with machinic life appear less as an esoteric scientific project on the periphery of the postindustrial landscape than as a manifestation in science of an *essential* tendency of the contemporary technical system as a whole. This tendency, I think, can best be described not as a becoming-organic, as Simondon puts it, but as a becoming-machinic, since it involves a transformation of our conception of the natural world as well. As I suggest below (and further elaborate in the book), our understanding of this becoming-machinic involves changes in our understanding of the nature and scope of computation in relation to dynamical systems and evolutionary processes.

The Computational Assemblage

The contemporary technical system, it is hardly necessary to point out, centers on the new technology of the computer; indeed, the computer's transformative power has left almost no sector of the Western world—in industry, communications, the sciences, medical and military technology, art, the entertainment industry, and consumer society—untouched. Cybernetics, artificial life, and robotics also develop within—in fact, owe their condition of possibility to—this new technical system. What sets them apart and makes them distinct is how they both instantiate and provoke reflection on various ways in which the computer, far from being a mere tool, functions as a new type of abstract machine that can be actualized in a number of different *computational assemblages*, a concept I develop to designate a particular conjunction of a computational mechanism and a correlated discourse. A computational assemblage thus comprises a material computational device set up or programmed to process information in specific ways together with a specific discourse that explains and evaluates its function, purpose, and significance. Thus the discourse of the computational assemblage consists not only of the technical codes and instructions for running computations on a specific material device or machine but also of any and all statements that embed these computations in a meaningful context. The abacus no less than the Turing machine (the conceptual forerunner of the modern computer) has its associated discourse.

Consider, for example, the discourse of early AI research, which in the late 1950s began to construct a top-down model of human intelligence based on the computer. Alan Turing inaugurated this approach when he worked out how human mental computations could be broken down into a sequence of steps that could be mechanically emulated.[17] This discourse was soon correlated with the operations of a specific type of digital computer, with a single one-step-at-a-time processor, separate memory, and control functions—in short, a von Neumann architecture.[18] Thinking, or cognition, was understood to be the manipulation of symbols concatenated according to specifiable syntactical rules, that is, a computer program. In these terms classic AI constituted a specific type of computational assemblage. Later its chief rival, artificial neural nets, which were modeled on the biological brain's networks of neurons—the behavior of which was partly nondeterministic and therefore probabilistic—would constitute a different type.[19] In fact, real and artificial neural nets, as well as other connectionist models, the immune system, and ALife programs constitute a group of related types that all rely on a similar computational mechanism—bottom-up, highly distributed parallel processing. Yet their respective discourses are directed toward different ends, making each one part of a distinctly different computational assemblage, to be analyzed and explored as such. This book is thus concerned with a family of related computational assemblages.

In their very plurality, computational assemblages give rise to new ways of thinking about the relationship between physical processes (most importantly, life processes) and computation, or information processing. For example, W. Ross Ashby, one of the foremost theorists of the cybernetic movement, understood the importance of the computer in relation to "life" and the complexity of dynamical systems in strikingly radical terms:

> In the past, when a writer discussed the topic [of the origin of life], he usually assumed that the generation of life was rare and peculiar, and he then tried to display some way that would enable this rare and peculiar event to occur. So he tried to display that there is *some* route from, say, carbon dioxide to amino acid, and thence to the protein, and so, through natural selection and evolution, to intelligent beings. I say that this looking for special conditions is quite wrong. The truth is the opposite—every dynamic system generates its own form of intelligent life, is self-organizing in this sense.... Why we have failed to recognize this fact is that until recently we have had no experience of systems of medium complexity; either they have been like the watch and the pendulum, and we have found their properties few and trivial, or they have been like the dog and the human being, and we have found their properties so rich and remarkable that we have thought them supernatural. Only in the last few years has the general-purpose computer given us a system rich enough to be interesting yet still simple enough to be understandable. With this machine as tutor we can now begin to think about systems that are simple enough to be comprehensible in detail yet also rich enough to be suggestive. With their aid we can see the truth of the statement that *every isolated determinate dynamic system obeying unchanging laws will develop "organisms" that are adapted to their "environments."*[20]

Although Ashby's statement may not have been fully intelligible to his colleagues, within about twenty years it would make a new kind of sense when several strands of innovative research began to consider computational theory and dynamical systems together.

The most important strand focused on the behavior of cellular automata (CA).[21] Very roughly, a cellular automaton is a checkerboard-like grid of cells that uniformly change their states in a series of discrete time steps. In the simplest case, each cell is either on or off, following the application of a simple set of preestablished rules. Each cell is a little computer: to determine its next state it takes its own present state and the states of its neighboring cells as input, applies rules, and computes its next state as output. What makes CA interesting is the unpredictable and often complex behavior that results from even the simplest rule set. Originally considered a rather uninteresting type of discrete mathematical system, in the 1980s CA began to be explored as complex (because nonlinear) dynamical systems. Since CA instantiate not simply a new type of computational assemblage but one of fundamental importance to the concerns of this book, it is worth dwelling for a moment on this historic turning point.[22]

The first important use of CA occurred in the late 1940s when, at the suggestion of the mathematician Stanley Ulam, John von Neumann decided to implement the logic of self-reproduction on a cellular automaton. However, CA research mostly languished in obscurity until the early

1980s, the province of a subfield of mathematics. The sole exception was John Conway's invention in the late 1960s of the Game of Life, which soon became the best known example of a CA. Because the game offers direct visual evidence of how simple rules can generate complex patterns, it sparked intense interest among scientists and computer programmers alike, and it continues to amuse and amaze. Indeed, certain of its configurations were soon proven to be computationally universal (the equivalent of Turing machines), meaning that they could be used to implement any finite algorithm and evaluate any computable function. The turning point in CA research came in the early 1980s. In a groundbreaking article published in 1983, Stephen Wolfram provided a theoretical foundation for the scientific (not just mathematical) study of CA as dynamical systems.[23] In the same year, Doyne Farmer, Tommaso Toffoli, and Wolfram organized the first interdisciplinary workshop on cellular automata, which turned out to be a landmark event in terms of the fertility and importance of the ideas discussed.[24] Wolfram presented a seminal demonstration of how the dynamic behavior of CA falls into four distinct universality classes. Norman Margolus took up the problem of reversible, information-preserving CA, and pointed to the possibility of a deep and underlying relationship between the laws of nature and computation. Gerard Vichniac explored analogies between CA and various physical systems and suggested ways in which the former could simulate the latter. Toffoli showed that CA simulations could provide an alternative to differential equations in the modeling of physics problems. Furthermore, in a second paper, Toffoli summarized his work on Cellular Automata Machine (CAM), a high-performance computer he had designed expressly for running CA. As he observes, "In CAM, one can actually *watch*, in real time, the evolution of a system under study."[25] Developing ideas based on CA, Danny Hillis also sketched a new architecture for a massively parallel-processing computer he called the Connection Machine. And, in a foundational paper for what would soon become known as ALife, Langton presented a cellular automaton much simpler than von Neumann's, in which informational structures or blocks of code could reproduce themselves in the form of colonies of digital loops.

The discovery that CA could serve as the basis for several new kinds of computational assemblage accounts for their contemporary importance and fecundity. For a CA is more than a parallel-processing device that simply provides an alternative to the concept of computation on which the von Neumann architecture is built. It is at once a collection or aggregate of information processors and a complex dynamical system. Although completely deterministic, its complex behavior results from many simple but simultaneous computations. In fact, it is not even computational in the common meaning of the term since it does not produce a numerical solution to a problem and then halt. On the contrary, it is meant to run continuously, thus producing ongoing dynamic behavior. Nor do its computations always and forever produce the same result. Conway's Game of Life made this plainly visible: although the individual cells uniformly apply the same simple set of rules to compute their next state, the global results seldom occur in the same sequence of configurations and are usually quite unpredictable. The states, therefore, cannot be computed in advance—one can only "run" the system and see what patterns of behavior emerge. Indeed, it was this capacity to generate unpredictable complexity on the basis of simple, deterministic rules that made the game seem "lifelike." But as Wolfram demonstrated, there are actually four different computational/dynamic regimes: one that halts after a reasonable number of computations, one that falls into a repetitive loop or periodic cycle, one that generates a chaotic, random mess, and one (the most complex) that produces persistent patterns that interact across the local spaces of the grid. Langton theorized that this last regime, which constitutes a phase transition located "at the edge of chaos," instantiates the most likely conditions in which information processing can take control over energy exchanges and thus in which life can gain a foothold and flourish.[26]

Narratives of Machinic Life

The example of cellular automata clearly demonstrates why it is much more useful to focus on specific computational assemblages—both the machines themselves and their constituent discourses—than simply to discuss the computer as a new technology that automates and transforms what existed before. While it is self-evident that the computer lies at the heart of the contemporary technical system, the latter actually consists of a multiplicity of different computational assemblages, each of which must be described and analyzed in its material and discursive specificity. At the same time, we should not ignore certain transformations and rearticulations that occur at the general level of the technical system. Specifically, the advent of the computer and the birth of machinic life mark a threshold in which the technical system is no longer solely engaged with the production of the means to sustain and enrich life but is increasingly directed toward its own autonomization and cybernetic reproduction. This seemingly inevitable tendency toward a form of technogenesis was first anticipated by Samuel Butler in his fictional narrative *Erewhon* (1872). Influenced by Darwin and acutely aware of the increasing pace of technological transformation, Butler explored the sense in which the human subject, beyond serving as the eyes and ears of machines, also functioned as their "reproductive machinery."[27] According to this seemingly inevitable logic, our human capacity as toolmakers (*homo faber*) has also made us the vehicle and means of realization for new forms of machinic life.

This strand of thinking has given rise to two conflicting cultural narratives, the adversarial and the symbiotic. According to the first, human beings will completely lose control of the technical system, as silicon life in the form of computing machines performs what Hans Moravec calls a "genetic take-over" from carbon life, thanks to the tremendous advantage the former possesses in information storage, processing speed, and turnover time in artificial evolution.[28] Since silicon-based machines will eventually increase their memory and intelligence and hence their complexity to scales far beyond the human, their human makers will inevitably find themselves surpassed by their own inventions. According to the second narrative, human beings will gradually merge with the technical system that defines and shapes the environment in a transformative symbiosis that will bring about and characterize the advent of the posthuman.[29] Just as "life" now appears to be an emergent property that arises from distributed and communicating networks rather than a singular property of certain stand-alone entities, so "the human" may come to be understood less as the defining property of a species or individual and more as an effect and value distributed throughout human-constructed environments, technologies, institutions, and social collectivities. The proliferation of machinic life, of course, can be marshaled as evidence supporting either of these two narratives.

Rather than engage directly with these two cultural narratives, this book focuses on their scientific and technological condition of possibility, that is, on the specific scientific achievements that underlie them. As I have already suggested, central to the book's subject matter is the dramatic unfolding of a new space, or relationship, between the metaphysics of natural or biological life and the relatively recent appearance of a new kind of technical object—the liminally lifelike machine or system. This space, however, is not defined by opposition and negation (*phusis* versus *technē*). Although the methods deployed in the technical field involve a mimicking of the natural, what results is not a duality of nature and artifice but the movement of evolution and becoming, relay and resonance, codefinition and codetermination of processes and singularities that constitute something different from both: the machinic phylum. To be sure, the unfolding of this new realm or space entails boundary breakdowns and transformations of age-old oppositions, events that spawn a multiplicity of overlapping and contradictory perspectives in a heady mix of scientific, cultural, and explicitly science fictional narratives. In other words, as machinic life emerges from within a biotechnical matrix seldom discussed as such, it is so entwined with other often contradictory narratives that its own singularity may not be fully discernible and comprehensible.

Consider the example of ALife, whose very possibility of scientific autonomy reflects this betweenness. On one side, in scientific publications and conference presentations (and especially grant applications), ALife is compelled to justify itself in relation to the knowledge claims of theoretical biology, to which it is in danger of becoming a mere adjunct; on the other, its experiments in simulated evolution are often seen as merely useful new computational strategies in the field of machine learning or as new software and/or methods in the development of evolutionary programming.[30] Inscribed in neither of these flanking discourses is the possibility of a potentially more powerful intrinsic narrative, to wit, that artificial life is actually producing a new kind of entity—at once *technical object* and simulated *collective subject*. Constituted of elements or agents that operate collectively as an emergent, self-organizing system, this new entity is not simply a prime instance of the theory of emergence, as its strictly scientific context suggests. It is also a form of artificial life that raises the possibility that terms like subject and object, *phusis* and *technē*, the natural and the artificial, are now obsolete. What counts instead is the mechanism of emergence itself, whatever the provenance of its constitutive agents. More specifically, the real questions are how global properties and behaviors emerge in a system from the interactions of computational "primitives" that behave according to simple rules and how these systems are enchained in dynamic hierarchies that allow complexity to build on complexity. Darwinian evolutionary theory necessarily enters the picture, but natural selection from this new perspective is understood to operate on entities already structured by self-organizing tendencies. In fact, in the wake of Kauffman's and Langton's work, evolution is seen as the mechanism that holds a system near the "edge of chaos," where it is most able to take advantage of or further benefit from varying combinations of both structure and near chaotic fluidity.[31] With this new understanding of Darwinian evolutionary theory before us, the lineaments of an underlying narrative begin to loom into view.

Revised in the light of the dynamics of self-organization and emergence, Darwinian theory assumes a role of fundamental importance in the study of complex adaptive systems—a new theoretical category designating emergent, self-organizing dynamical systems that evolve and adapt over time. Examples include natural ecologies, economies, brains, the immune system, many artificial life virtual worlds, and possibly the Internet. While evolutionary biology is divided by debate over whether or not evolution by natural selection is the primary factor in the increase of biological complexity (Stephen Jay Gould, for example, has argued that contingency and accident are more important),[32] many systems provide direct evidence that, in the words of John Holland, "adaptation builds complexity." Holland describes New York City as a complex adaptive system because, in and through a multiplicity of interacting agents and material flows, it "retains both a short-term and long-term coherence, despite diversity, change, and lack of central direction."[33] Much of Holland's recent work is devoted to understanding the special dynamics of such systems. In Echo, his model of a complex adaptive system, agents migrate from site to site in a simulated landscape, taking in resources and interacting in three basic ways (combat, trade, and mating) according to the values inscribed in their "tag" and "condition" chromosomes. When agents mate, new mixes of these chromosomes are passed to offspring, and "fitter" agents evolve. Amazingly, highly beneficial collective or aggregate behaviors emerge that are not programmed into the individual agents. These behaviors include "arms races," the formation of "metazoans" (connected communities of agents), and the specialization of functions within a group.[34] The occurrence of such highly adaptive behavior in the natural world is common of course; but that it should also occur in an artificial world with artificial agents should be cause for new thinking.

The computer simulation of such agent-based systems has been one of the signal achievements of contemporary science. Yet the deeper theoretical significance of complex adaptive systems stems not simply from the novel simulations deployed to study them but from the fact that these systems are found on both sides of the nature/culture divide, thus suggesting that this age-old boundary may actually prevent us from perceiving certain fundamental patterns of organized being. Indeed, a primary intention of artificial life is not simply to problematize such boundaries and conventional conceptual frames but to discover such patterns from new vantage points by multiplying the examples of life. Even so, one significant current in ALife research asserts that complexity (or complex adaptive systems) rather than "life" (and thus the opposition to nonlife) is the conceptually more fruitful framework. In this vein Thomas Ray explicitly reverses the modeling relationship: "The objective is not to create a digital model of organic life, but rather to use organic life as a model on which to base our better design of digital evolution."[35] Similarly Mark Bedau defines life as a property of an evolving "supplely adaptive system" as a whole rather than as what distinguishes a particular individual entity.[36] This definition follows from and extends Langton's contention that life should not be defined by the material medium in which it is instantiated but by the dynamics of its behavior.

Meanwhile, artificial life experiments continue to respond to the challenge that true, open-ended evolution of the biosphere may not be possible in artificial (specifically, computer-generated) systems. Whereas all of the components of biological life-forms interact and are susceptible to mutation, change, and reorganization, in computer simulations the underlying hardware and most of the time the code are unalterably set by the experimenter, who thus limits in advance the kind and amount of change that can occur in the system. Although current research is determined to overcome this limit, we may be witnessing the end of a first phase in official ALife research, which thus far has been based primarily on small-scale, computer-generated "closed-world" systems. In any event, the need to develop other approaches is clearly evident. Thomas Ray created one such closed system (Tierra) and set up a second, more ambitious version on the Internet. In another example, which amounts to an inversion of the official ALife agenda established by Langton, David Ackley is attempting to build a computer or "living computational system" following principles characteristic of living systems. And on yet another research track, efforts to create artificial protocells, and thus a viable form of "wetlife," have recently made astonishing strides.

Almost from its inception, ALife research has had a cross-fertilizing influence on contemporary robotics. That influence is also apparent in the closely related fields of animats (the construction of artificial animals) and the development of autonomous software agents.[37] Complicating this story was the appearance (almost simultaneously with ALife) of what was called the new AI, which generally meant a wholesale rejection of classic symbolic AI and the full embrace of a dynamical systems perspective. The central figure in the new AI is Rodney Brooks, who inaugurated a new approach to constructing autonomous mobile robots from the bottom-up, based on what he called subsumption architecture. But while following a bottom-up approach similar to that of ALife, contemporary robotics distrusted simulation and believed that the deficiencies of ALife could be overcome by building autonomous robots that can successfully operate in the noisy physical world. At the same time, the development of the neural net controllers needed to make these robots function came to depend on simulation and the deployment of evolutionary programming techniques. Thus, on all fronts, the further development of artificial life-forms (including mobile robots and autonomous software agents) continues to require computational assemblages that can simulate Darwinian evolution and provide an environment in which artificial evolution can occur. A remarkable success in this regard was achieved by Karl Sims with his computer-generated "virtual creatures" environment, in which (as in nature but not yet in physical robotics) neural net controllers (i.e., a nervous system) and creature morphology were made to evolve together.[38] In fact, the attempt to use evolutionary programming techniques to evolve both controllers and robot morphologies for physical robots now defines the new field of evolutionary robotics, which is considered in chapter 7.

Lamarckian Evolution or Becoming Machinic

Perhaps not surprisingly, the current renovation of Darwinian theory—some would argue it is more a deepening than a revision—has been accompanied by a revival of interest in Lamarck's theory that *acquired* traits are passed down to subsequent generations through hereditary mechanisms.[39] Indeed, at first glance a Lamarckian model would seem to be more directly applicable to the evolutionary tendencies of machines and technical systems. As John Ziman frames it, the transformation of an evocative metaphor like "technological evolution" into a well-formed model requires several steps.[40] The first is to address the problem posed by several "disanalogies," foremost among which is that technological innovation exhibits certain Lamarckian features normally forbidden in modern biology. For Ziman, however, the real question is not Darwin or Lamarck but whether or not modern technology as a process guided by design and explicit human intention can be reconciled with evolution, "which both Darwin and Lamarck explained as a process through which complex adaptive systems emerge *in the absence* of design." "We may well agree that technological change is driven by variation and selection," he continues, "but these are clearly not 'blind' or 'natural.'"[41]

Yet despite these reservations, Ziman believes that an evolutionary model can incorporate the factors of human intentionality, since human cognition is itself the product of natural selection and takes place, as he puts it, "in lower level neural events whose causes might as well be considered random for all that we can find out about them" (7). Thus the process as a whole can be said to operate blindly. Actually, the process need not even be blind in the way that mutations or recombinations of molecules in DNA are blind; rather, all that is required is that "there should be a stochastic element in what is actually produced, chosen and put to the test of use" (7). Given that there are no universally agreed upon criteria that determine which technological innovations are selected and that "artifacts with similar purposes may be designed to very different specifications and chosen for very different reasons," Ziman concludes that "there is usually enough diversity and *relatively* blind variation in a population of technological entities to sustain an evolutionary process" (7). Finally, in a not altogether unanticipated move, he suggests that instead of lumping technology and biology together we should treat them as "distinct entities of a larger genus of *complex systems*." Essentially this means that instead of worrying about whether evolutionary processes conform to strictly Darwinian or neo-Darwinian principles, we should be exploring the properties of "a more general *selectionist* paradigm" (11). The most compelling exemplification of selectionism in action, Ziman finds, is ALife.[42]

The question of whether the evolution of artificial life should be considered in Lamarckian rather than Darwinian terms was raised early in ALife research, most notably by J. Doyne Farmer and Alletta d'A. Belin in "Artificial Life: The Coming Evolution," a speculative essay directed toward the future of artificial life.[43] Actually, Herbert Spencer's concept of evolution rather than Darwin's frames the discussion. As Farmer and Belin explain, for Spencer "evolution is a process giving rise to increasing differentiation (specialization of functions) and integration (mutual interdependence and coordination of function of the structurally differentiated parts)" (832). It is thus the dominant force driving "the spontaneous formation of structure in the universe," from rocks and stars to biological species and social organization. In these terms evolution entails a theory of organization that opposes it to disorder or entropy while also anticipating contemporary theories of self-organization. Given the fundamental importance of self-organization to much of contemporary science, it has become essential "to understand why nature has an inexorable tendency to organize itself, and to discover the laws under which this process operates" (833).

Within this larger framework artificial life signals a momentous change in the way evolution takes place. Following the spontaneous formation of structure through processes of self-organization (which leads to the origins of life), biological reproduction becomes the means by which information and patterns from the past are communicated to the future. Moreover, with Darwinian evolution (random mutations and natural selection), incremental changes are introduced that produce structures of greater variety and adaptability. With the advent of human culture a great speed-up occurs through Lamarckian evolution, since changes in the form of "acquired characteristics" can now be transmitted directly to the future rather than only through *genetic* information. The invention of the computer is another benchmark, since it allows a much more efficient storing of information and the performing of certain cognitive functions that heretofore only humans could perform. But with artificial life it becomes possible "for Lamarckian evolution to act on *the material composition of the organisms themselves*" (834). More specifically, with computer-generated life-forms the genome can be manipulated directly, thus making possible not only the genetic engineering of humans by humans but a "*symbiotic* Lamarckian evolution, in which one species modifies the genome of another, genetically engineering it for the mutual advantage of both" (835). Finally, it will be possible to render or transfer this control of the genome to the products of human technology, producing self-modifying, autonomous tools with increasingly higher levels of intelligence. (As shown in chapter 7, this tendency is already evident in contemporary robotics.) "Assuming that artificial life forms become dominant in the far future," Farmer and Belin conclude, "this transition to Lamarckian evolution of hardware will enact another major change in the global rate of evolution.... The distinction between the artificial and the natural will disappear" (835).

Presented in these broad and sweeping terms, Farmer and Belin's narrative resonates with several familiar cosmological narratives. Examples include Henri Bergson's *Creative Evolution* (1907) and Pierre Teilhard de Chardin's *The Phenomenon of Man* (1955), where "life" and "intelligence" respectively are understood to be forces for creative change that bring about the adaptation of the universe itself as they continually spread outward. Human beings are simply one vehicle by means of which they disseminate and proliferate. A recent version of this narrative can be seen in Harold J. Morowitz's *The Emergence of Everything*, which sketches in twenty-eight instances of emergence the origin of the physical universe, the origin of life, and the origin of human mind.[44] Yet none of these narratives of "becoming" envision the extension of life and intelligence through the propagation of human-constructed machines that replicate and evolve in complexity. In order to pursue this scenario, and in particular to pose the question of how nature itself could be caught up in a "becoming machinic," I turn to Gilles Deleuze and Félix Guattari's theory of becoming, which despite its philosophical rather than scientific impetus exhibits notable similarities to Farmer and Belin's narrative of a symbiotic Lamarckian evolution.

Initially the theory seems directly opposed to evolution, at least to evolution by descent and filiation.[45] Although Deleuze and Guattari mention the relationship of the orchid and the bee, certain supposed "alliances" with viruses, and "transversal communications between heterogeneous populations," the biological realm remains largely outside their concerns precisely because of its very capacity to reproduce and self-propagate "naturally," that is, along lines of family descent and filiation. Against the natural mode of propagation they extol alliance, monstrous couplings, symbiosis, "side-communication" and "contagion," and above all those doubly deterritorializing relays they call "becomings." A primary example comes from mythic tales of sorcerers and shamans who enter into strange and unholy relationships with animals in order to acquire their powers, but as Deleuze and Guattari emphasize, these instances of becoming-animal do not involve playing or imitating the animal. Either to imitate the other or remain what you are, they say, is a false alternative. What is involved, rather, is the formation of a "block"—hence they speak of "blocks of becoming"—constituted by alliance, communication, relay, and symbiosis. Since the block "runs its own line 'between' the terms in play and beneath assignable relations," the outcome cannot be reduced to familiar oppositions like "appearing" versus "being" and "mirroring" versus "producing." More to the point, a becoming-animal "always involves a pack, a band, a population, a peopling, in short, a multiplicity" (*A Thousand Plateaus*, 239). Thus to enter into a block of becoming is to enter a decentered network of communication and relay that will lead to becoming someone or something other than who you are, though not through imitation or identification. In fact, Deleuze and Guattari speak of "a-parallel evolution," where two or more previously distinct entities relay one another into an unknown future in which neither will remain the same nor the two become more alike. In certain respects, this dynamic process resembles that of biological coevolution, in which distinct species or populations are pulled into either an escalating arms race or a symbiosis. Or, to take a more pertinent example, there are some species known to survive as a cloud, or quasi, species when exposed to a high mutation rate.[46] No single organism contains the entire genome for the species; rather, a range of genomes exists in what ALife scientist Chris Adami has described as a "mutational cloud."[47] This is neither the evolution nor the disintegration of a species, but appears to be an instance of a becoming-symbiotic.

For Deleuze and Guattari, the act of imitation serves only as a mask or cover, behind which something more subtle and secret (i.e., imperceptible) can happen. If we follow their idea that becoming-animal is not a mimicking of an animal but an entering into a dynamic relationship of relay and aparallel evolution with certain animal traits, it becomes possible to theorize how becoming-machinic is a force or vector that, under the guise of imitation, is directing and shaping not only ALife experiments and contemporary robotics but much of the new technology transforming contemporary life. The rapid innovation and evolution of computer technology and the changes brought about as a result—what the popular media refer to as "the digital tidal wave"—are part of this larger movement. These developments, and the constellation of dynamic processes driving them, cannot be understood simply as the fabrication and widespread usage of a new set of tools. Unlike the telescope and microscope, which extended a natural human capacity, the computer is a machinic assemblage of an altogether different order, one that transforms the very terms of human communication and conceptualization. In Heideggerian terms, it sets into motion a new and different "worlding of the world," one that has brought forth a machinic reconception of nature itself.

The assumption that physical processes are rule-governed and therefore simply or complexly computational is a central aspect of this new worlding.[48] Computationalism, as this assumption may be loosely called, includes the metaphorical version as well: all physical processes can be viewed or understood as computations. One widely accepted example is the view that evolution itself is simply a vast computational process, a slow but never-ending search through a constantly changing "real" fitness landscape for better-adapted forms. No doubt the most extreme version of computationalism has been advanced by Edward Fredkin, who believes that all physical processes are programs running on a cosmic cellular automaton; nature itself, in short, is a computer. Fredkin argues that subatomic behavior is well described, but not *explained*, by quantum mechanics and that a more fundamental science, which he calls digital mechanics, based on computational mechanisms, will one day supply a deeper and more complete account.[49] A comparable version of computationalism centered on a notion of "computational equivalence" across a diverse range of phenomena has been advanced by Stephen Wolfram in *A New Kind of Science*.[50] Whether warranted or not, this conceptualization of computation as a process that generates nature itself is a strong—perhaps ultimate—expression of the becoming-machinic of the world.

More than anything else, our acculturated reliance on relationships of mimesis and representation still makes it difficult to comprehend becoming-machinic in its own terms. The example of ALife is again illustrative. Because ALife appears to abstract from and mimic living systems, we tend to understand its meaning and function in terms of a model, representation, or simulation of life. In this perspective the experimenter simply constructs a particular kind of complex object that reduces and objectifies a natural process in a simulacrum. From a Deleuzian perspective, on the other hand, the human subject appears not in the image of a godlike creator but as a participant in a larger process of becoming-machinic. This process is not fully explicable anthropomorphically, either as a natural process of growth or as a human process of construction. Rather, it is a dynamic self-assembly that draws both the human and the natural into new forms of interaction and relay among a multiplicity of material, historical, and evolutionary forces. As a result, the human environment is becoming a field in which an increasingly self-determined and self-generating technology continues natural evolution by other means.

As participants in a block of becoming composed of both natural and artificial life forms and traits, human subjects do not thereby become more machinelike; nor do artificial life-forms become more human. Instead, as new relays and networks of transversal communications begin to cohere, boundaries rupture and are newly articulated, energy and image are redistributed, and new assemblages form in which human being is redefined. As human cognitive capacities increasingly function within and by means of environmentally distributed technologies and networks, these capacities will necessarily be further augmented by new relationships with information machines. However, by creating conditions and methods by which machines themselves can become autonomous, self-organizing, and self-reproducing, human beings change not only the environment but the way they constitute and enact themselves, thus reshaping their own self-image.

In certain respects this narrative of becoming-machinic does not differ much from Farmer and Belin's narrative of Lamarckian symbiosis. But whereas the latter focuses almost exclusively on the alteration of the genome in both humans and artificial life forms and the great speedup in the "global rate of evolution," becoming-machinic leaves open the question of exactly what new kinds of assemblages human beings will enter into and become part of.[51] To be sure, at this level of generalization it may not be possible or feasible to gauge the differences between directed evolution (with all its unforeseen consequences) and a becoming-machinic guided by a logic of relay and coevolution. Yet both demand that we think beyond the protocols of mimesis and representation toward new and hybrid forms of self-organized being. It is often said that biological entities evolve "on their own," whereas the evolution of intelligent machines requires human intervention. But this may be only an initial stage; once intelligent machines are both fully autonomous and self-reproducing they will be subject to the full force of the evolutionary dynamic as commonly understood. Within this larger trajectory what will become paramount is how both evolutionary change and the ongoing experiments of artificial life will produce and instantiate new learning algorithms. The latter will involve not only pattern recognition, adaptability, and the augmentation of information-processing capabilities but also new search spaces created by new forms of machinic intelligence, as the natural and social environment is increasingly pervaded by machinic forms. Thus far the computer has been one of many means by which new learning algorithms are developed and implemented. However, if the next step beyond artificial evolution within computers (ALife) and by means of computers (contemporary robotics) is to be computers that can evolve by themselves, then the new learning algorithms will have to be instantiated in the computer's very structure. In other words, a new kind of computational assemblage will have to be built to fit the learning algorithms, rather than the other way around.[52] In effect, the computer itself will have to become a complex adaptive system, able to learn and adapt because it can change the rules of its operating system not only for different computational tasks but as the environment itself changes. This would bring to its fullest realization the project first adumbrated in the cybernetic vision of a new kind of complex machine.

I FROM CYBERNETICS TO MACHINIC PHILOSOPHY

1 Cybernetics and the New Complexity of Machines

Cybernetics is not merely another branch of science. It is an Intellectual Revolution that rivals in importance the earlier Industrial Revolution.
—Isaac Asimov

Before the mid-twentieth century, the very idea that human beings might be able to construct machines more complex than themselves could only be regarded as a dream or cultural fantasy. This changed in the 1940s and '50s, when many scientists and mathematicians began to think in innovative ways about what makes behavior complex, whether in humans, animals, or machines. One of the scientists, John von Neumann, often referred to what he called the "complexity barrier," which prevented current machines or automata from following the path of evolution toward the self-reproduction of ever more complex machines. Others, more commonly, thought of complexity in relation to the highly adaptive behavior of living organisms. Many of these scientists were directly involved in the advent of cybernetics and information theory, a moment that should now be considered essential to the history of our present, rather than a merely interesting episode in the history of technology and science. For, contrary to widespread belief, cybernetics was not simply a short-lived, neomechanist attempt to explain all purposeful behavior—whether that of humans, animals, or machines—as the sending and receiving of messages in a feedback circuit. Rather, it formed the historical nexus out of which the information networks and computational assemblages that constitute the infrastructure of the postindustrial world first developed, spawning new technologies and intellectual disciplines we now take for granted. Equally important, it laid the grounds for some of the most advanced and novel research in science today.

Historically, cybernetics originated in a synthesis of control theory and statistical information theory in the aftermath of the Second World War, its primary objective being to understand intelligent behavior in both animals and machines. The movement's launching pad was a series of interdisciplinary meetings called the Macy Conferences, to which a diverse assortment of scientists, mathematicians, engineers, psychiatrists, anthropologists, and social scientists were invited to discuss "the Feedback Mechanisms and Circular Causal Systems in Biology and the Social Sciences." In all, ten conferences were held in New York City from 1946 to 1953. (Unfortunately, only the transactions of Conferences 6–10 were published.) The movement's chief spokesman, Norbert Wiener, explains that the term "cybernetics" (from the Greek *kybernetes*, for steersman) was introduced in 1947 "by the group of scientists about Dr. Rosenblueth and myself" to designate a new field centered on problems in "communication, control, and statistical mechanics, whether in the machine or in living tissue."[1] The new science of cybernetics, he continues, extends to "computing machines and the nervous system" as well as to the brain as a "self-organizing system." After the publication of Wiener's book, *Cybernetics; or, Control and Communication in the Animal and Machine* (1948), Macy Conference participants decided to incorporate the term into the title of the sixth and all subsequent conferences.

Producing not only new theories but new kinds of machines, cybernetics was not simply or solely an intellectual movement. During the war many of those who would later form the movement's core worked on new weapons systems, radar, and the rapid calculating machines that would eventually lead to the electronic, stored-program computer.[2] But just as significantly, many others were trained in neurophysiology and the behavior of living organisms, which gave the movement a double perspective and strong interdisciplinary flavor. The movement's conceptual center, however, was defined by a new way of thinking about machines. The machines, or automata, that most interested the cyberneticists were those that were self-regulating and that maintained their stability and autonomy through feedback loops with the environment. Since living organisms characteristically exhibit this capacity, a conjoint study of their behavior was both necessary and inevitable.

In several luminous pages in *Cybernetics*, Wiener signals a critical shift in how science understands the living organism. In the nineteenth century it was understood above all as "a heat engine, burning glucose or glycogen or starch, fats, and proteins into carbon dioxide, water and urea" (41). Wiener and his contemporaries realized, however, that "the [organism's] body is very far from a conservative system, and that its component parts work in an environment where the available power is much less limited than we have taken it to be." In fact, the nervous system and organs responsible for the body's regulation require very little energy. As Wiener puts it, "The bookkeeping which is most essential to describe their function is not one of energy." This "bookkeeping," rather, is accomplished by regulating the passage of information, like a vacuum tube does in an electronic circuit, from the body's sense organs to effectors that perform actions in the environment. Accordingly, "the newer study of automata, whether in the metal or in the flesh, is a branch of communication engineering, and its cardinal notions are those of message, amount of disturbance or 'noise'—a term taken over from the telephone engineer—quantity of information, coding technique, and so on" (42). As a consequence, the "theory of sensitive automata," by which Wiener means automata that receive sensory stimulation from the environment and transfer that information through the equivalent of a nervous system to effectors that perform actions, must be a statistical one:

> We are scarcely ever interested in the performance of a communication-engineering machine for a single input. To function adequately, it must give a satisfactory performance for a whole class of inputs, and this means a statistically satisfactory performance for the class of input which it is statistically expected to receive. Thus its theory belongs to the Gibbsian statistical mechanics rather than to the classical Newtonian mechanics. (44)

From this last point Wiener concludes that "the modern automaton exists in the same sort of Bergsonian time as the living organism." The statement is more far-reaching than perhaps Wiener intended, for Bergsonian time is also the time of becoming and "creative evolution." In any event, while the vitalist-mechanist controversy that Bergson attempted to resolve might have had a certain validity for an earlier, less technologically developed period, from Wiener's cybernetic point of view it had already been relegated "to the limbo of badly posed questions" (44).

Wiener's valorization of information feedback circuits was reinforced by Claude Shannon's quantitative theory of information, published in the same year as Wiener's *Cybernetics*.[3] Though intended to overcome engineering problems in electronic communications—specifically to reduce noise in telephone lines—Shannon's was a formal theory that could be applied to the communication of information in any medium. Like Wiener, Shannon defined information in statistical terms; his formula for computing information, in fact, was based directly on Ludwig Boltzmann's famous formula for computing the entropy, or amount of randomness, of a thermodynamic system. Given the uncertainty of molecular states, Boltzmann proposed a measure based on their statistical distribution and even thought of our incomplete knowledge of these states as "missing information." For Shannon, the uncertainty of a message stems from the freedom of choice in the selection of a particular message (or set of symbols constituting a message). The greater the number of possible messages, the greater the uncertainty and hence the greater the information value of a single selected message or symbol. But whereas for Shannon information measures this uncertainty, or entropy, for Wiener it measures a gain in certainty; information, therefore, he considered to be a measure of negative entropy, or "negentropy."[4] Wiener's positive definition, as well as his emphasis on continuous (analog) rather than discrete (digital) modes of information transfer, may have reflected his greater interest in living organisms. In *The Human Use of Human Beings: Cybernetics and Society*, a popular version of his theory, he explains how the process of cybernetic feedback made possible systems and forms of organization that ran counter to nature's statistical tendency to disorder and increasing entropy.[5] Only in these pockets of negentropy, he argued, could something like biological life arise.

Yet Shannon was certainly not indifferent to the cybernetic interest in making machines that could model or instantiate the behavior of living organisms. At the 8th Macy Conference in 1951 he presented an electromechanical "mouse" that could learn to find its way through a maze.[6] The mouse would first explore the maze, and then, after one complete exploration, run through it perfectly. The maze consisted of twenty-five squares separated by moveable partitions; the goal was a pin that could be inserted into a jack in any square. The mouse was a small vehicle with two motors enabling it to move in any of four directions (north, south, east, west). The motors were wired to a "sensing finger" and set of relays that allowed the mouse to "explore" the maze and to "remember" its path through it. From a starting point it would enter a square and then move forward. If it bumped into a wall, it would move back to the center and move forward in a second direction (counterclockwise by ninety degrees from the first) until it passed to another square. It would then visit each square systematically, doing the same until it arrived at the goal. A set of relays would lock into memory each of the successful directions. When the mouse arrived at the goal a bell would ring and a light turn on, but it would also switch from "exploratory strategy" to "goal strategy." When repositioned at the starting point or at any square it had visited, the mouse would now go directly to the goal; if placed in a square it had not visited, it would explore until it found one, then move directly to the goal. In short, it had *learned* the maze.

While no one could or did confuse the behavior of the "maze-solving machine" with that of a real mouse, the similarities between the two were uncanny. In fact, most of the machines built by the cyberneticists exhibited behavior that, if witnessed in living organisms, would be deemed intelligent, adaptive, or illustrative of learning. Hence discussion at the Macy Conferences often revolved around questions of whether these machines were models or mere simulations, in the pejorative sense of giving only the appearance of something.[7] And yet, hovering in the air was a tacit or groping sense that it was really a matter of a new kind of machine that transcended this opposition, just as the opposition between causal and teleological behavior is transcended in the cybernetic notion of circular causality. Indeed, a seminal paper instigating the Macy Conferences had made this very argument. In "Behavior, Purpose, and Teleology," Arthur Rosenblueth, Norbert Wiener, and J. Bigelow proposed a classification of behavior based on how an entity (whether object or organism) produces changes in the environment.[8] One of the categories includes entities (both animals and machines) able to engage in goal-directed actions through feedback mechanisms. Traditionally, such actions had been considered outside the bounds of scientific study, since to explain an action in relation to a goal meant explaining it in terms of an event that had not yet happened, as if the cause could somehow come after the effect. The authors, therefore, proposed a different kind of model, one that substitutes a type of "circular causality" based on feedback for the usual cause-and-effect relationship, or that of stimulus followed by response. According to the new model, part of the output for an action taken by the entity is returned to the entity as input for the next action in a continuous circuit of auto-regulation, hence transcending the supposed opposition between causal and teleological explanation.

Cybernetic discourse, moreover, tended to speak of machines in terms of living organisms and living organisms in terms of machines. There was in fact an assumed or implicit agreement that the two differed only in the complexity of their respective organization. The unspoken—and perhaps unspeakable—objective directly follows: to bridge the gap between the organic and the inorganic, the natural and the artificial. The cyberneticists' use of the word *automaton*—or, more often, the plural *automata*—also points in this direction. Conventionally, of course, the term designates a self-moving machine, often a mechanical figure or contrivance meant to convey the illusion of autonomy. The cyberneticists, however, speak of natural *or* artificial automata, of automata "in the metal or in the flesh," as Wiener puts it in the passage quoted above. Does this transgressive breaching of natural boundaries confirm J.-C. Beaune's thesis that the modern automaton is the philosophical machine par excellence, concretizing in its fundamental ambiguity the fear and anguish just under the surface of the industrial era's belief in a progressive rationality?[9] The difficulty here is that the terms no longer seem applicable, for cybernetics makes the modern automaton central not only to its own development but to sciences yet to be born, within which the automaton's fuller realization will transform (or abolish) the oppositions to which it owes its existence.

Conceptualizing the New Machine

In considering Wiener and Shannon, we have only broached the familiar part of cybernetics, understood as a theory of control and self-regulation achieved through the sending and receiving of information in feedback circuits. Note that feedback and self-regulation, though implicated from the outset with purposeful, intelligent, adaptive behavior—and later with memory and learning—are always understood by the cyberneticists in terms of pure physical embodiment and performance, not symbol making and representation.[10] This understanding is particularly evident in W. Ross Ashby's rigorous reframing of cybernetics, which begins with the question, What are all the possible behaviors of any given machine? As a consequence, the terms of his approach are dictated not by information theory but by dynamical systems theory. Indeed, within this wider frame, cybernetics becomes a more ambitious undertaking: it is not simply about feedback but constitutes a theory of machines, as Ashby states at the beginning of his *Introduction to Cybernetics*:

> Cybernetics ... is a "theory of machines," but it treats, not things but *ways of behaving*. It does not ask "what *is* this thing?" but "*what does it do*?" ... It takes as its subject-matter the domain of "all possible machines," and is only secondarily interested if informed that some of them have not yet been made, either by Man or by Nature. What cybernetics offers is the framework on which all individual machines may be ordered, related and understood.[11]

Yet Ashby, no less than Wiener, was also interested in how machines can model and thereby help us understand the behavior of living organisms. Indeed, his "homeostat machine" (discussed below) was inspired by the question of how an organism adapts to its environment. However, in *Design for a Brain*, as in all his published research, Ashby proceeds from the assumption that the organism *is*—or must be treated as—a machine.[12] At the outset he notes that the brain resembles a machine "in its dependence on chemical reactions, in its dependence on the integrity of anatomical paths, and in the precision and determinateness with which its component parts act on one another." Yet psychologists and biologists have confirmed "with full objectivity the layman's conviction that the living organism behaves typically in a purposeful and adaptive way" (1). Ashby's objective is to reconcile these apparently opposed perspectives, essentially by demonstrating exactly how a machine can be adaptive. To do this, he will make "extensive use of a method hitherto little used in machines" (10)—by which he obviously means feedback—and will define the organism and environment as a coupled dynamical system. Thus he defines the environment as "*those variables whose changes affect the organism, and those variables which are changed by the organism's behavior*" (36, author's emphasis). His homeostat machine allowed him to experiment precisely with this coupling of two distinct systems.

Although Ashby takes the radical view that *every* self-organizing dynamical system or machine generates a form of "life" adapted to its environment, his approach is entirely consistent with the cybernetic program more generally. Essentially, cybernetics proposed not only a new conceptualization of the machine in terms of information theory and dynamical systems theory but also an understanding of "life," or living organisms, as a more complex instance of this conceptualization rather than as a different order of being or ontology. Hence the complexity of life is not attributed to some ineffable, mystical force, as in vitalism, nor is it reduced to a determinate mechanical linkage of cause and effect, as in Descartes's understanding of animals as complicated machines. Rather, what we usually find in actual cybernetic research is the assumption that some aspect of a living organism's behavior can be accounted for by a mechanism or mechanisms that can be modeled by a machine. For example, in *An Introduction to Cybernetics*, Ashby shows first how the somewhat unpredictable behavior of an insect that lives in and about a shallow pond, hopping to and fro among water, bank, and pebble, can illustrate a machine in which the state transitions correspond to those of a stochastic transformation (165). But then he shows how this same machine, which he now calls a "Markovian machine," is capable of great complexity (225–235).[13] And this is typical: a machine that serves as the model often turns out to be capable, under certain conditions, of exceedingly complex behavior. Finally, this complex behavior, which often illuminates the actual behavior of living organisms, becomes interesting in and of itself, even though the early cyberneticists lacked a coherent framework for discussing it in these terms.

	↓	E	
		1	2
A	1	2	1
	2	1	1

Figure 1.1
Processor P.

Before turning to specific examples, it may be instructive to illustrate Ashby's relatively abstract (because generalized) approach to machines. As he defines it, a "machine is that which behaves in a machinelike way, namely, that its internal state, and the state of its surroundings, defines uniquely the next state it will go to" (*An Introduction to Cybernetics*, 25). Consider a simple example—a "processor" P.[14] Figure 1.1 summarizes all of P's possible behaviors, given a specific input A and a specific state of the environment E.

Thus (following the downward arrow), if the input A has the value 1 and the environment E is in a state indicated by the number 1, then the machine moves to a state indicated by the number 2; alternatively, if the environment is in the state indicated by the number 2, then the machine will remain in its initial state. If the input A has the value 2 (again reading downward from the arrow), then the machine will remain in its initial state regardless of whether the environment is in state 1 or 2. This simple mapping of all the machine's possible states as a function of its input and the environmental state tells us two things. First, we don't need to know what internal mechanism brings about P's change of state (it is treated as a "black box"). All that matters is how its behavior changes as a consequence of changes in two variables, its input and the state of the environment. (This also means that P's behavior can be easily expressed as a mathematical function.) Second, P does not operate in isolation; its behavior only makes sense in relation to the input it receives and the environment to which it is "coupled."

We could easily make P's behavior more complicated simply by increasing the number of variables. But we could also make its behavior *more complex*. Let us connect a large number of Ps in a gridlike array, such that the output of one is connected to the input of two others, and so on throughout the array. Of course, the environmental state E would also have to be redefined. One possibility would be to make its value a summation of the neighboring P states for all Ps within a "certain neighborhood" (e.g., not more than one P away in any direction). This would mean that the state-transition table (for this is what the table above is called) would be more complicated, but not unmanageably so. Let us make one further modification. Whenever a P moves to a state 2, a small light turns on, and it remains on as long as P remains in that state. What would be the result of this new setup? First, as soon as an initial input stimulus is given to the array, lights would begin to flash on and off randomly. Eventually, a stable pattern of lights might appear, indicating that the array as a whole had found a stable configuration. Next, we try a second initial input, different from the first. Would we see a repetition of the results of the first stimulus? In fact, there is no way to know in advance; we would simply have to try it and see. This means that we now have a machine—we could call it a P-array—whose behavior is completely deterministic (since the state of every individual P is still determined by two variables, input and environmental state) but is completely unpredictable. It is unpredictable simply because the changing variables in all the "P-neighborhoods" are now influencing one another in such a tangle of nonlinear feedback circuits that there is no way to compute the outcome in real time.

Aside from its illustrative value, does the P-array have any use in the world, or is it just an interesting artifact? As a machine, it can always be connected to another machine. Let's assume that the P-array exhibits a number of stable patterns. Each stable pattern could be made to trigger a specific actuator that accomplishes something. We would now have as many different actuators as there are stable patterns, making the P-array a very flexible switchboard. It is flexible because a pattern, rather than a digital on-off switch, triggers an actuator. Thus its reliability does not depend upon the perfect functioning of every P. Since there are many Ps, the array as a whole possesses a built-in redundancy. It is also flexible in that similar or even partially incomplete patterns may evoke the same response (triggering a single actuator), or multiple patterns may evoke multiple simultaneous responses (triggering multiple actuators). These flexible capacities begin to make the P-array resemble a brain, or at least a mechanism for producing intelligent, purposeful behavior. And the P-array also raises a number of fruitful questions: What are the precise dynamics that lead it to stabilize, or self-organize into specific patterns? What if it never stabilizes? Could a mechanism be added that would automatically change the individual P settings until it did? If a specific input leads to an action that damages or destroys the P-array, could another mechanism inhibit the execution of this action? Are there other mechanisms that would enable the P-array to learn and adapt to changes in its environment? These are the kinds of questions that the cyberneticists would typically ask.

The most fruitful aspect of the cybernetic heritage was thus to have created a conceptual space in which a new theory of machines could be elaborated, often in relation to or involving a fresh examination of the behavior of living organisms. This encouraged new reciprocal understandings of the realms of the artificial and the natural in a series of relays and reversals of perspective. As a consequence, and with irreversible force, the boundary lines soon began to dissolve, and humanity eventually found itself in the present era of the posthuman cyborg and the proliferation of what I call machinic life. Before proceeding to show how several "species" of this new form of life arose in tandem and relay with efforts to break the complexity barrier—this will be the central subject of later chapters—I want to examine the theories and machine constructions set forth by three of the original cyberneticists: John von Neumann, W. Ross Ashby, and W. Grey Walter. These specific examples will lead to a discussion of the concept of self-organization in Ashby and Heinz von Foerster and thus to the end of the first phase of cybernetics as an historical movement. I conclude the chapter with a reflection on the imaginary constructions of Valentino Braitenberg, whom I (somewhat playfully) consider the last cyberneticist. While each of these instances exhibits a particular aspect of complexity, when taken together they provide a historical backdrop for developments I take up in subsequent chapters.

Von Neumann's Self-Reproducing Automata

John von Neumann is the first theoretician of complexity in the sense that will be developed here. Those familiar with his scientific contributions—to the mathematics of quantum physics, to game theory and economics, to the implosion device for triggering the first atomic bomb, and to the architecture of the first all-purpose electronic computer—won't be surprised to learn that from the late 1940s to the end of his life in 1957 von Neumann devoted enormous attention to automata theory. In fact, before von Neumann's work, automata theory as such could hardly be said to have existed. Specifically, before a formal theory of computation was initiated in the 1930s by the mathematicians Alan Turing, Alonso Church, and Stephen Kleene and the new "thinking machines" began to be built in the 1940s and '50s, the construction of automata had been a distinctly marginal concern, a playful sideline activity associated more with toys than scientific models. One thinks, for example, of Jacques Vaucanson's mechanical duck and other wind-up devices that enchanted the European courts of the eighteenth century, or the Swiss cuckoo clocks that amused the middle class. As was typical of von Neumann, his interest in automata was predicated not simply on their new importance, now that extremely fast computing machines were the order of the day, but more centrally on the their lack of a theory as such, a general and logical theory, as he put it. Indeed, from the theoretical point of view that von Neumann envisioned, automata theory offered the possibility of giving a rigorous and fertile formulation to questions that cybernetics and information theory made it possible to ask for the first time. How, for example, can the differences that underlie the logic of organization in biological as opposed to artificial entities be used to build more reliable machines? More specifically, how can unreliable components be organized to become highly reliable for a machine or automaton as a whole? And what are the conditions that would enable simple automata—understood as information-processing machines that exhibit self-regulation in interaction with the environment—to produce more complex automata?

In providing a framework within which these and similar questions could be addressed, von Neumann established the intellectual value and usefulness of a general theory.[15] Here I consider only a few singular details, and mostly in reference to von Neumann's preferred two examples: the human brain, or nervous system, and the computer, which represented the most interesting examples of natural and artificial automata, respectively. Far from considering them separately, von Neumann made frequent comparisons. At the outset of "The General and Logical Theory of Automata" he states that "some regularities which we observe in the organization of the former may be quite instructive in our thinking and planning of the latter; and conversely, a good deal of our experiences and difficulties with artificial automata can be to some extent projected on our interpretations of natural organisms."[16] But while often noting differences in computational speed, number of processing units, material composition, and so forth, he was mostly concerned with the "logic" underlying the organization of these two kinds of automata. In his short book *The Computer and the Brain* (published posthumously), he suggests that the human brain, like most biological organisms, is a "mixed" machine, both logical and statistical, which functions according to both digital and analog mechanisms.[17] He emphasizes not only that the "message-system used in the nervous system [is] of an essentially *statistical* character" (79) but also that the "language of the brain [is] not the language of mathematics" (80). He concludes, therefore, that "when we talk mathematics, we may be discussing a *secondary* language, built on the *primary* language truly used by the nervous systems" (82). This sensitivity to the limits of specific languages—in particular, that of formal logic—becomes critical when confronting complexity.

Von Neumann knew from Warren McCulloch and Walter Pitts's pathbreaking essay, "A Logical Calculus of the Ideas Immanent in Nervous Activity," that neurons, or nerve cells, are always connected with other neurons in "neural nets."[18] Assuming that individual neurons function as digital on-off switches for "nervous [i.e., electrical] impulses," McCulloch and Pitts show that these neural nets can be understood as switching networks that perform logical and arithmetic calculations, and that indeed these natural "logic machines" are equivalent to Turing machines (35), hence computationally universal.[19] In this way the authors pinpoint the underlying neurophysiological mechanism that makes logic and calculation possible. Methodologically, however, McCulloch and Pitts treat the neuron as a black box and simply axiomatize its functioning. While highly appreciative of this axiomatization, von Neumann was skeptical nonetheless that it reflected the way the brain actually works. A brain built according to these principles which could carry out all the required functions "would turn out to be much larger than the one that we actually possess" and may even "prove to be too large to fit into the physical universe," he asserted ("General and Logical Theory," 33–34). In short, while granting that McCulloch and Pitts had proved logically that any behavior unambiguously describable in words could be computed or emulated by a neural net, he also believed, somewhat disconcertingly, that the type of "logic" these nets deployed was too limited to account for the behavior exhibited by more complex automata.

The explanation von Neumann brought into play centered on a very specific notion of complexity. For simple automata, he thought, it is easier to describe the behavior itself than exactly how this behavior is produced or effectuated. Take, for example, Vaucanson's duck. Any observer could describe its behavior and attest that it was capable of doing things like waddle around and take food into its bill. Yet it would be very difficult for that same observer to describe the complicated structure—the mechanical interactions of specific gears and springs—by means of which these activities were accomplished. Even so, von Neumann postulated that above a certain threshold of complexity the description of the structure would be simpler than a description of the behavior. In discussing McCulloch and Pitts's theory, he gives the example of seeing visual analogies. Given the number and kind of visual analogies that are perceptible, the task of describing them seems endless. The feeling thus arises that it may be "futile to look for a precise logical concept, that is, for a precise verbal description of 'visual analogy.'" It is therefore possible, he concludes, that "the connection pattern of the visual brain itself is the simplest logical expression of this principle" ("General and Logical Theory," 24). The application to McCulloch and Pitts's theory is clear: there is an equivalence between the logical principles and their embodiment in a neural network. In simpler cases this means that the principles would supply a simplified expression of the network. However, in cases of "extreme complexity" the reverse might be true: the network itself would be the simplest expression (or source) of an indescribably complex behavior.[20]

Reasoning in this manner, von Neumann came to believe that the theory of automata demanded a new type of logic, essentially different from the formal, combinatorial logic of mathematics. In his introduction to von Neumann's book, *Theory of Self-Reproducing Automata*, Arthur Burks enumerates several of its general features, and points to areas where von Neumann expected to find it. First, the logic of automata would have to be continuous rather than discrete, analytical rather than combinatorial. Second, it would have to be a "probabilistic logic which would handle component malfunction as an essential and integral part of automata operation."[21] And third, it would most likely have to draw on the resources of thermodynamics and information theory. Von Neumann himself had arrived at a first formulation of these features:

> In fact, there are numerous indications to make us believe that this new system of formal logic will move closer to another discipline which has been little linked in the past with logic. This is thermodynamics, primarily in the form it was received in Boltzmann, and is that part of theoretical physics which comes nearest in some of its aspects to manipulating and measuring information. Its techniques are indeed much more analytical than combinatorial. ("General and Logical Theory," 17)

Not surprisingly, it was the ever-present example of *natural* automata that pushed von Neumann in this direction. This is most directly evident in his concern with "reliability" in computing devices. By his own calculations, the neurons in the brain are some 5,000 times slower than the vacuum tubes used as switching devices in the first electronic calculators, yet they are far more reliable. This is simply because they are far more numerous, and their connections more complicated. Although large numbers of neurons die daily, except in extraordinary conditions our nervous system continues to function normally. Obviously there is a high redundancy, but the system also evinces a high degree of organization. Quite reasonably, von Neumann believed that the "logic" of this organization would throw light on the problem of how to enhance the reliability of computer components, and he broached the problem in many of his lectures and published writings. In *Theory of Self-Reproducing Automata*, for example, he notes that there is a high degree of error tolerance in natural organisms. This flexibility, he speculates, probably requires an "ability of the automaton to watch itself and reorganize itself." It is this high degree of autonomy of parts that allows for a system in which "several organs [are] each capable of taking control in an emergency" (73).

A long paper entitled "Probabilistic Logics and the Synthesis of Reliable Organisms from Unreliable Components" contains von Neumann's most sustained attempt to consider the technical details of the problem of reliability.[22] He begins by defining the automaton as a black box with a finite number of inputs and outputs but restricts the brunt of his considerations to the operational logic of a single-output automaton. Although he refers—somewhat curiously—to these automata as "organs," much of what he says now seems recognizably close to neural net theory. For example, the multiple input lines can have either inhibitory or excitatory values and thus define a threshold "firing" function. He then shows that any single output organ can be replaced by a network built up from organs providing three logical operations (**a** and **b**, **a** or **b**, **a** and **not-b**), or from "majority organs" built on a different set of logical primitives: (**a** and **b**) or (**a** and **c**) or (**b** and **c**). From these majority organs automata can be constructed that exhibit simple memory, simple scaling (counting by twos), and simple learning (stimulus **a** is always followed by stimulus **b**). Turning to the problem of error, he introduces the idea of "multiplexing," that is, of carrying a single message simultaneously on multiple lines, and demonstrates statistically that by using large bundles of lines any degree of reliability for a circuit can be insured. Summarily, what von Neumann outlines here represents a groping attempt to develop an artificial version of the kind of parallel information processing found in natural automata like the brain and thus an alternative to the serial, one-step-at-a-time kind of processing that he himself had proposed for the computing machine EDVAC and that is still referred to as a von Neumann architecture.

His work on automata culminates in the *Theory of Self-Reproducing Automata*, a collection of lectures, notes, and construction plans for a self-reproducing automaton that Von Neumann never completed in his lifetime. These construction plans are described in chapter 4. Here it need only be noted that complexity is considered in relation to biological evolution. In the "Fifth Lecture" von Neumann notes that living organisms are so inherently complicated and improbable that it is a miracle that they appeared at all. The only thing mitigating the effect of this miracle is that they reproduce themselves. But he quickly adds that this process is "actually one better than self-reproduction, for organisms appear to have gotten more elaborate in the course of time" (78). Thus, considered phylogenetically, organisms must be said to have the ability to produce something more complicated than themselves. However, in the case of artificial automata, we are led, at least initially, to the opposite conclusion. An automaton A that can produce automaton B would need a complete description of B in addition to detailed instructions for its construction. Automaton A, therefore, will necessarily be more complicated than B, which will appear degenerative by comparison. There is, he believes, a way around this dilemma (later described in his theory of self-reproduction). The issue of degeneration, nevertheless, leads him to posit a threshold: "There is a minimum number of parts below which complication is degenerative, in the sense that if one automaton makes another the second is less complex than the first, but above which it is possible for an automaton to construct other automata of equal or higher complexity" (80). The exact number depends on how the parts are defined, and von Neumann wisely suggests that it probably cannot be determined in the absence of "some critical examples" (he provides two, discussed in chapter 4). He is certain, however, that

> there is . . . this completely decisive property of complexity, that there exists a critical size below which the process of synthesis is degenerative, but above which the phenomenon of synthesis, if properly arranged, can become explosive, in other words, where syntheses of automata can proceed in such a manner that each automaton will produce other automata which are more complex and of higher potentialities than itself. (80)

Although this "explosive" process has yet to be fully achieved, it now informs the agenda of the new science of ALife.

Ross Ashby's Homeostat

Among the participants in the cybernetic movement, W. Ross Ashby was perhaps closest to von Neumann in appreciating the fundamental importance of complexity. An English psychiatrist turned cyberneticist, Ashby has been unjustly represented as a peripheral latecomer, no doubt because he participated only in the next-to-last, 9th Macy Conference, held in March 1952.[23] On this occasion, however, he made two presentations. The first, a discussion of his homeostat machine, deeply engaged the other participants and immediately validated his reputation as an important theorist. The second, a short exchange on whether or not a mechanical chess player could be taught to outplay its human designer, was of obvious theoretical significance for the yet-to-be-established discipline of artificial intelligence. In both presentations we see the earmarks of his typical approach—to consider the machine, whether existing or not, as a set of specific actions in relation to or as a function of a field of possible actions. This approach, rigorously elaborated in *Design for a Brain* and *An Introduction to Cybernetics*, substantiates his understanding of cybernetics as a comprehensive theory of the machine, one that "envisages a set of possibilities much wider than the actual, and then asks us why the particular case should conform to its usual particular restriction" (*An Introduction*, 3). In these terms Shannon's information theory (which always deals with a *set* of possibilities) is introduced, which will also enable cybernetics to treat complex systems (4–6). Eventually this approach leads Ashby to define the concept of self-organization, making him, according to most accounts, one of the initiators of the second phase of the cybernetic movement.[24]

In Ashby's writings we encounter something of a paradox. *Design for a Brain*, for example, is resolutely mechanistic and deterministic in its approach to adaptive behavior, which is understood as a problem in dynamical systems theory. Yet Ashby always displays resourceful ingenuity in his specific explanations. This is mostly due to his uncanny understanding of how dynamical systems actually work, as when he shows how a complex system like the brain forms a larger, coupled system with the environment. Indeed, as discussed in chapter 7, this application of a dynamical systems perspective has been "rediscovered" and deployed by contemporary roboticists. For Ashby, the "joining" of brain and environment provides the key to adaptive behavior, which turns out to mean the maintenance of stability by keeping the variables of the organism as machine within acceptable limits. In a not quite tongue-in-cheek example, he reminds us that "civilized behavior" depends first on keeping the air temperature in an enclosed space within acceptable limits.

The problem Ashby addresses in his first Macy Conference presentation is "how the organism manages to establish homeostasis in the larger sense, how the learning organism manages to organize its cerebral neuronic equipment so that, however unusual a new environment may be, it can learn to take appropriate action."[25] In other words, when changes in the environment occur, an organism must adapt itself to the new conditions in order to survive. If the temperature drops below a certain level, for example, a rabbit must find shelter and hibernate, or it will freeze to death. In actuality, when we consider both organism and environment as two interacting parts of a larger system, it is not possible or even necessary to include all the variables of their possible interaction, only those that directly effect the stability of the organism. Thus if the organism remains stable throughout a wide range of values of an environmental variable, the latter need not be considered. Ashby calls such a variable a null-function; he also distinguishes among a full-function, which varies all the time, a part-function, which except for certain time intervals remains constant, and a step-function, which, like a relay, changes dramatically when a certain threshold is reached. If fluctuations of an environmental variable produce changes in the organism, the variable will obviously have to be considered as an essential part of the system. When the system becomes unstable, the organism for its part must adapt by some mechanism of corrective feedback and reestablish stability—meaning that the values of the variables must be returned to within an acceptable range. What interests Ashby is not so much the mechanism by which this adaptation is achieved (presumably it is some kind of neural network) as how this mechanism can be modeled by a machine.

At the beginning of the presentation, before Ashby has gone very far, discussion erupts with questions about how the variables of the environment and the organism are to be modeled. There is, nevertheless, general agreement that the environment consists of different kinds of variables, some of which (alone or in combination) produce an observable effect on the animal or organism that it must somehow counter or adjust to. In Ashby's approach, the environment—designated as *E*—is also a transducer, or operator, in the sense that "it converts whatever action comes from the organism into some effect that goes back to the organism" (74). The brain of the organism must therefore act as an inverse operator E^{-1} capable of reacting in such a way that the environmental disturbance is followed by an action that returns the organism to the proper values of its own variables. What kind of mechanism can do this? Although he never says so, Ashby is clearly looking for a way to go beyond a fixed repertoire of stimulus-response reactions, that is, beyond the resources of behavioral psychology:

> The fundamental problem is one of organization, of finding the appropriate switching pattern. Clearly, the instructions for what is appropriate must come, ultimately, from the environment, for what is right for one environment may be wrong for another. The problem is how the information from the environment can be used to adjust the switching pattern. What the organism needs is a system or method, which, if followed blindly, will almost always result in the switching pattern changing from "inappropriate" to "appropriate." I have given reasons for thinking that there is only one way in which this can be done.[26] The switching must be arranged, at first, at random, and then there must be corrective feedback from the essential variables into the main network, such that if any essential variable goes outside its proper limits, a random, disruptive effect is to be thrown into the network. I believe that this method is practical with biological material and is also effective, in the sense that it will always tend automatically to find an inverse operator, an E^{-1}. (76)

In order to explore these issues experimentally, Ashby built a machine he called the homeostat, described in his presentation (see fig. 1.2). The machine consists of four units connected to one another. Each unit contains an electromagnetic circuit connected to a needle that indicates a

Figure 1.2
The homeostat. W. Ross Ashby, *Design for a Brain* (London: Chapman & Hall, 1972; orig. pub. 1952), 101.

range of deviations from the vertical (zero) position. A feedback circuit attempts to keep the unit within a normal operating range, defined as a needle position of plus or minus forty-five degrees. If a disturbance pushes the needle past forty-five degrees, it trips a relay that causes a device called the "uniselector" to reconfigure the unit's entire circuit according to a completely different set of values, both in terms of the polarity of the voltage and the resistances of the circuit. These new values are taken from a table of random numbers; if a particular unit goes unstable, its entire configuration of settings are randomly changed. Moreover, since the output from each unit is connected to every other unit, the effect of this change is propagated to the other units and thus throughout the entire system. As Ashby notes, multiple connections among the unit give the machine as a whole over three hundred thousand combinatorial possibilities (95). In many or even most cases the new setting causes one or more of the other units to go unstable, thus causing it or them to jump randomly to another configuration. The process repeats itself in all four units until the overall system finds a stable configuration.

Ashby points out that if the uniselectors in some of the units are "locked," they can be regarded as the environment, while the remaining units can be regarded as the "brain" struggling to control changes in the environment by searching randomly for a stable combination of the configurations in all the units, that is, for the system as a whole. Not surprisingly, many of the conference participants voiced difficulties in seeing how this randomized mechanism models the organism's adaptation to changing variables in the environment. As Julian Bigelow remarks, "It may be a beautiful replica of something, but heaven only knows what" (95). Before considering what it is exactly that the homeostat does model, an essential feature of the machine must be described in more detail, since it underlies what may be the most important aspect of Ashby's presentation.

In response to Bigelow's questions about the machine's feedback loops, Ashby explains that the homeostat "is really a machine within a machine." This is necessary because it must deal with two kinds of variables. First, there is the "continuously fluctuating type" to which the machine responds with small corrective movements. These corrections of deviations from the normal state Ashby compares to the small movements made by an airplane's automatic pilot, or to "the trip made by a rat in a cage when, being thirsty, it goes to the water bottle and has a drink" (96). These actions entail no learning, and no change from one form of behavior to another occurs. However, such a change would indeed occur if the design of the automatic pilot were altered. This is what the homeostat actually does, since it is capable of making both kinds of changes:

> What happens is that the resistances on the uniselectors are fixed and constant, temporarily. On this basis, the feedbacks can show, by the movement of the needles, whether the whole is stable or unstable. The changes at this stage are continuous and correspond to the continuous fluctuations of the automatic pilot. Then comes, perhaps, the other change; if the resistances make the feedbacks wrong, making the whole unstable, the uniselector moves to a new position and stops there. (This would correspond to making a change in the design of the pilot.) Then the continuous changes occur again, testing whether the new pattern of feedbacks is satisfactory. It is clearly essential, in principle, that the resistances that determine the feedbacks should change as step-functions; they must change sharply, and then they must stay constant while the small fluctuations test whether the feedbacks they provide are satisfactory. All design of machinery must go in stages: make a model, test it, change the design, test again, make a further change, test again, and so on. The homeostat does just that. (96)

As Ashby emphasizes, the homeostat thus performs two activities. On the one hand it behaves like a "properly connected thermostat," reacting to disturbances and by negative feedback restoring itself to its optimal position. On the other hand, when it is unable to restore itself, it changes from one set of feedbacks, which it has found to be unstable, to another set. This second type of change is of a different order than the first and differs also in the means by which the machine "converts itself from an unstable system to a stable system" (98). In *Design for a Brain* (chapter 7) Ashby calls such a system an "ultrastable system."

In the exchanges among the conference participants the question of what the homeostat can be said to model remains uppermost. In retrospect, what is valuable about this discussion is how it hints at an incipient forking in what we might call the dominant discursive framework, for it is apparent that most of the interlocutors simply assume that the value of the machine resides in its capacity as a model. Julian Bigelow expresses this position most directly: "Sir, in what way do you think of the random discovery of an equilibrium by this machine as comparable to a learning process?" To which Ashby responds, "I don't think it matters. Your opinion is as good as mine." Why Ashby should say this is not apparent, although he quickly adds that, as the machine's inventor, he is not "going to stick up for it and say I think it is homologous" (103). We might take these statements to suggest that for Ashby two things are involved: first, the fact that the machine's capacity to model a particular natural process does not exhaust its interest; and second, with no evident or agreed upon understanding of how the process of learning works and what it entails, the group is not yet equipped to assess the machine's value in these terms (i.e., it is still a matter of opinion).

As expected, the discussion evolves along the second track, and with fruitful results. Specifically, as the group moves toward a more precise conception of learning, Ashby admits that his machine possesses a "serious fault": "If you disconnect the environment and give it a second environment, and then bring the first environment back again, its memory of the first environment is totally lost" (104). Ralph W. Gerard remarks that this makes the homeostat very similar to the electromechanical mouse that Shannon had presented at the 8th Macy Conference—after one complete exploration of a maze, the mouse would run through it perfectly. While all were agreed that this ability made it a learning device, Warren McCulloch confirms Gerard's point: if put into a second maze, the mouse would quickly learn it, but at the same time forget the first maze. In both cases, these machines with limited memory thus raise questions about learning and how it should be defined. The session concludes with an unresolved discussion of the virtues and limitations of what Walter Pitts calls "random machines" in comparison with the behavior of real animals.

To pursue another way of thinking about Ashby's machine, let us return to his assertion that "all design of machinery must go in stages: make a model, test it, change the design, test again, make a further change, test again, and so on. The homeostat does just that" (96). Viewed in terms of what it does, the homeostat is simply a machine that adapts to changing environmental conditions by repeatedly changing and testing its own design until it reaches a state of equilibrium. This is the action it performs as a material device. Curiously, its capacity for what Ashby calls, in *Design for a Brain*, "self-reorganization" (107) is what enables it both to realize its purpose and to remain utterly unpredictable from one moment to the next. Thus although we might know what it does and how it does it, the behavior of this automaton would continually surprise us.

Since there were very few if any machines that could do anything like this at the time, we may well wonder why the participants could not appreciate the machine in these terms. Why did they have to look for its justification in a model? The simple answer, of course, is that the assumption of a model is explicit in Ashby's presentation, and operative in *Design for a Brain* as well. Even so, a careful examination of what Ashby actually says reveals an almost systematic reluctance to press the analogy. Indeed, one could argue that for Ashby there is no analogy: the brain, like the homeostat, is simply a material switching device, connected through sensors and effectuators with the forces of the environment. It does not "represent" the world but provides a complex, dynamic way of engaging it. What Ashby insists upon more than anything in the presentation is what the homeostat actually does and is capable of doing. At the same time, we certainly cannot say that the other participants were wrong in trying to understand the homeostat's behavior in relation to a living organism's capacity to adapt. How then should we understand these divergent ways of characterizing Ashby's machine?

In several essays that extend to the early phase of cybernetics the argument of his book, *The Mangle of Practice: Time, Agency, and Science* (1995), Andrew Pickering suggests an intriguing framework for understanding this divergence and conflict of views. He observes that "traditionally, science studies has operated in what I called the *representational idiom*, meaning that it has taken for granted that science is, above all, about representing the world, mapping it, producing articulated knowledge of it."[27] Thus science studies is essentially "a venture in *epistemology*." Pickering finds, however, that this approach is inadequate to the "analysis of [scientific] practice" and argues therefore that "we need to move towards *ontology* and what I call the *performative idiom*—a decentred perspective that is concerned with agency—doing things in the world—and with the emergent interplay of human and material agency" (1). Cybernetics, and particularly the work of the English cyberneticists, Pickering now realizes, "is all about this shift from epistemology to ontology, from representation to performativity, agency and emergence, not in the analysis of science but within the body of science itself" (2). In these terms, the ambiguity in Ashby's discourse and the confusion among the Macy participants makes perfect sense: both Ashby and his interlocutors are caught up in a moment of transition from one discursive framework to another, contradictorily viewing the homeostat both as a model according to the representational idiom and, according to the performative idiom, an ontologically new kind of a machine capable of surprisingly complex behavior. As Pickering notes, in relation to industrial machines typical of its day the homeostat can be said to possess "a kind of agency—it did things in the world that sprang, as it were, from inside itself, rather than having to be fully specified from outside in advance" (4). It is precisely this new form of agency that makes comparisons between the new cybernetic machines and living organisms inevitable, while also obscuring the singular ontology of these new machines.

In later chapters I explicitly characterize ALife, contemporary robotics, and even artificial intelligence in performative terms, arguing that in different ways—both conceptually and technologically—they instantiate a new kind of science that produces the very objects that they are (purportedly) only studying. However, it did not occur to me while writing these chapters that this shift could be said to have occurred earlier, in the machines constructed by the first cyberneticists. But rather than pursue this important argument here, in the terms Pickering adumbrates, I want to consider another instance of a cybernetic machine, the mobile tortoises constructed by W. Grey Walter.[28] Like his compatriot Ashby, Grey Walter has generally been considered a marginal figure in the early cybernetics movement. And, as in the case of Ashby, this neglect is entirely unjustified. Indeed, in so far as these cybernetic tortoises exhibited a new kind of machinic complexity, the rehabilitation of Grey Walter's role in the movement is necessary if there is to be a proper assessment of the complexity and richness of cybernetics itself.

Grey Walter's Tortoises

Even if Grey Walter had never involved himself in cybernetics, he would still deserve a notable place in biomedical engineering for his pioneering achievements in the development of electroencephalography (EEG) and in modern technology for his work on radar. Both before and after the Second World War he made many important discoveries in how to measure and interpret the oscillating electrical fields generated by the brain: he located the source of the alpha rhythm (8–12 Hz) in the occipital lobe; he discovered delta waves (1–2 Hz) and developed a method of using them to locate brain tumors and foci of brain damage; and he built the first device—basically an ink-writing oscillograph—used to register the frequency of an EEG trace. He also developed a method of measuring what is called the readiness potential in human subjects, which permits an observer to predict a subject's response about a half to one second before the subject is aware of any intention to act. As Walter J. Freeman notes, this cerebral phenomenon can be interpreted as evidence "that intentional actions are initiated before awareness of such actions emerges, and that consciousness is involved in judging the values of actions rather than in the execution of them."[29] Another device Walter constructed, which he called the "toposcope," allowed him to observe the amplitude and phase differences of alpha rhythms as they change over time, providing a means of doing time series analysis of alpha activity. Using what Freeman beautifully describes as "cinemas of an array of 22 oscilloscopes," the toposcope "visualized the spread of alpha waves across the surface of the brain in ways resembling the ebb and flow of tidal waves around the earth" (2). Walter's hypothesis, which remains controversial though not yet superseded, is that alpha activity, which is only observed when the subject is at rest with his or her eyes closed and disappears with the onset of any focused activity, is actually a "scanning" by the brain in search of local centers of activity: once it locates a "target" in the cortex, it stops. During the war, Walter also made a crucially important contribution to the development of radar technology by helping to develop a scanning mechanism known as the "plan position indicator." We are all familiar with this type of radar screen, which is still commonly used on ships, submarines, and in air traffic control towers: on the screen an electron beam shaped like the spoke of a wheel sweeps counterclockwise at the screen's refresh rate. With each sweep the "target" appears as a bright spot of light, its position and direction of movement clearly displayed.

Given these interests and accomplishments, it is hardly surprising that in the years after the war Grey Walter should turn his attention to the construction of devices that imitate or model goal-seeking and scanning activities. As he recalls in his book *The Living Brain*, the war coupled these two activities in the form of guided missiles and radar detection.[30] The combination of goal-seeking and scanning, he reasoned, would yield "the essential mechanical conception of a working model that would behave like a very simple animal." This conception, moreover, would test his theory that it is not so much the "multiplicity of units [that] is ... responsible for the elaboration of cerebral functions as the richness of their interconnections" (125). In two articles published in *Scientific American* he calls his constructions a new genus of "mechanical tortoises," and provides not only details of their construction but analysis of their complex behavior. In the first, "An Imitation of Life," he describes Elmer and Elsie, two examples of the genus *Machina speculatrix*, and in the second, "A Machine That Learns," a new species, *Machina docilis*, formed by "grafting" an electronic circuit called the Conditioned Reflex Analogue (CORA) onto *M. speculatrix*. This added device allows *M. docilis* to learn new behaviors (*docilis* means "easily taught"), as well as to forget them if they are not reinforced.[31]

These creatures were simply constructed, with three wheels in a tricycle arrangement, two motors for steering and motive power, a light and bump censor, an electronic circuit and two batteries (see fig. 1.3). A plastic shell fit over the chassis—hence their resemblance to tortoises. Amazingly, Grey Walter discovered how to connect these simple elements in ways that produced efficient but complex and unpredictable behavior. Be-

Figure 1.3
Grey Walter's tortoise. Owen Holland, "Grey Walter: The Pioneer of Real Artificial Life," in *Artificial Life V* (Cambridge, Mass.: MIT Press, 1997), 36. Photograph courtesy of Owen Holland.

cause the light censor was connected to the steering mechanism, the tortoise would move out into the environment in cycloid spirals scanning for a "target" light; when a light source was located, the steering mechanism was altered so that the tortoise could home in on it more directly. When the tortoise arrived at a short distance from the light, or when the light reached a specific intensity, a feedback circuit would cause the tortoise to back away rapidly. If there were a second light source, the tortoise would scurry back and forth between the two. When its batteries weakened below a certain level, it would respond to light in yet another way, returning to its "hutch" to recharge its batteries. This more complex behavior was accomplished by means of an indicator light on its shell that turned on or off depending on whether its motor was running. A connection with another part of the circuit caused it to plug into or be released from the battery recharger in the hutch. This indicator light in turn resulted in more complex behavior. If the tortoise encountered a mirror, its indicator light would flash on and off as its motor turned on and off, causing it to "flicker and jig at its reflection," as Walter put it, in a manner suggesting a capacity for self-recognition. Moreover, if it encountered another tortoise, as when Elmer and Elsie would meet, both tortoises would enter into a complicated dance of "mutual oscillation, leading finally to a stately retreat." If it encountered any obstacle during these various activities, a ring bump censor would trigger an amplifier and several relays that blocked the light censor circuit and transformed the tortoise's gait "into a succession of butts, withdrawals and sidesteps until the interference is either pushed aside or circumvented" ("An Imitation of Life," 45). Since these oscillations persisted for a full second, the tortoise could free itself and move clear of the obstacle.

Unexpectedly autonomous, self-regulating, and unpredictable, *M. speculatrix* was thus capable of "exploratory, speculative behavior," as Grey Walter intended its species name to suggest. When he added the CORA circuit to *M. speculatrix*, the machine's capacity to learn a conditioned reflex further increased its already remarkable behavioral repertoire, and a second generation of tortoises, *M. docilis*, came into being. In one sense this capacity was simply a mechanical equivalent of Pavlov's famous experiment: a dog salivates when food is placed before it; if a bell is also rung at the same time, after a number of repetitions the dog will salivate at the sound of the ringing bell, even in the absence of food. In the case of *M. docilis*, these conditioned reflexes were accomplished with light, touch, and sound censors, and of course the necessary additional feedback circuit (reproduced in "The Machine That Learns"). In several experiments a whistle sound was made to replace the stimulus to the light and touch censors. As a result, if *M. docilis* approached an obstacle and "heard" a warning whistle it would immediately stop and withdraw.

This conditioning led to several interesting complications. For example, noise from the tortoise's motors often interfered with its reception of the whistle sound. One solution was to alter the circuit so that the sound switched off the motors momentarily, producing a "freezing" effect analogous to the way some animals play possum when they hear a strange noise. But this response interfered with the process of conditioned learning—in the instance, for example, where sound comes to "mean" light—and had to be inhibited. Grey Walter saw here an example of how an "instinctive" effect would have to be suppressed in order to bring about a positive conditioning. Furthermore, by adding a second learning circuit, it became very easy to produce conflicts and interferences that amounted to what Walter called "experimental neurosis." In one instance, when stimulated simultaneously with sound and light, a tortoise became incapable of reentering its hutch when its batteries ran low. In fact, experiments with multiple learning circuits led Walter to predict a weakness or limit in elaborate systems: "Extreme plasticity cannot be gained without some loss of stability." Specifically, the more learning circuits or paths of association, the more unstable the system as a whole. Generalizing to the human condition, Walter concludes that it is "no wonder that the incidence of neuropsychiatric complaints marches with intellectual attainment and social complexity" ("A Machine That Learns," 63).

In *The Living Brain* and his articles Grey Walter establishes his interest in constructing machines that are lifelike in the basic sense of being purposeful, independent, and spontaneous ("A Machine That Learns," 45). *M. speculatrix*, he asserts, is "designed to illustrate ... the uncertainty, randomness, free will or independence so strikingly absent in most well-designed machines" (44); and *M. docilis* "behaves astonishingly like an animal" (*The Living Brain*, 179). What is most striking about this lifelike behavior, however, is the extreme economy of means by which it is generated. Grey Walter himself was acutely aware of the relationship between simplicity of means and complexity of results in his work. In fact, his explanatory comments adumbrate a rudimentary principle of what might be called a "behavioral design philosophy." As already mentioned, constructing a mechanical device that combined goal seeking and scanning "held promise of demonstrating, or at least testing the validity of, the theory that [a] multiplicity of units is not so much responsible for the elaboration of cerebral functions, as the richness of their interconnection" (125). In "An Imitation of Life" he is more explicit:

> The number of components in the device [*M. speculatrix*] was deliberately restricted to two in order to discover what degree of complexity of behavior and independence could be achieved with the smallest number of elements connected in a system providing the greatest possible number of interconnections. From the theoretical standpoint two elements equivalent to circuits in the nervous system can exist in six modes; if one is called A and the other B, we can distinguish A, B, A + B, A → B, B → A, A ⇔ B as possible dynamic forms. To indicate the variety of behavior possible for even so simple a system as this, one need only mention that six elements would be more than enough to form a system which would provide a new pattern every tenth of a second for 280 years—four times the human lifetime of 70 years! It is unlikely that the number of perceptible functional elements of the human brain is anything like the total number of nerve cells; it is more likely to be of the order of 1,000. But even if it were only 10, this number of elements could provide enough variety for a lifetime of experience for all the men who ever lived or will be born if mankind survives a thousand million years. (44)

While the numbers may not seem exactly right, the basic idea is clear: connecting simple elements in multiple ways generates complexity. As

Grey Walter acknowledges, this approach allows him simply to black box the daunting internal intricacy of the biological brain. But it also allows him to construct tortoises that exhibit both purpose and unpredictability.

In retrospect, it appears that the tortoises were the first true autonomous robots as well as the first serious attempt to produce real artificial life.[32] No one can attest to the first part of the claim better than the roboticist Rodney Brooks, who founded behavior-based robotics in the late 1980s. Devoting several pages to Grey Walter's constructions in his recent book, *Flesh and Machines: How Robots Will Change Us*, Brooks is most impressed by how often Walter's robots exhibit emergent behavior, "where multiple behaviors couple with the environment to produce behaviors that are more than simple strings or suppositions of the elementary behaviors."[33] Brooks attributes this complexity to the "nonlinear coupling" of different elements with the environment, as when the light and motor function together in one way under certain conditions and in another way under others. Owen Holland points out a similar instance when he observes that "Grey Walter's architecture responded to sensory input—the [tortoise] shell being touched—by changing the pattern of interconnection between its neuron-level elements to produce a fundamentally different circuit—an oscillator rather than a two stage amplifier" ("Grey Walter," 41). This "rich exploitation of interconnectivity" not only underlies the construction of the tortoise but is also found in natural creatures. Holland notes, for example, that "in the crustacean stomato-gastric system . . . stimuli external to the network modulate connectivity to produce altered networks with radically different characteristics" (41). In this instance, however, Walter does not imitate or model nature. One might say, rather, that Walter pursues his design principle at the material level of the specific components and mechanisms he works with, and that nature often does the same—hence a parallelism rather than a form of biomimeticism.

While the focus thus far has been on how the theoretical work and material constructions of von Neumann, Ashby, and Walter produce complexity, it has also been apparent that in this body of work the boundary line between nature and artificial machines—living and nonliving matter—is no longer well defined or rigidly fixed. How should we understand this boundary loosening and complication? Andrew Pickering argues that the nature-machine opposition is monstrously broached by cybernetics and that we need to shift our perspective from a representational and mimetic understanding of these machines to a performative one—to think about ontology rather than epistemology. But would this simply replace one dualism with another? Pickering claims that Grey Walter himself "recognized that the tortoises could be seen as performative technological artifacts as well as models of the brain" ("The Tortoise against Modernity," 9). As an adaptive system, the tortoise would thus assume its place within the history of technology, specifically of servomechanisms and feedback devices, within which its technical innovations would be assessed. And similarly for his theory of the brain. However, Pickering's summary description of this theory is made in wholly technical, artifactual terms:

> The tortoise's brain (the capacitors, relays and tubes), like the homeostat, was a performative and embodied one, a brain continuous, as it were, with the tortoise's sensory and motor apparatus. The brain functioned as a switchboard between the motors and sensors, and not, importantly, as a hierarchical controller running the show from above. In the absence of the sensors or the motors, the tortoise's brain was just a handful of inert components, having no interesting properties in themselves. ("The Tortoise against Modernity," 6)

In Grey Walter's model of the brain, in other words, agency is fully embodied in a material set of parts and connections. Yet what is missing from this account is the necessary emphasis on the *complexity* of these connections. For it is precisely this complexity of connection that makes "a handful of inert components" yield behavior that is interesting in itself. In these terms, Pickering is surely correct: this isn't simply another dualism. But what exactly is it? We have adjectives and nouns ("performative," "ontology," "machine") but no name for this unifying view, which combines strands of mechanics and materialism with a new and still incomplete account of what makes the machine behave in a lifelike manner. Perhaps the cyberneticists' failure to name this new and innovative relationship of machine to natural living systems explains the ease with which the movement slipped out of view, despite its revolutionary significance. In any event, it is tempting to think that the concept that might have made the difference is self-organization, which arose within cybernetics but remained on its conceptual periphery.

Self-Organizing Machines

First introduced by Ross Ashby in a short article published in 1947, the concept of self-organization has come to enjoy a rich provenance and wide range of applications in contemporary science.[34] In the absence of a specific context, the term usually designates a system that spontaneously—that is, without external guidance or control—moves from a random, or less unorganized, state to one that exhibits a more orderly pattern of behavior. In 1947 the concept had little or no effect and Ashby doesn't bother to mention it in either of his subsequent books. However, things had changed by the late 1950s, when the idea of self-organization was taken up by Heinz von Foerster as a focus of theoretical activities at the University of Illinois's Biological Computer Laboratory, which von Foerster founded and directed. Under his auspices self-organization became the unifying theme of three conferences.[35] From this point on, however, the term *self-organizing* becomes more difficult to track, as it begins to appear in a number of distinct fields—physics, biology, general systems theory—where its various elaborations take on a life of their own no longer unified under the common framework of cybernetics. Indeed, by this point cybernetics itself could be said to have dissolved as a coherent movement. It partially reformed at the Biological Computer Laboratory (Ashby himself was there from 1969 to 1970), albeit with fewer participants and a less broadly conceived set of ideas. The new agenda, which is often taken to mark the beginning of the cybernetic movement's second phase, focuses on self-organization, self-referential systems, and the role of the observer.[36]

Ashby's first paper on self-organization is simply a mathematical demonstration that a self-organizing system or machine is possible. Doubts arise from the apparent contradiction: how can the system be both "(a) a strictly determinate physico-chemical system and (b) ... undergo 'self-induced' internal reorganizations resulting in changes of behavior" ("Principles of the Self-Organizing Dynamic System," 125). If the change comes from within, then its organization cannot be described as a set of states determined by functions that define this organization. Simply put, you can't have a function that both defines and is changed by the state of the organization. If S defines a set of functions (f_1, f_2, f_3, etc.) but f_3 can also change the organization of S, then S is a function of f_3, which renders the nomenclature illogical. On the other hand, if the change comes from without, then the system is no longer *self*-organizing. Ashby's solution is really a logico-mathematical version of the "machine within a machine" embodied in his homeostat. Basically, the system or machine will have to contain two distinct organizations, "each of which is absolute [i.e., completely determinate] if considered by itself" (128). What connects them is a single step-function of time with two values. Assuming that there are finite intervals of time between the change from one value to the other, then a spontaneous change of organization can occur. In other words, during a first period of time the system has one organization, and during a second it has another. This implies—although Ashby doesn't consider this aspect—that inasmuch as time becomes an internal determinant of the system's organization, it is no longer a Newtonian machine but "lives" in Bergsonian time.

In his second paper, where the context is a rigorous and searching examination of the concept of organization itself, Ashby distinguishes two meanings of the term "self-organizing system."[37] First, a system can be said to be self-organizing if it encompasses parts that are separate and independent and that then join. In a strict sense, however, this means that the system is simply "changing from unorganized to organized." Second, the term means that the system is "changing from a bad organization to a good one." Deploying an argument similar to the one used in his first paper, Ashby asserts that "no machine can be self-organizing in this sense" (267). How then does he account for the fact that "the homeostat rearranges its own wiring" and would therefore seem to be an instance of a self-organizing system? Curiously, instead of repeating his earlier "machine within a machine" argument, he now separates the system into two parts, S and $\alpha(t)$, the latter being a function of time with two values, one before and the other after the change in organization. In effect, the notion of "self" is "enlarged to include this variable α" by making the latter a separate machine coupled to S. As Ashby puts it, "Thus the appearance of being 'self-organizing' can be given only by the machine S being coupled to another machine (of one part).... Then the part S can be 'self-organizing' within the whole $S + \alpha$" (269). Paradoxically, then, Ashby both argues for the logical impossibility of self-organization and spells out the terms by which this impossibility can be overcome.

Commenting on Ashby's essay, the physicist Cosma Shalizi remarks that this is not what most people have in mind when they speak about self-organization.[38] Rather, he suggests, what Ashby is really getting at is how there can be a "selection of states" by the organization of the system. Shalizi then adds that "a system would be self-organizing if it takes a flat, even distribution of states into a peaked, non-uniform one." In other words, the entropy of a self-organizing system would have to decrease. Written over forty years after Ashby, Shalizi's rephrasing of the problem in these terms now strikes us as self-evident. We may be surprised, therefore, to discover that this is precisely the approach that Heinz von Foerster takes in his essay, "On Self-Organizing Systems and Their Environments."[39] For von Foerster, a self-organizing system is one whose "internal order" increases over time. This immediately raises the double problem of how this order is to be measured and how the boundary between the system and the environment is to be defined and located. The second problem is provisionally solved by defining the boundary "at any instant of time as being the envelope of that region in space which shows the desired increase in order" (36). For measuring order, von Foerster finds that Claude Shannon's definition of "redundancy" in a communication system is "tailor-made." In Shannon's formula,

$$R = 1 - H/H_m,$$

where R is the measure of redundancy and H/H_m the ratio of the entropy H of an information source to its maximum value H_m. Accordingly, if the system is in a state of maximum disorder (i.e., H is equal to H_m), then R is equal to zero—there is no redundancy and therefore no order. If, however, the elements in the system are arranged in a such a way that, "given one element, the position of all other elements are determined" (37), then the system's entropy H (which is really the degree of uncertainty about these elements) vanishes to zero. R then becomes unity, indicating perfect order. Summarily, "Since the entropy of the system is dependent upon the probability distribution of the elements to be found in certain distinguishable states, it is clear that [in a self-organizing system] this probability distribution must change such that H is reduced" (38).

The formula thus leads to a simple criterion: if the system is self-organizing, then the rate of change of R should be positive (i.e., $\delta R/\delta t > 0$). To apply the formula, however, R must be computed for both the system and the environment, since their respective entropies are coupled. Since there are several different ways for the system's entropy to decrease in relation to the entropy of the environment, von Foerster refers to the agent responsible for the changes in the former as the "internal demon" and the agent responsible for changes in the latter as the "external demon." These two demons work interdependently, in terms of both their efforts and results. For the system to be self-organizing, the criterion that must be satisfied is now given by the formula shown in figure 1.4. This criterion, von Foerster asserts, is not at all difficult to fulfill.

Reflecting further on self-organizing systems, von Foerster considers Erwin Schrödinger's observations about order in the latter's book *What Is Life?* (1944). Schrödinger is particularly struck by the high degree of order exhibited by the genes, or what he calls the "hereditary code-scripts," despite their exposure to the relatively high heat of "thermal agitation." This leads him to remark that there are two mechanisms that produce order: the first, a statistical mechanism producing "order from disorder," is "the magnificent order of exact physical law coming forth from atomic

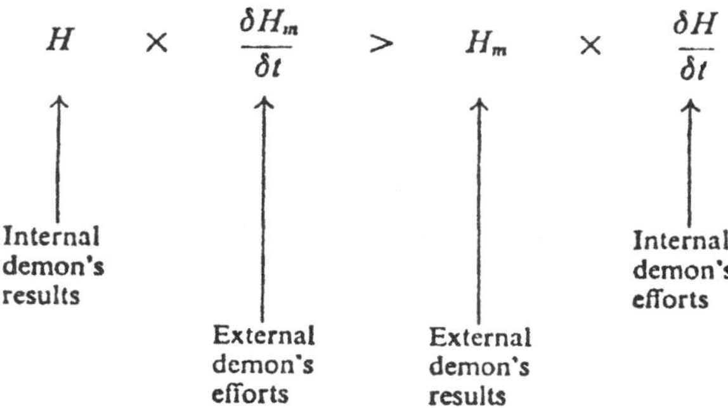

Figure 1.4
Von Foerster's formula. Heinz von Foerster, "On Self-Organizing Systems and Their Environments," in *Observing Systems* (Salinas, Calif.: Intersystems Publications, 1984), 13.

and molecular disorder"; the second, a less familiar mechanism that produces "order from order," holds "the real clue to the understanding of life," since "what an organism feeds upon is negative entropy" (42–43). In von Foerster's view, however, self-organizing systems may also provide another clue. The principle he now proposes, though it may sound like Schrödinger's mechanism of "order from disorder," is actually quite different, and von Foerster calls it "the order from noise" principle. Thus he states, "In my restaurant self-organizing systems feed not only upon order, they also find noise on the menu." He concludes with what he describes as a "trivial ... but amusing example" (43): if one repeatedly shakes a box filled with cubes, each of which is magnetized on one of its faces, instead of a series of random assemblages a highly intricate structure will eventually emerge.

Given the fruitfulness of the idea that a complex order can emerge from a system's exposure to "noise" or other disturbances, von Foerster's illustration can only seem disappointing.[40] Or rather, viewed in the light of the sea changes that the concept of self-organization would undergo over the next twenty years, von Foerster's proposal—and some would say the same about Ashby's theorizing—can have at most an anticipatory value. These sea changes followed from the interaction and relay of two series of developments. On the one hand, Ilya Prigogine (chemistry), Hermann Haken (physics), and Manfred Eigen (biology) made groundbreaking empirical discoveries of self-organizing systems, in which instabilities resulting from the amplification of positive feedback loops spontaneously create more complex forms of organization. On the other hand, Steven Smale, René Thom, Benoit Mandelbrot, and Mitchell Feigenbaum, to name a few, made discoveries in topology and nonlinear mathematics that led to a complete revamping of dynamical systems theory. This story, which involves the discovery of how nonlinear factors can produce deterministically chaotic systems, is now fairly familiar. One simple but telling difference this sea change has made in dynamical systems theory is that the concept of the *attractor* has replaced notions like stability and equilibrium. As a system moves toward increasing instability it may reach a point where two sets of values for its variables—and hence two different states—are equally viable. (In its phase portrait as a dynamical system this is referred to as a bifurcation.) But which state will it "choose"? There is no way of knowing since the system's behavior has become indeterminate at this point. The very presence of a bifurcation means that the system is falling under the influence of a different attractor and thus undergoing a dynamical change as a whole. These discoveries and the conceptual tools developed to describe them will be discussed in greater depth in chapter 3. The main point here is that they entirely transformed the conceptual landscape on which cybernetics had arisen.

The Last Cyberneticist

Unfortunately, the sea change that eventually gave birth to chaos theory, or more precisely, nonlinear dynamical systems theory, also tended to eclipse recognition of the degree to which cybernetics functioned as a kind of historical a priori, or condition of possibility, clearing the ground and providing a necessary initial framework for future developments, among which I would include AI and ALife, computer simulations of complex adaptive systems, and subsumption architecture in contemporary robotics. What then happened to cybernetics, and why is it not part of the general curriculum in scientific training? Kevin Kelly has proposed three theories.[41] First, the birth of AI in 1956 drew away most of the funding supplied by both the government and the university and thus also the graduate students. Second, "cybernetics was a victim of batch-mode computing" (453), and the cyberneticists never had easy, real-time access to computers; unfortunately, the mainframes then available were guarded by a priesthood not especially receptive to their new brand of science. In fact, when relatively easy access did come (especially with desktop computers in the late 1970s), the discoveries of chaos theory and artificial life soon followed. Third, von Foerster shifted attention to "observing systems" and "the cybernetics of cybernetics."[42] In Kelly's view, including the observer in the system as part of a larger metasystem proved useful for family therapy and sociologists interested in the effects of observing systems, but it also meant that the main constituency of cybernetics came to consist of therapists, sociologists, and political scientists. As a result, he concludes, cybernetics "died of dry rot" (454).

While the full story is of course more complicated, there is some truth in Kelly's first two theories,[43] which are really two aspects of the same theory. Because there was not yet a clear-cut separation between hardware and software, the cyberneticists understood computers as self-controlling computational devices made up of switching devices and feedback circuits rather than multitasking machines capable of running different programs. To be sure, they were fully cognizant of how the mathematical and logical properties of these new machines made them unlike previous machines. The work of von Neumann in particular was essential for the development of this machine from a high-speed calculator to a stored-program computer and hence a universal symbol processor.[44] Yet on the whole the cyberneticists did not participate in the shift in interest from machine to program that would characterize early AI.

Significantly, an early sign of this shift could be glimpsed at the conference on self-organizing systems at which von Foerster presented "On Self-Organizing Systems and their Environments." At this same conference, A. Newell, J. C. Shaw, and H. A. Simon presented "A Variety of Intelligent Learning in a General Problem Solver." The GPS, as it was called, was a computer program that incorporated heuristic strategies based on means-ends analysis to solve problems. It was a further development of their earlier program, the Logic Theorist, which could generate original proofs of theorems in symbolic logic. When first presented at the Summer Dartmouth Conference in 1956, which is generally taken to be the official founding moment of AI as a scientific discipline, the Logic Theorist was perceived to be an undeniably persuasive exhibit of a "thinking machine." Moreover, with their notions of information-processing psychology and a "physical symbol system," Newell and Simon provided the theoretical foundations of early AI. Armed with the belief that all cognitive processes could be simulated by a computer, they dismissed cybernetic machines as irrelevant:

> Although the tortoises [of Grey Walter] aroused considerable interest and have been further developed by other investigators, they appear no longer to be in the main line of evolution of psychological simulations. The interesting properties they exhibited could be rather easily simulated by digital computers, and the digital simulation lent itself to greater elaboration than did the analogue devices.[45]

Whether or not the emergent behavior of Walter's tortoises could be so easily simulated might be contested, but there is no doubt that Newell and Simon's claims for the proper approach to constructing artificial intelligence constituted a clear rupture and turning point. These claims will be considered in detail in chapter 6; meanwhile, it is enough to note that for the nascent field of AI, cybernetic machines were found wanting simply because they did not manipulate symbol structures in a computer program. In other words, they did not control their own behavior by means of abstract, "disembodied" representations. On the contrary, in the terms that the contemporary roboticist Rodney Brooks later valorizes, these cybernetic machines were fully embodied and situated in the world. Thus those who pursued AI research from the late 1950s through the 1970s would have found Claude Shannon's design for a chess-playing program and Arthur Samuel's checker-playing program far more interesting.[46]

This point is especially important in the present cultural context. Particularly in the humanities, when cybernetics is spoken of at all, it is asserted or simply assumed that as a historical movement cybernetics is responsible for the view that information is somehow disembodied, in the sense that it exists independently of any particular material substrate. This view, I claim, is misleading, and will not survive an attentive consideration of the constructions and published writings of the first-generation cyberneticists. As we have seen, for at least three of the major participants in the movement, information theory was neither central nor at all understood in this sense. Thus the mischaracterization seems partly due to the nearly exclusive attention usually given to Wiener and Shannon, whose abstract mathematical models can easily be taken out of context.[47] Even so, Shannon makes it clear that information is a measurable, statistical property of the symbols that make up a message. These symbols always exist in a material medium, whether as ink on paper or voltage differences in an electronic circuit. To claim, therefore, that information is disembodied makes no more sense than to say that the temperature of a gas is disembodied. Both, rather, are abstracted from the entities they are meant to measure, like all mathematical functions that refer to properties of material elements. Because the measure of information transmitted by or stored in a particular set of symbols is derived from a statistical correlation between this set and all those symbols that *might have been selected instead* does not mean that information exists independently of the material nature of the symbols; it only means that its measurement is not a function of the latter. Nor is the measurement of information a function of the meaning of the symbols, an aspect of Shannon's theory that leads

Warren Weaver (in his introduction) to a somewhat defensive explanation. Yet this is also something of a false problem. Lines and dots on a piece of paper are meaningless until it is realized that they *might be* a source of information because they could indicate where the treasure is buried. While it is true that a random or meaningless concatenation of symbols still contains a measurable amount of information, information acquires its status and value as information only because there is an assumed correlation between a message composed from a set of discrete symbols and physical events and processes in the world (i.e., the symbols in themselves are not meaningless). For some readers, nevertheless, the abstract measure of these correlations seems to have been misunderstood as the hypostatization of information itself.

As I have tried to show, this received view of cybernetics—that is, its reduction to an abstract, disembodied understanding of information—is easily countered by considering the new machines and rigorous theoretical approaches that von Neumann, Ashby, and Walter actually produced. However, while it is *not* true that information is disembodied for the cyberneticists, it *is* true that in early AI cognitive processes *are* disembodied, in the specific sense that it was believed that these processes could be modeled and understood independently of any material substrate. This view is predicated on the realization that the computer is a new type of abstract machine, defined by its form or organization and functionality rather than by the substance out of which it is made. It is hardly surprising then that the early practitioners of AI saw both the human brain and the digital computer as roughly equivalent material instantiations of information- or symbol-processing machines. More specifically, in their claim that it is the symbol system—not the material substrate—that really matters, Alan Newell and Herbert Simon essentially reduced artificial intelligence to software and reinstalled a Cartesian duality that cybernetics—at least at its best moments—had entirely transcended. This new Cartesian dualism is not one of matter and mind but of matter and psychology, now conceived of as symbol processing. Hence the birth of early AI was also the birth of cognitive psychology.[48]

In order to convey a very different understanding of psychology—one much closer to that of the original cyberneticists, I conclude this chapter with a brief look at Valentino Braitenberg's *Vehicles: Experiments in Synthetic Psychology*. In the same whimsical but serious tone that animates the book, Braitenberg himself may be thought of as "the last cyberneticist." What is the basis for this presumptive and perhaps ahistorical categorization? Like several of the original cyberneticists, Braitenberg is an accomplished neuroscientist with a creative interest in constructing mobile machines. But since he comes several decades later, he can assume a playful and self-conscious attitude toward the toy world that he invites the reader to cocreate. *Vehicles* is thus really a thought experiment about how simple machines can be fashioned to produce complex behavior and how we will be tempted "to use psychological language in describing their behavior."[49] Starting with the simplest vehicle imaginable, we learn to "build" a series of increasingly more capable machines. At the same time, the behavior of these machines cannot help but evoke certain affective states, like fear, aggression, love, foresight, egotism, and optimism. Since we know that these affective states are not *in* the vehicles themselves but only in our perception of their behavior, what Braitenberg initially proposes is simply an "interesting educational game" (2). Yet things are not quite so simple. After progressing through the imaginative construction of fourteen of these vehicles, we are treated to eight complex drawings of various kinds of machinic arrangements.[50] But then, in an extended section called "Biological Notes on the Vehicles," Braitenberg concludes by reconsidering many of the vehicles in light of current brain research. However, even within this play of perspectives, what is most striking (and important) is the autonomy Braitenberg grants these machines. Because of what they can do, they are treated as interesting in and of themselves, an approach unavailable to the early cyberneticists.

Let us dwell for a moment on these vehicles—all simple mechanical constructions made of wheels, motors, sensors, wires, and threshold devices. Consisting of only a single wheel, motor, and sensor, Vehicle 1 moves directly toward a source of stimulation (say, light) unless deflected by friction. Vehicles 2, 3, and 4 are made by simply adding more motors, wheels, and sensors and crossing the connecting wires, enabling movement to be stimulated and/or inhibited in various ways. Vehicle 3, for example, will approach a light source and then turn away at a certain proximity. With the further addition of threshold devices, which can be connected in series, parallel, or in networks, primitive "brain" functions become possible. In Vehicle 5, for example, different devices (a red light and a bell) can activate each other in a feedback circuit, providing a kind of memory. With additional threshold devices (simple neural networks), subsequent vehicles acquire more complex cognitive functions, such as the ability to recognize movement, shape, and bilateral symmetry as well as a generalized response to color.

In Braitenberg's presentation two basic ideas come into play. The first is what he calls "the law of uphill analysis and downhill invention" (20).

Essentially this means that building vehicles that work and do things—especially things that are unplanned—is usually easier than analyzing from external observation the internal structures that make this behavior possible. This is because, as Braitenberg explains, induction is slow and requires a search for solutions. However, from the present vantage point it is clear that the actual complex behavior of his vehicles results from their often unpredictable interactions with the environment. This behavior is complex because it is emergent in the sense developed by contemporary dynamical systems theory, to be explored in some detail in chapter 5. Let it suffice here to observe that Braitenberg introduces a jump in complexity simply by wiring threshold devices in feedback circuits that produce associative connections. Vehicle 12 is constructed in this manner, and as a result Braitenberg claims that it exhibits "FREE WILL" (68, author's emphasis). What he means, simply, is that above a certain number of active elements and cross-connections in his vehicle's brain, its behavior becomes unpredictable to a human observer, even though that behavior is completely determined. This unpredictability is also true for individual human brains, Braitenberg adds, and serves as the basis for our own pride in our assumed autonomy.

The second idea is captured in the chapter title for Vehicle 6: "Selection, the Impersonal Engineer." Suppose, Braitenberg proposes, that we put the entire collection of vehicles on a large table. From among the circulating vehicles we begin to pick up one at a time and copy it, then put both model and copy back on the table. Meanwhile some of the vehicles will have fallen onto the floor, where they remain (they will not be copied). Importantly, the copying will have to be done quickly, in order to replace as soon as possible the faulty or inadequate or unlucky vehicles. There won't be enough time to test the copies or to make sure the wiring is correct. Hence errors will creep into the copying process. Some, perhaps most, will result in dysfunctional or barely functional vehicles. But other errors will act as "germs for improvement," particularly when "we pick up one vehicle as a model for one part of the brain and then by mistake pick up another vehicle as a model for another part of the brain" (27). While improbable at first, in the long run the lucky successes that arise from these mistakes will have a much greater chance of being reproduced. They will also be more resistant to analysis: "The wiring that produces their behavior may be so complicated and involved that we will never be able to isolate a simple scheme" (28). In short, Vehicle 6 is produced by "unconscious engineering," which is clearly a mimicking of the process of reproduction, copy errors, and selection we recognize as Darwinian evolution. Moreover, whereas analysis "will necessarily produce the feeling of a *mysterious* supernatural hand guiding the creation," we ourselves will have seen how "this simple trick" can produce machines that are "beautiful, marvelous, and shrewd" (28).

Braitenberg's *Vehicles* thus occupies a Janus-faced position: it both looks back to the cybernetic tradition, particularly to the ideas and machines of von Neumann, Ashby, and Walter, and anticipates the bottom-up approach and emphasis on the dynamics of emergence that will characterize ALife and behavior-based robotics in the 1980s and '90s. Although Braitenberg makes no attempt to copy or reproduce behavior typical of organic life, in many ways these vehicles exhibit a lifelike complexity similar to that of the cybernetic machines examined above. Not surprisingly, his design philosophy as well relies on a multiplicity of connections among simple mechanical elements. This too is a vital part of the cybernetic heritage and accounts for the emphasis on concrete embodiment and performativity rather than symbol processing and representation (i.e., the symbol-based approach of early AI). Not only is it appropriate to speak of his vehicles as autonomous agents, but his seriously playful deployment of evolution as a strategy to produce more complex vehicles anticipates the latest and perhaps most important conceptual turn in contemporary robotics.

2 The In-Mixing of Machines: Cybernetics and Psychoanalysis

Perhaps, really, what we are seeing is a gradual merging of the general nature of human activity and function into the activity and function of what we humans have built and surrounded ourselves with.
—Philip K. Dick, "The Android and the Human"

Among the many debates that characterized the Macy Conferences, which publicly launched the cybernetic movement, none were more heated and acrimonious than those generated by psychoanalysis. Although many of the participants were trained in psychiatry, neurophysiology, and psychology, only Lawrence Kubie was a practicing psychoanalyst. Trained in neurophysiology, he had "converted" to psychoanalysis at midcareer. In 1930, during the first part of his career, he had published a highly influential paper in *Brain* suggesting that the central nervous system could be pictured "as a place in which, under certain conditions and in certain areas, excitation waves move along pathways which ultimately return them to their starting points."[1] Later in the 1930s these reverberating circuits of neurons were experimentally verified and studied by Lorente de Nó, another conference participant. This research, in turn, informed the neural net theory of Warren McCulloch and Walter Pitts, whose essay "A Logical Calculus of the Ideas Immanent in Nervous Activity" was seminal to the formation of cybernetics.[2] In 1941, in the second phase of his career, Kubie postulated a connection between certain reverberating closed circuits in these nets and the compulsive behavior known in Freudian theory as *Wiederholungszwang*, or repetition compulsion.[3] By the time of the Macy Conferences, however, Kubie had come to believe that these circular neuronal paths were the physiological substrate of a behavior that could not be explained in these terms alone. Having become an orthodox Freudian analyst, he understood neurotic behavior to be the outward symbolic expression of unconscious fears and desires.

It is hardly surprising, then, that at the Macy Conferences Kubie assumed the mantle of representing the psychoanalytic point of view, delivering papers entitled "Neurotic Potential and Human Adaptation," "The Relation of Symbolic Function in Language Formation and in Neurosis," and "The Place of Emotions in the Feedback Concept." From the outset, however, Kubie's presentations drew strong opposition, and he was put on the defensive by hard-line experimentalists who sneered at the unscientific status of the Freudian unconscious and relentlessly questioned the subjective tenor of psychoanalytic interpretation. In a vitriolic attack on Freud circulated just before the 9th Macy Conference, Warren McCulloch argued that, rather than account for the data collected in observations of human behavior, psychoanalysis creates its own, self-justifying data.[4] Despite Kubie's patient and sustained efforts to bring psychoanalysis into dialogue with the cybernetic perspective, the overall result was a greater exposure of what Heims calls "the problematic nature of psychoanalysis as science" (*The Cybernetics Group*, 125). Yet this failure does not appear to have been a foregone conclusion. Indeed, at the outset of the conferences the most powerful spokesperson for the cybernetic point of view, Norbert Wiener, made it clear that he had no essential objection to psychoanalysis but simply believed that it needed to be rewritten in the language of information, communication, and feedback (126).

It can almost be said that this is what the French psychoanalyst Jacques Lacan accomplishes in his second seminar. Among the most scintillating intellectual events in Paris in the 1950s, Lacan's yearly scheduled seminars were the site of a methodical revolution in psychoanalytic theory. Having devoted the first year's seminar to Freud's papers on technique, Lacan took up "The Ego in Freud's Theory and in the Technique of Psycho-analysis" in the second.[5] The importance of this second seminar stems from the fact that there Lacan developed the distinction between the imaginary and the symbolic registers of experience he had proposed in his first seminar and that henceforth would be central not only to his definition of the ego (*le moi*) but to the entire framework of his thought. Yet what is most surprising about the seminar is the attention and significance Lacan granted to cybernetics and information theory. Indeed, on June 22, 1955, a week before its last yearly meeting, Lacan presented a paper to the Société française de psychanalyse entitled "Psychoanalysis and Cybernetics, or on the Nature of Language." But how do we account for this interest? And what does cybernetics have to do with Freudian psychoanalysis and Lacan's innovative transformations of it? As we'll see, the cybernetic concept of the machine and the digital language of information theory led Lacan to believe that the world of the symbolic is the world of the machine. His encounter with cybernetics thus produced a moment of crossing between two distinct discourse networks, with unexpected consequences that bear on the relationship between language and machines, the symbolic and the real. Moreover, and of special importance here, Lacan understood the symbolic function as a particular kind of computational assemblage that made human behavior meaningful.

Machines and the Symbolic Order

It is not at all certain how much Lacan knew about the Macy Conferences. He certainly knew about Kubie's work as well as the follow-up research done by John Z. Young.[6] It is also likely that Lacan discussed cybernetics and perhaps the conferences with Roman Jakobson, the Russian-born linguist whom Lacan had met in 1950 and whose work was a major influence.[7] Jakobson taught at MIT, where both Norbert Wiener and Claude Shannon were professors, attended the 5th Macy Conference, and later published an article about information theory and language. In addition to these direct connections, there was widespread interest in cybernetics among French intellectuals after the war. In Paris, a detailed and penetrating review of Wiener's *Cybernetics; or, Control and Communication in the Animal and the Machine* appeared in *Le Monde* on December 28, 1948. Written by the Dominican friar Père Dubarle, the review so impressed Wiener that he took up several of its arguments in *The Human Use of Human Beings*, a popular version of his earlier book. In 1950, at the invitation of Benoit Mandelbrot, Wiener himself gave a well-publicized lecture at the Collège de France. In 1953 Pierre de Latil's *La pensée artificielle* appeared, followed in 1954 by Raymond Ruyer's *La cybernétique et l'origine de l'information*, which Lacan disparages in his seminar. This first wave of European reaction to cybernetics also included Martin Heidegger's wholesale dismissal of it as the latest form of calculative thinking. However, while revering Heidegger as the most important contemporary philosopher in Europe, Lacan did not share Heidegger's belief that cybernetics was destined to "replace philosophy" and come to "determine and guide" the new sciences.[8] Nor did Lacan view cybernetics as a particularly American strain of thought. Rather, as he asserts in his lecture "Psychoanalysis and Cybernetics," cybernetics was a new kind of "conjectural science" that for the first time made it possible to understand the autonomy of symbolic processes.[9] In these terms, the subjectivity that so troubled the experimentalists at the Macy Conferences becomes in Lacan a matter of subject position, of where the subject finds himself or herself in a predetermined structure. However, this structure should not be understood simply or exclusively as another instance of French structuralism, for Lacan's second seminar makes formal automata theory at least equally pertinent. If the movement of a symbol dictates the correlation between a place in a structure and a state of the subject, this is because the symbolic order itself operates as a machine—a new kind of machine that cybernetics first brings to light.

Lacan's second seminar of 1954–1955, as its title suggests, is overtly concerned with how the ego in psychoanalytic theory should be defined and understood. In proposing a new theory of the workings of the unconscious and the determinations of desire, Freud had brought about a fundamental decentering of the subject in relation to the self, or ego, and thus inaugurated a new stage in the history of Western subjectivity. It is a new stage, Lacan insists, because the modern ego as theorized by Freud did not exist for either Socrates or Descartes, although it is anticipated in La Rochefoucauld's notion of *amour-propre* (self-love). Yet the modern sense of the ego brought about by Freud's "Copernican revolution" was by no means secure or well understood. Even in psychoanalytic theory, Lacan argues, what Freud meant by the ego is often confused with consciousness, or, more egregiously, the ego is sometimes made substantial and even "autonomous" through a process Lacan calls entification. In this context Lacan redefines the ego by distinguishing between the symbolic and the imaginary orders and by introducing a notion of the unconscious as the "discourse of the other." While he claims only to "read" Freud, that is, to make explicit what Freud leaves implicit, it can be argued that these innovations bring psychoanalytic theory to a more precise degree of conceptualization.

The term *cybernetics*, along with subsidiary notions like feedback, the circuit, and the message as information, enters Lacan's seminar rather casually.[10] What is most important, in the early stages of the seminar, is the cybernetic concept of the machine, which will come to throw new light on fundamental psychoanalytic concepts. Lacan introduces this general idea when he begins to discuss the symbolic universe, to which, he insists, the machine is closely related. At the same time, he continues to use the term *machine* in several senses. He states, for example, that "the machine is much freer than the animal," which is really a "jammed machine," where "certain parameters are no longer capable of variation."

He continues: "It is inasmuch as, compared to the animal, we are machines, that is to say something decomposed [*décomposé*], that we possess greater freedom, in the sense in which freedom means the multiplicity of possible choices." Most importantly, he states, "The meaning of the machine is in the process of complete transformation, for us all, whether or not you have opened a book on cybernetics" (31). But suppose we *had* opened such a book: what new notion of the machine would we have found there?

As we saw in the previous chapter, we would have discovered that the cyberneticists were mainly interested in machines that exhibit feedback and that, by means of internal control mechanisms like the thermostat on the modern heater, regulate their own functioning. We would also have noticed references to the mechanical automata that so fascinated the courts of eighteenth-century Europe, like the Jaquet-Droz drawing dolls and Jacques Vaucanson's duck. By means of clocklike mechanisms, for example, the latter could not only waddle but simulate eating. But eventually we would have encountered a very different kind of machine, one that existed initially only as a thought experiment. Or rather, because it was a computational machine, the functioning of which did not depend on any particular form of material embodiment, it could be said to exist in a very singular way. This new and revolutionary concept of the machine first appears in Alan Turing's foundational paper of 1936, in which he addresses the problem of computability—whether a number or function can be computed—and thus the larger question of whether a mathematical problem can be solved.[11]

Turing proposed that if the problem can be expressed as an algorithm, or a precise set of formal instructions for arriving at the specified solution, then it can be computed mechanically by a machine. The question then becomes whether or not the machine will halt with a finite answer. This machine, as described by Turing, came to be called a Turing machine. It consists of three parts: a reading/writing "head," an infinitely long "tape" divided into squares that passes through the head, and a table of instructions that would tell the head what to do as a function of what it reads on the tape and the machine's current state. Specifically, the head would scan the tape square by square; depending on the head's current state and whether a mark was present or absent in a particular square, it would enter a mark, erase a mark, or leave the square blank, then move to another square, either to the left or right. Since at any moment the reading/writing head could only be in one of a finite number of internal states defined by a table of instructions (now known as its state-transition table), it was considered to be a finite-state machine, or automaton. (A familiar example of a finite-state automaton is a two-way traffic light set to flash green, yellow, and red in a sequence that allows traffic to pass safely at an intersection.) With this simple device two things could be accomplished. Data could be entered in the form of a string of symbols—for example, binary numbers (one or zero) could be encoded as the presence or absence of the mark; and operations could be performed according to the table of instructions given to the head, as in: if no mark and the machine is in state 1, enter a mark and move to the square on the left and move to state 2; if a mark, move to the square on the right and remain in state 1. These instructions and the memory constituted by the tape allow the head to manipulate the marks or symbols in a variety of ways, thereby performing mathematical operations. However, what makes this finite-state machine a Turing machine is its auxiliary memory—the infinite tape, which it can access in either direction—for it is this memory capacity that enables it to perform a range of different computations.[12] A simple finite-state machine, for example, cannot multiply large numbers, because it has no way to store and bring forward the results of previous stages of calculations as it advances. From this simple fact we can grasp the importance of memory—not only how much but from where (i.e., what state) it can be accessed—in defining a machine's computational capacity.

Turing's thesis, subsequently accepted by virtually all mathematicians, states that every computation expressible as an algorithm or every determinate procedure in a formal system has its equivalent in a Turing machine. More important, Turing further postulated the existence of a universal machine (now known as a universal Turing machine), which could emulate the behavior of any specific Turing machine. A universal computing machine would therefore be one that, given any table of instructions that defined a Turing machine, could carry out those instructions; it would, in short, be programmable. Turing's ultimate purpose, we should recall, was to prove that there is no way to determine in advance whether certain numbers *are* computable, that is, whether the machine set up to compute them will ever halt. Invented as part of the proof, his notion of the Turing machine would eventually provide a formal basis for the modern computer, in which different sets of instructions, or programs—for computation, data processing, sending and receiving data, and so on—allow the same machine to do a variety of tasks. This capacity makes the computer a fundamentally new *type* of machine, defined by a logical and functional rather than a material structure. It is an abstract, second-order machine in which the logical form of many different kinds of machines is abstracted and made equivalent to a set of algorithms. Although today's desktop computers are usually made of silicon and copper wire encased in plastic and metal, in principle they could be constructed out of a wide variety of materials. As abstract machines, their functions are not defined by the specific behavior of the materials from which they are constructed; rather, this behavior is used to physically instantiate a symbol system with its own independent rules or syntax.

Although we might not have gleaned all of this from simply perusing a book on cybernetics, we could not have missed the fact that the first electronic computers—often called "thinking machines" in the popular press—were constructed in the years just preceding Lacan's seminar: ENIAC in 1945, followed by EDVAC, MANIAC, ILLIAC, and JOHNIAC (after von Neumann), then SAGE in 1949, and the highly popularized UNIVAC in 1951. What was striking about these machines, in addition to their capacity for rapid calculation, was that they automated—that is, rendered in self-regulating mechanistic procedures—the operations of formal systems in which arbitrarily chosen symbols could be combined according to rules of composition—a syntax —to produce more complex operations. Because these machines thus automated the "laws of thought" in a series of logical and combinatorial operations (heuristic search strategies and pattern recognition procedures would come later), these symbol-processing machines were unlike any machines seen before.[13] The only thing remotely comparable was the clock, another autonomous machine that when widely introduced into Europe completely restructured human behavior. And just as the clock—or so we can imagine Lacan thinking—is a time-marking machine that can be found not only on our walls and wrists but also in our bodies, institutions, and exchanges both economic and informational, so too these logic machines must inhabit and traverse us in unnoticed ways, giving structure and meaning to what we all too casually call life.

In his second seminar Lacan works his way toward a formulation of precisely this import. Meanwhile, throughout the early sessions, he keeps open the common understanding of the machine, while also alluding to this new formal one. The results can sometimes be confusing, as when he sets up an opposition between perceptual consciousness and the ego and then opposes both to the realm of the symbolic and the world of machines. Not long after the passage cited above, where Lacan refers to both animals and humans as machines, he offers a materialist definition of consciousness, suggesting that it is simply a reflection, like a mountain on the surface of a lake, easily emulated by a camera and photoelectric cell. Since consciousness is contingent on "the existence of our eyes and ears" (48), it is ostensibly not a phenomenon of the ego. Moreover, the reflections of consciousness occur in an inner space of phenomenal images where things appear to be centered and that Lacan calls the imaginary order. But there is another order, he asserts, one of play and exchange that begins with the circulation of speech, a symbolic realm that "has nothing to do with consciousness" (49) and where "man is de-centered." Since it is with "this same play, this same world that the machine is built" (47), the world of the symbol is also the world of the machine:

> Speech is first and foremost that object of exchange whereby we are recognized, and because you have said the password, we don't break each other's necks, etc. That is how the circulation of speech begins, and it swells to the point of constituting the world of the symbol which makes algebraic calculations possible. The machine is the structure detached from the activity of the subject. The symbolic world is the world of the machine. Then we have the question as to what, in this world, constitutes the being of the subject. (47)

This, in a nutshell, is the central question of the seminar as well as the terms in which it will be addressed. To make fully intelligible his claim that "the symbolic world is the world of the machine," Lacan will define the symbolic order as a new and fundamental concept, redefine the unconscious as the "discourse of the other," and relegate the ego to the order of the imaginary, all while claiming to remain true to Freud's original intent. Key to Lacan's revision of Freud are the three differential orders, or registers, of experience that he calls the symbolic, the imaginary, and the real. These three orders, moreover, are not simply oppositional but always "in-mixed" with one another. Indeed, it is precisely the in-mixing [*immixtion*] of these different registers that generates both problems and complexity for human beings.

From the outset Lacan associates the ego with the imaginary order. In his earlier essay on the mirror stage (the gist of which he repeats in the seminar), he explained how the ego is formed on the basis of a specular image of the body. Before this stage the infant can only experience itself in isolated fragments, as a morcelated body without clear boundaries and who finds unity and wholeness only in the "other." However, when the infant beholds itself in the mirror, it realizes that *that reflected other* is me. Yet it is an ambiguous experience—this mirrored "hey, that's me"—since from then on the infant experiences itself as a unity, but a unity that comes from without, that is, from an "other." It is, Lacan says, an alienated unity. This (mis)recognition of one's bounded and unified image is soon followed by other imperfect identifications. As a result, Lacan jokingly puts it, the "ego is like the superimposition of various coats borrowed from what I would call the bric-à-brac of its props department" (155).

For Lacan, then, this ego is not the subject. In relation to the subject the ego is an imaginary construction, a specular object, or mirage. As such, it serves a crucial function, as the support of imaginary identification(s). But the ego is not consciousness either. The field of perceptual consciousness does not provide a unity for the ego; rather, the ego is precisely what "the immediacy of sensation is in tension with" (50). Lacan implies that while consciousness itself is a mere physical phenomenon, in humans it brings about the illusion of an agent responsible for it: thus there is not simply "a seeing" but an "I who sees" that remains constant. Nevertheless, although an imaginary construct, the ego is the locus of no less real libidinal investments. The relationship of one ego to another, for example, is characterized as one of oppositional duality, necessarily defined by aggressive conflict and rivalry. This relationship Lacan compares to the encounter of two machines, each one jammed on the image of the other.[14] An unjamming can occur only through the intervention of a legislating or mediating function, which can perform the service of a "symbolic regulation." Tellingly, it will have to be another machine, inserted into the first two, one that speaks "commandingly" but in "the voice of no one."

We already know—Lacan has repeated it several times—that this mediating function is served by the order of the symbolic. It is the order that the anthropologist Claude Lévi-Strauss invokes when he describes the "elementary structures of kinship"; it is also what Freud was seeking in *Beyond the Pleasure Principle*. Reading the latter, Lacan distinguishes between two kinds of repetition. In the first, a restitution of an equilibrium is achieved by a homeostasis, or regulation, of energies coming from the external world. It is the kind of mechanism that enables the gratification of pleasure and that Lacan will later extend to adaptation and instinct in animal behavior (pp. 86–87 and 322–323 in the seminar). This kind of setup is usually modeled as a type of machine involving the exchange of energy, and therefore subject to entropy and the laws of thermodynamics. But beyond this setup there is a compulsion to repeat that is inexplicable in these terms and that Freud associated with the death drive. Lacan now makes a daring move. If we consider cybernetics, he says, we find another kind of machine—a second type of mechanism—in which the circulation of information works against entropy (and here Lacan is simply paraphrasing Norbert Wiener). Lacan then asks:

> What is a message inside a machine? Something which proceeds by opening and not opening, the way an electronic lamp does, by yes and no. It's something articulated, of the same order as the fundamental oppositions of the symbolic order. At any given moment, this something which turns has to, or doesn't, come back into play. It is always ready to give a reply, and be completed by this selfsame act of replying, that is to say by ceasing to function as an isolated and closed circuit, by entering into the general run of things. Now this comes very close to what we can conceive of as *Zwang*, the compulsion to repeat. (89)

Here, in the difference between a mechanical or energy-driven machine and the information machine, Lacan finds the means to clarify and further conceptualize Freud's distinction between restitution and repetition in the subject's psychic economy.

Let us consider this distinction more closely. For Lacan, the subject's compulsion to repeat is precisely how the unconscious reveals itself: as a form of "insistence." More specifically, the unconscious "insists" as the "discourse of the other":

> This discourse of the other is not the discourse of the abstract other, of the other in the dyad, of my correspondent, nor even of my slave, it is the discourse of the circuit in which I am integrated. I am one of its links. It is the discourse of my father for instance, in so far as my father made mistakes which I am condemned to reproduce—that's what we call the super-ego. I am condemned to reproduce them because I am obliged to pick up again the discourse he bequeathed to me, not simply because I am his son, but because one can't stop the chain of discourse, and it is precisely my duty to transmit it in its aberrant form to someone else. (89)

Now, if the discourse of the other is the circuit in which I am integrated, then the ego can only be a point of resistance, and Lacan will declare that that is exactly what it is—an obstacle, interposition, or filter. If there were no resistance from the ego, he states, "the effects of communication at the level of the unconscious would not be noticeable" (120). By way of analogy, he mentions the electronic tube, with its anode and cathode, between which a third element (like the imaginary ego) regulates the passage of electrical current.

But here it might be objected that this whole appeal to cybernetics, information machines, electronics, and so on, is *merely* one of analogy—all the more so since during this part of the seminar Lacan also speaks of Freud's "two completely different structurations of human experience," not in relation to machines but in relation to the distinction Søren Kierkegaard makes between reminiscence and repetition in his short book *Repetition*. It is along the paths of the lost object, never to be recovered and for which the modern subject never ceases to generate substitutes, Lacan says, that the modern subject's experience is structured. Yet while this theory of the lost object may account for the structuring of the object world, the world of the symbolic must have a higher priority. This follows from the fact that in order for the symbolic order to serve a mediating function, it must exist at a higher logical level than the world of objects and images. Early in the seminar Lacan emphasized the "autonomy of the symbolic" (37), claiming that the symbolic order, while it "has its beginnings elsewhere than in the human order ... intervenes at every moment and at every stage of the [latter's] existence" (29). But up to this point he has only argued that this order manifests itself ("insists") in the subject's compulsion to repeat and that it reveals itself in the circuit of discourse in which the subject is integrated. He has yet to show exactly how this order emerges and intervenes concretely in human reality. This he accomplishes through a reading of Edgar Allan Poe's short story, "The Purloined Letter." It is also where he begins to make good on his claim that the world of the symbolic is the world of the machine.

The Machination of the Subject

The importance Lacan attributed to his reading of Poe's story is indicated by the fact that he revised it and made it the portal to his *Écrits*.[15] Although Lacan's reading has often been commented upon, what is rarely acknowledged and has never been adequately discussed is the relationship of his reading to information machines and cybernetic themes, a context that is clear and unmistakable in the *Seminar* but not so evident in the version published in *Écrits*. Simply to provide and underscore this context will require a somewhat detailed elaboration, since it bears on the process most fundamental to Lacan's argument: the encoding of the real in the symbolic order.

In chapter 15 of the seminar, "Odd or Even? Beyond Intersubjectivity," Lacan asks his interlocutors to consider an anecdote within the Poe story in which the detective Dupin tells the prefect of the police about a certain eight-year-old boy he once knew who was a whiz at winning marbles in the game of odd or even. In this simple game, one boy would hold up his closed fist and the other would try to guess whether it held an odd or even number of marbles; the winner would receive one of the loser's marbles. The point of the anecdote is that this particular boy was phenomenally successful because he had a strategy: to identify his own thinking with that of his opponent in order to calculate the latter's most likely next move. But as Lacan shows, there are only two "moments" to this strategy: in the first the opponent is naive, in a second as smart as the boy himself. Beyond this position there is no third, except to play randomly ("like an idiot") (181), that is, to return to the first position. The only way out of or "beyond" this dilemma is to play against a machine, which offers no possibility of (imaginary) intersubjective identification; one is "from the start forced to take the path of language, of the possible combinatory of the machine" (181). This path follows an emergent logical order, in contrast with the imaginary relation to the other, which is predicated on an experience that necessarily "fades away." Later we shall consider an instance of what it is like to play against a machine, a very sophisticated modern computer, but here let us follow Lacan's development of this difference.[16]

Lacan begins by asking two of his interlocutors to play this game of "odd or even" and to record the wins as pluses and the losses as minuses. He then makes two points: first, the game is only meaningful as a sequence (winning one game doesn't mean anything) and therefore we have to remember (i.e., record) the results; second, in the sequence of wins and losses a theoretical distinction emerges. If one player begins to win repeatedly he or she will appear to be unnaturally lucky, and the odds against his/her continued winning will seem to increase. But this means that we are "no longer in the domain of the real but in that of the symbolic," since in the domain of the real each player has an equal chance of winning or losing for each game played, no matter what the past results. Our subjective reaction is not merely an illusion, however, since the introduction of a simple set of symbols like pluses and minuses to record even a random sequence necessarily gives rise to an emergent order. As Lacan puts it: "Anything from the real can always come out [*N'importe quoi de réel peut toujours sortir*]. But once the symbolic chain is constituted, as soon as you introduce a certain significant unity, in the form of unities of succession, what comes out can no longer be just anything" (193). In short, the very recording of random events gives rise to a rudimentary form of order, since it allows the formation of units and hence the emergence of a syntax governing their possible sequences of succession.

To illustrate, Lacan sorts the pluses and minuses into three groups, according to all of their possible combinations (fig. 2.1). With these groupings, only certain sequences are possible. For example, a 1 (+++

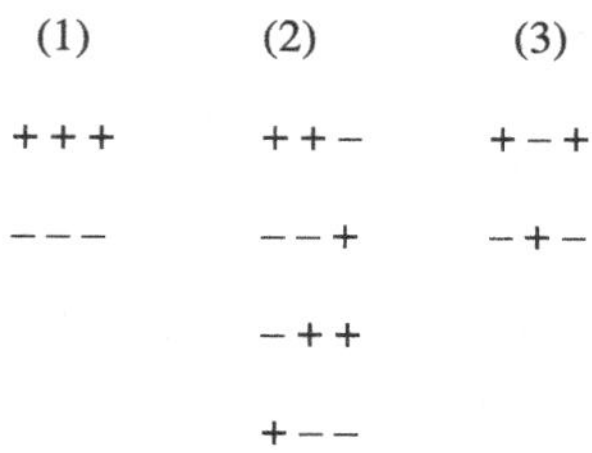

Figure 2.1
Lacan's pluses and minuses.

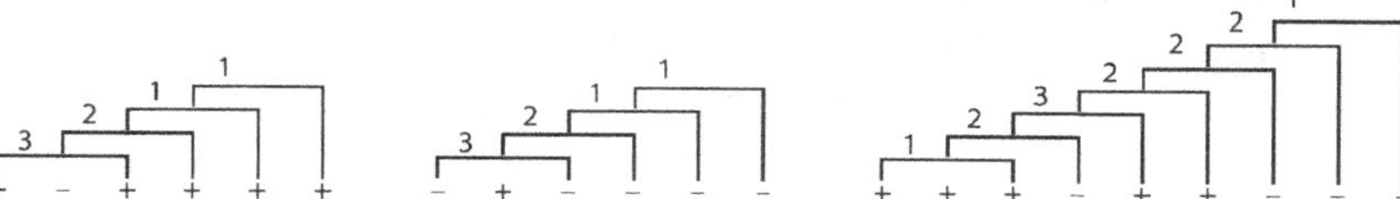

Figure 2.2
Plus and minus sequences. Image by Lucas Beeler.

Passing from 1 to 2 → β

Passing from 2 to 2 → ý

[...]

Shift back from 2 to 1 → ∂

Shift back from 2 to 3 → ∂ (193)

Figure 2.3
Lacan's group intervals.

or −−−) will never be able to follow directly after a 3 (+−+ or −+−), because in the transition a 2 will appear (fig. 2.2). Lacan then notes that other significant unities can be constituted from these "laws," "representing the intervals between these groups" (193) (fig. 2.3).

Lacan provides similar but more elaborate illustrations in *Écrits* (41–61), yet even these simple series of pluses and minuses, which initially may be random, indicate how only certain orderings or sequences of integrations are possible. Constituting a rudimentary syntax, they inscribe a form of memory that operates with the force of a "law."

For Lacan, this demonstration establishes two things. First is the autonomy and self-organization of the symbolic order: "From the start and independently of any attachment to some supposedly causal bond," he says, "the symbol already plays, and produces by itself, its necessities, its structures, its organizations" (193). Second is the fact that within this play of the symbol, the subject will always find his or her place:

> By itself, the play of the symbol represents and organizes, independently of the peculiarities of its human support, this something which is called a subject. The human subject doesn't foment this game, he takes his place in it, and plays the role of the little *pluses* and *minuses* in it. He is himself an element in this chain which, as soon as it is unwound, organizes itself in accordance with laws. Hence the subject is always on several levels, caught up in crisscrossing networks. (192–193)

From here Lacan proceeds to his reading of Poe's "The Purloined Letter," which illustrates the operations of this symbolic order. Specifically, he shows how the plot inscribes three subject positions, or positions that subjects can come to occupy. They are marked and defined according to a relative state of blindness regarding the presence or absence of a compromising letter that the queen receives (unbeknownst to the king) and the minister purloins, by surreptitiously substituting for it another letter. The king sees nothing, the queen sees this and takes advantage of it, and the minister does the same to her in turn. After the police secretly but futilely search the minister's apartment, the detective Dupin is brought in and recovers the letter by deploying the same strategy against the minister that the latter had used against the queen. In this concatenation, the story thus registers a redistribution or "step-wise displacement" (203) from a first sequence of three subject positions, occupied by the king, queen, and minister, to a second, occupied by the police, minister, and detective.[17]

What is at stake in this part of the seminar can now be summarized. Lacan introduces the game of odd and even in order to illustrate the limit of its play in the oscillation of intersubjective relations: in a first moment, I (as ego) assume the position of the other in order to determine how he or she will play; in a second moment, I realize that he or she (another ego) is doing the same with me, that is, that I am an other for this (imaginary) other. The problem is how to pass "beyond" into "a completely different register from that of imaginary subjectivity" (181), which cannot be accessed by way of identification. As Lacan never tires of repeating, this realm "beyond" is precisely the unconscious, "beyond the pleasure principle, beyond relations, rational motivations, beyond feelings, beyond anything to which we can [willingly] accede" (188).[18] However, having established that this realm beyond subjectivity is defined by the symbolic order, Lacan has yet to explain how the symbolic order *is* also the world of the machine. To do this, he must show how the "symbolic emerges into the real," that is, how the symbolic order itself arises and functions as a machine.

Briefly put, it is the symbolic order's encoding of the real in numbers ("it ties the real to a sequence," as Lacan later puts it), that allows the recording and integration of data (which he calls memory, while cautioning us to distinguish between memory and remembering), which in turn gives rise to a syntax of different combinatorial possibilities.[19] The resulting "machine"—it is actually a finite-state automaton—is not created by human beings but appears to emerge spontaneously when discrete digital marks are used for counting. This machine, moreover, does not require the intervention of human consciousness in order to function. In Poe's story the circulation of the letter reveals the three positions a subject may come to occupy in a determined field of social relations, but without any explicit awareness of this structuration on the subject's part. Each position corresponds to a specific prevailing "state"—the minister's "feminization," for example—regardless of his or her personal psychology or individual disposition. As Lacan asserts in *Écrits*, "In its origin subjectivity has no relationship to the real, but [only] to a syntax that engenders in it a signifying mark" (50). In Poe's story the possession of or desire for the letter is precisely the signifying mark, and the letter's movement reveals the syntax that engenders the subject positions that make subjectivity a possibility. The story is thus appropriately read as a finite-state diagram or transition-state table of the symbolic order's functioning, in terms of three positions and three corresponding states. Repetition of the sequence (king, queen, minister; prefect of police, minister, detective) is necessary in order to mark out and underscore these positions, as well as the possible transitions from one position or state to another.

Computational Media: A New Discourse Network

Again the objection might be raised that we've heard Lacan's definitions before, as well as his interpretation of the Poe story, but without this all-important emphasis on machines. Are they really necessary to his argument, or are they just fashionable, provocative metaphors, even for Lacan himself? After all, Lacanian theory has pretty much ignored Lacan's interest in cybernetics, with no apparent loss of completeness or intelligibility. So, even if we agree that "cybernetics clearly highlights ... the radical difference between the symbolic and the imaginary orders" (306), as Lacan asserts, we may wonder to what extent his introduction of machines is a handy illustration of his theory that doesn't add anything essential.

To my knowledge, only the German media theorist Friedrich Kittler has addressed the question of the *necessity* of cybernetics to Lacan's theory. In *Gramophone, Film, Typewriter*, Kittler considers both Freud and Lacan in relation to modern technical media and understands Lacan's methodological distinctions between the real, the imaginary, and the symbolic as "the theory (or merely a historical effect)" of the differentiations brought about by technical media around the beginning of the twentieth century, when nature became a series of data flows that "stopped not writing themselves."[20] In a later essay Kittler argues more fully that the differences between Freud's psychoanalytic theory and Lacan's rewriting of it reflect the differences in the operating standards of information machines and technical media in their respective epochs. In constructing his model of the psychic apparatus, "Freud's materialism reasoned only as far as the information machines of his epoch—no more, no less."[21] According to the scientific imperatives to which Freud willingly bent himself, ineffable emanations of spirit had to be replaced by systems of neurons that differ (and defer) according to separable functions, in this case recording (memory) and transmission of data (perceptual consciousness). In Freud's time, however, storage functions could only be conceived of on the model of the engram, which includes not only the graphic inscriptions theorized by Jacques Derrida but also the grooves on Edison's newly invented phonograph. Significantly, for both his "case studies" and lectures Freud relies on his own "phonographic memory," as he emphasizes on several occasions. In medial terms, psychoanalysis ("the talking cure") is actually a form of "telephony," a communication circuit between patient and analyst in which the former's unconscious is transformed into sound or speech and then back into the unconscious. As Kittler puts it, "Because mouths and ears have become electro-acoustical transducers, the [analytic] session remains a simulated long distance call between two psychic apparatuses."[22] In fact, a telephone cable had been laid in Freud's house, but not in the consulting room, in 1895. Yet Freud doesn't limit himself to the phonographic. In *The Interpretation of Dreams* the transmission medium is optical, a cameralike apparatus that converts latent dream thoughts into a system of conscious perception, the virtual images of which Lacan would understand as cinema. In constructing his model of the psychic apparatus, Freud thus implemented all storage and transmission media available at the time: print, phonograph, telephone, and cinema, though the last term never appears in his writing.

As suggested by the titles of his own works—*Écrits*, the *Seminars*, *Television*, "Radiophonie"—Lacan grasped both the importance of these media for Freudian theory and the extent to which the foundations of psychoanalysis stood at the beginning of a new era, characterized by the technical differentiation of media and the end of the print monopoly. For Kittler, this explains why Lacan's triple register of the real, the imaginary, and the symbolic corresponds to the separations of technical media, that is, to gramophone, film, and typewriter, respectively. Yet, as Kittler insists, the Lacanian symbolic corresponds not only to the linguistic signifiers inscribed mechanically by the typewriter but to the entire domain of computation. Hence the final chapter of *Gramophone, Film, Typewriter* moves from a study of the typewriter to the German cryptographic machine Enigma and the Allied efforts (led by Alan Turing) to decipher its encoded commands to the German military, efforts that directly contributed to the development of the modern computer. The latter differs from "a several-ton version of Remington's special typewriter with calculating machine" (258) because it includes "conditional jumps" in its programmed instruction set. (These "jumps" are conditional branchings like IF-THEN sequences.) Through these and other feedback loops, the computer itself becomes a subject:

> Computers operating on IF-THEN commands are therefore machine subjects. . . . Not for nothing was [Konrad] Zuse "frankly nervous" about his algorithmic golems and their "halting problem." Not for nothing did the Henschel Works or the Ministry of Aviation assign the development of cruise missiles to these golems. On all fronts, from top-secret cryptoanalysis to the most spectacular future weapons offensive, the Second World War devolved from humans and soldiers to machine subjects. (259)

All of which makes cybernetics—the theory of self-guidance and feedback loops—"a theory of the Second World War" (259).

Does this mean that Lacan's writings are at bottom a theory of the machinic subject? Kittler doesn't pose the question. He merely insists that when Lacan utters the word "machine" we hear the word "computer." Lacan's machines are really information machines, simple ones at that, since all they do is count, store, and integrate codings of binary numbers (1s and 0s). Yet these basic operations underlie all that a modern digital computer does. Since the computer, or rather its conceptual forerunner, Alan Turing's universal machine of 1936, is the most important technological invention between Freud and Lacan, Lacan's rewriting of Freud should be understood as an attempt to redefine the psychic apparatus according to contemporary conditions of mediality. In these terms Lacan can be said to implement a more fully functional model of the psychic apparatus, since it now includes the most up-to-date media of information storage, transmission, *and* computation.

In this perspective the presumed "presence" of Turing and technical media should not be understood to operate as an instance of technological determinism. Rather, in pointing out that Lacan's refashioning of the psychic apparatus is necessarily built on and therefore reflects contemporary conditions of mediality, conditions that presuppose a theory of computation (i.e., the concept of a universal Turing machine), Kittler makes explicit both the technical and discursive conditions of possibility of Lacan's discourse. Put differently, the interlacing of Lacan's discourse with the new discourse of cybernetics and information theory means that Lacan participates in a specific "discourse network."[23] But here an anachronism appears to enter Kittler's historical scheme, for he situates both Freud and Lacan in the discourse network of 1900, which is marked by the emergence of technical media and the discourses of psychophysics and psychoanalysis. However, Lacan's discourse, though closely tied to Freud, depends on a different set of historical a priori, namely, the universal Turing machine and the digital information machines that operate through cybernetic feedback mechanisms. It would seem more reasonable therefore to argue that Lacan's revision of Freudian theory, at least in its early stages, draws on and participates in a *new* discourse network, one that emerges in the aftermath of the Second World War and that has subsequently become our own. In this new discourse network psychophysics is replaced by the computational paradigm and psychoanalysis by cognitive science. Since the computer is a universal symbol manipulator that can simulate any computational device, the new discourse network is not defined by any particular set of specific machines but rather by networks of computational assemblages. Whereas in the earlier discourse network nature became a series of data flows that "stopped not writing themselves," here we might say that all dynamical systems, including living systems, have become a series of data flows that "stopped not computing themselves."

Following Turing, computational theory was further developed by Alonso Church, Emile Post, Stephen Kleene, and, from very different perspectives, McCulloch and Pitts (neural net theory), and John von Neumann (automata theory). However, the new discourse network would be consolidated only with the advent and full flowering of cybernetics and information theory, following discussions of information, feedback, and circular causality at the Macy Conferences and the construction of the stored-program computer. As organisms and molecules began to be viewed as information storage and retrieval systems and DNA as a coded program, this discourse network would come to include molecular biology and genetics. With Noam Chomsky's work (discussed below), linguistics would be added as well. By the end of the 1950s, in fact, a specific set of concerns, assumptions, and languages had coalesced across a web of scientific and technological connections that included the university and scientific establishment, the military, and a nascent communications industry. And as later chapters suggest, subsequently developed computational assemblages in artificial intelligence and cognitive science, distributed emergent computation in Artificial Life and the new sciences of complexity, as well as evolutionary computation in robotics would become vectors for this discourse network's further spread, consolidation, and refinement.

To be sure, the concept of a discourse network needs to be more precisely defined, both in terms of historical limits (a *terminus ab quo* and *terminus ad quem*) and the material/informational practices that make each network unique. But rather than elaborate a detailed definition here, I want to demonstrate further how deeply Lacan's theory—at its initial stage at least—is embedded in the discourse of cybernetics and information theory. The anachronism in his historical scheme notwithstanding, I begin with Kittler's passing remark that Claude Shannon

> calculated the probability of every single letter in the English language, and from these calculations produced a beautiful gibberish. [He] then went on to take into account the transition probabilities between two letters, that is, digraphs, and the gibberish began to sound a bit more like English. Finally, through the use of tetragrams (not to be confused with the names of God) there arose that "impression of comprehension" which so loves to hallucinate sense from nonsense. Lacan's analysis of Poe works with precisely these types of transition probabilities, the major mathematical discovery of Markoff and Post.[24]

Offered without comment, Kittler's observation that Lacan "works with" the same types of "transition probabilities" essential to cybernetics and information theory leads to a whole web of discursive linkages. To trace several out will further clarify Lacan's conceptualization of the role of information machines in the functioning of the symbolic order, and specifically in relation to language. The degree to which these linkages and discursive affiliations instantiate a new discourse network and the light they throw on Lacan's theory will then appear to be two sides of the same conceptual nexus.

Language and Finite-State Automata

It should first be noted that Lacan himself mentions the Markov chains to which Kittler draws our attention (see *Écrits*, 51). A Markov chain is a special instance of a discrete-time stochastic process, which Claude Shannon introduces into his *Mathematical Theory of Information* in order to treat the sending of a message in information theoretical terms:

> We can think of a discrete source as generating the message, symbol by symbol. It will choose successive symbols according to certain probabilities depending, in general, on preceding choices as well as particular symbols. . . . A physical system, or a mathematical model of such a system which produces such a sequence of symbols governed by a set of probabilities, is known as a stochastic process. We may consider a discrete source, therefore, to be represented by a stochastic process. Conversely, any stochastic process which produces a discrete sequence of symbols chosen from a finite set may be considered a discrete source.[25]

Note that sending (and receiving) messages are thus to be viewed as stochastic—not deterministic—processes and that even the constraints on the message (like syntax in language) are treated as probabilities.

Shannon goes on to explain how the statistical structure of several examples of discrete information sources can be described "by a set of transition possibilities $p_i(j)$, the probability that letter i is followed by letter j" (41). By increasing the transition probabilities to higher order approximations of English—for two, three, and four letters (digram, trigram, and tetragram, respectively), as in Kittler's example—the symbols from the discrete source begin to concatenate into strings that constitute recognizable words. Stochastic processes of this type, Shannon continues, are known as discrete Markov processes:

> The general case can be described as follows: There exist a finite number of "states" of a system; $S_1, S_2, \ldots S_n$. In addition there is a set of transition possibilities, $p_i(j)$, the probability that if the system is in state S_i it will go to state S_j. To make this Markoff process into an information source we need only assume that a letter is produced for each transition from one state to another. (45)

What Shannon doesn't make clear here is the specific property of the Markov stochastic process, or chain, namely, that the probability of the system's next state depends solely on it current state, therefore making its previous states irrelevant for predicting subsequent states. In effect, the transition matrix that calculates the probabilities for each change of state is "without memory." While this property may seem strangely inappropriate for describing language, it is actually what allows for a high degree of indeterminacy and freedom while not giving up structure and predictability.

The reason for these technical details will soon become apparent, but it should already be evident why Markov chains might be of interest to Lacan. If the "discourse of the other" is a discrete information source (sending messages from the unconscious), then it might indeed exhibit the character of a Markov process. While at first this might seem unlikely, in fact G. A. Miller had introduced Markov modeling into certain areas of psychology in 1952.[26] We should also recall that Ross Ashby discussed Markov chains and "Markovian machines" in *An Introduction to Cybernetics* (1956).[27] Moreover, in the 1950s Markov modeling was applied successfully to animal learning; in the 1960s, to human concept learning.[28] Lacan's particular interest, it can be inferred, was twofold. First, and more generally, without some psychic mechanism of "return" to earlier states, the process of association in the analytic session would never lead anywhere. (Freud had simply denied that there was any such thing as chance in the unconscious.) But could such "returns" reveal a pattern or even articulate a subjective structure? Could Markov chains, in revealing a pattern of probabilities, provide a model for understanding the "discourse of the other," which is neither random nor simply determined and therefore not easily predictable? If the effect of a Markov chain is not unlike the kind of emergent order Lacan had described with pluses and minuses, then it would seem much more likely. This leads us to the second aspect of Lacan's interest. We have already seen how a finite automaton describes the symbolic order's functioning in his reading of Poe's story. But it turns out that machines that produce languages in this manner are considered mathematically as finite-state Markov processes.

This convergence—particularly visible in the addenda to the seminar on "The Purloined Letter" published in *Écrits* (44–61) but omitted from the English version—further emphasizes the heretofore unrecognized extent to which Lacan's formulations resonate with "formal language theory," which was developing more or less contemporaneously.[29] This new branch of mathematical theory devoted itself to the study of relationships between languages and machines, or theories of grammar on one side and finite-state machines on the other. The connection with Lacan's thought is unmistakable, though it has been missed by scholars so preoccupied with the influence of structural linguistics (i.e., Saussure and

Jakobson) that, like the police in Poe's story, they have conducted their search strictly according to a priori assumptions. Not surprisingly, recontextualizing Lacan's seminar in relation to the mathematical formalisms of information theory and formal language theory produces a different perspective and a different set of consequences.

It becomes clear that, at least in the early stages of his thought, Lacan considered the working of the symbolic order not only in terms analogous to those of a cybernetic circuit but precisely as a circuit that operates as a finite-state automaton. To be sure, the two examples he provides—the counting system for the game of odd and even and the syntactical permutations marked by the purloined letter's passage—are very rudimentary. However, in the addenda in *Écrits* (see 48, 56–57), he includes more complex circular and directional graphs. Like the simpler examples, these graphs are intended to show how the operation of a "primordial symbol" can constitute a structure linking "memory to a law." The fact that Lacan's circular and directional graphs closely resemble the transition-state graphs found in textbooks on computation and automata theory clearly establishes the extent to which he was thinking of the symbolic order explicitly in terms of finite-state automata. (These graphs indicate schematically the sequence of possible transitions from state to state for a particular machine.) Figures 2.4, 2.5, and 2.6—two of Lacan's graphs followed by a transition diagram from a classic textbook on computation—should dispel any doubts.[30]

It may be useful to "read" and explain one of these graphs. Figure 2.6, for example, shows all the possible transitions for a finite automaton with four states (q_0, q_1, q_2, and q_3) that takes a string of 1s and 0s as input. Starting in its initial state (q_0), if the first digit is 1, the automaton moves

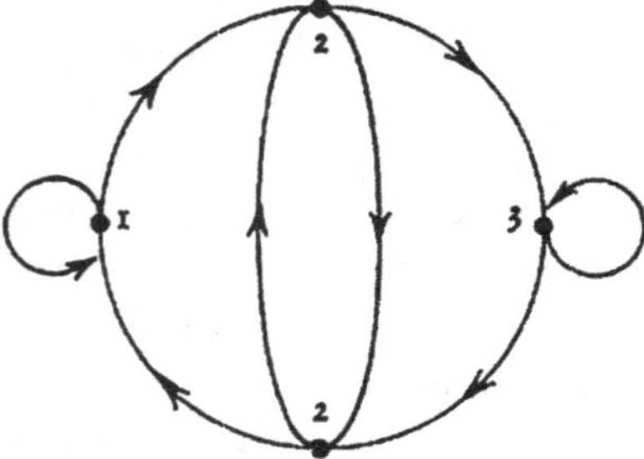

Figure 2.4
Lacan's first directed graph. Jacques Lacan, "Le séminaire sur 'La lettre volée,'" in *Écrits* (Paris: Éditions du Seuil, 1966), 48.

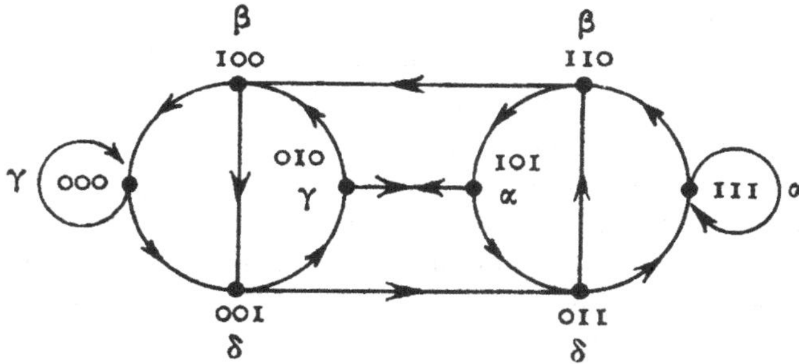

Figure 2.5
Lacan's second directed graph. Jacques Lacan, "Le séminaire sur 'La lettre volée,'" in *Écrits* (Paris: Éditions du Seuil, 1966), 57.

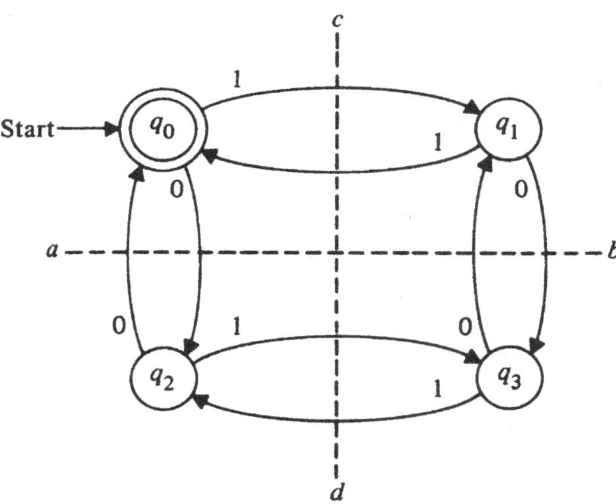

Figure 2.6
Transition diagram of a finite automaton. John E. Hopcroft and Jeffrey D. Ullman, *Introduction to Automata Theory, Languages and Computation* (Reading, Mass.: Addison-Wesley, 1979), 16.

to the q_1 state; if 0, it moves to q_2. Once in the q_1 state, if the next digit is 1, it moves back to the initial state; if a 0, it moves to q_3. If, however, it is at q_2 and the next digit is 1, it moves to q_3; if it is 0 it moves back to the initial state. Continuing in this manner, we can see that certain input strings will advance the automaton through all of its states, either clockwise or counterclockwise. Some, however, will not. The input 1101, for example, will leave the machine hanging in state q_3. This sequence of 0s and 1s, therefore, is not an "acceptable" input. Although it can be proved mathematically, simple trial and error will confirm that this finite automaton will accept *all* strings of 1s and 0s in which the number of both is even, since then control of the automaton will start at the initial state q_0 and be returned to it. This property allows the machine to be used, for example, to check a symbol string for parity.

Another consequence of the influence of automata theory on Lacan bears on the relationship between the symbolic order and natural language, a relationship that Lacan addresses in cybernetic terms in his paper presented to the Société française de psychanalyse. Before turning to the paper, it should be noted that in the seminar Lacan never identifies the symbolic order with language itself; rather, he understands the symbolic order as operating within or by means of language, in and by means of specific circuits of discourse in which signs of recognition or exchange are passed or not passed. Importantly, the operations of the symbolic order are never confused with or reduced to the operations of language. The idea that the distinction between the two must be rigorously preserved may have come from several sources, including number theory and Lévi-Strauss's work on structure and symbolic function in primitive societies. In any case, it is clear that the autonomy of the symbolic function—the key theme of the seminar—required a new conceptual framework for its full elucidation, and Lacan found it in the discourse of cybernetics and the new information machines. Indeed, it can be inferred that it was Lacan's familiarity with finite-state automata that enabled him to understand that simple information machines were not unlike simple restricted languages with a limited set of functions, in contrast to natural languages, whose full expressive powers enable them to be used for multiple purposes. Lacan presumably concluded that the workings of the symbolic order could be fully described by the grammar of a finite-state automaton, whereas natural language required a higher and more powerful grammar.[31]

We owe the scientific demonstration of this insight to Noam Chomsky, who begins *Syntactic Structures* (1957) from within the same newly emerging context of information processing and formal language theory on which Lacan implicitly draws and to which Chomsky had already contributed groundbreaking work on different types of formal grammars.[32] Chomsky asks what sort of grammar is necessary to generate all the sequences of morphemes that constitute grammatical English sentences—and only these. The "familiar communication theoretic model of language," by which he means Shannon's model in *The Mathematical Theory of Communication*, suggests a way to find the answer:

> Suppose that we have a machine that can be in any one of a finite number of different internal states, and suppose that this machine switches from one state to another by producing a certain symbol (let us say, an English word). One of these states is an *initial state*; another is a *final state*. Suppose that the machine begins in the initial state, runs through a sequence of states (producing a word with each transition), and ends in the final state. Then we call the sequence of words that has been produced a "sentence." Each such machine thus defines a certain language; namely, the set of sentences that can be produced in this way. Any language that can be produced by a machine of this sort we call a *finite state language*; and we can call the machine itself a *finite state grammar*.[33]

Chomsky then supplies a state-transition diagram for the two sentences "the man comes" and "the men come" (figure 2.7). Each node in the diagram corresponds to an internal state of the machine. The sequence of possible state-transitions ensures that if the singular subject *man* is chosen, then so is the singular verb form *comes*. Each state-transition thus limits the choice of the succeeding word or morpheme, and the sequence of possible transitions from one state of the machine to another determines a grammar. Of course, to adopt this conception of language entails our viewing the speaker as a type of machine, at least as a subject with this type of machine in his or her head, and Chomsky indisputably does (*Syntactic Structures*, 20).

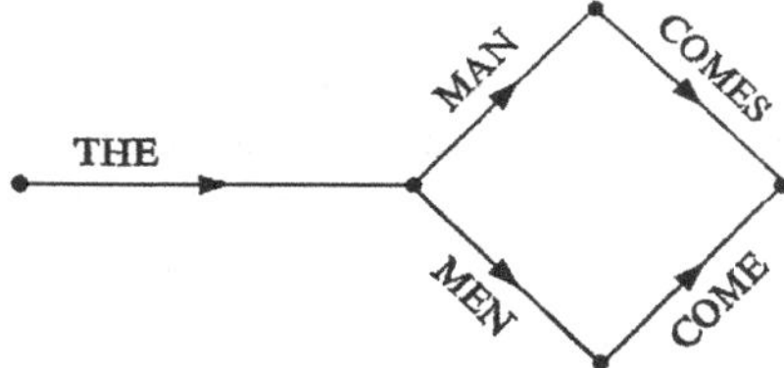

Figure 2.7
Chomsky's state-transition diagram for the two sentences "the man comes" and "the men come." Noam Chomsky, *Syntactic Structures* (The Hague: Mouton, 1957), 18–19.

Machines that generate languages in this manner are known as finite-state Markov processes, which constitute the simplest type of finite-state grammar that can generate an infinite number of sentences. But Chomsky demonstrates that this type is not adequate for generating English sentences, which require a more powerful kind of grammar—specifically, a phrase-structure grammar that allows the embedding of subordinate phrases. However, this type too turns out to be inadequate, and Chomsky is forced to conceptualize a transformational grammar that will allow one type of phrase structure to be transformed into another according to "re-write rules," which then yields the linguistic theory for which he is justly famous.

In addition to making explicit the difference between a relatively simple finite-state grammar and the more complex grammar of natural languages like English—a difference, I have argued, that is tacitly operative in Lacan's theory of the symbolic order as a finite-state grammar or automaton—Chomsky shows very clearly how formal theories of linguistic grammar are closely related to information machines. In "Three Models for the Description of Language," which provides the theoretical basis for *Syntactic Structures*, Chomsky conceives of language as either a finite or infinite set of sentences, each of finite length, all constructed from a finite alphabet of symbols. As the first step in the analysis of a language, he assumes that all sentences can be or already have been represented by a finite system of representation; in fact, they are represented as messages, in the information theoretical sense. As a matter of nomenclature, then: "If A is an alphabet, we shall say that anything formed by concatenating the symbols of A is a *string* in A. By a *grammar* of the language L we mean a device of some sort that produces all of the strings that are sentences of L and only these" (106). (Here "device" simply means a machine or automaton for generating symbol strings.) Different types of automata (or grammars) are distinguished by the type or configuration of symbol string they can recognize and generate. For a machine this is determined by the transition rules that define how the machine changes from one state to another after receiving a given symbol string as input. (As we saw above, certain symbol strings are not acceptable, since they would leave the machine hanging in an intermediary state rather than moving it to a final state or returning it to a previous state.) As in Chomsky's example above, in defining an automaton's possible sequence of states, these transition rules articulate a corresponding syntax, or grammar. More complex machines have a larger set of possible transition states, as well as a memory that allows them to transform or rewrite many more types of symbol strings into new or equivalent strings. They can therefore recognize and generate symbol strings with a more complex grammar.

In demonstrating that a finite-state automaton generating Markov strings is inadequate for modeling English grammar and that more complex types of automata like Turing machines are necessary for modeling natural and other "higher" types of language, Chomsky made an essential contribution not only to linguistics and formal language study but also to the theory of computation. In 1959 he proposed the following classification scheme of languages (or grammars) and their corresponding automata:

regular languages	finite automata
context-free languages	pushdown automata
context-sensitive languages	linear bounded automata
recursively enumerable languages	Turing machines[34]

Usually referred to the as Chomsky hierarchy, the scheme indicates an order of correlational complexity: as the grammar increases in complexity, so does the computational power of the corresponding automaton; conversely, as the computational power increases, so does the complexity of the possible orders, or grammar, of the symbol strings that it can recognize and generate. Building on this and related research, the field of automata theory deals with such topics as: "(1) Given a grammar, what is the simplest structure of a machine which will examine input strings and determine which ones are sentences in the language specified by the grammar? (2) Given a machine, find a grammar which describes the set of strings 'accepted' by the machine."[35] While these kinds of problems are obviously far more technical than the questions posed by Lacan and indicate a fairly advanced stage of automata theory, they nevertheless provide both the fullest context for and an unforeseen confirmation of his intuition that the "world of the symbolic is the world of the machine."

A Conjectural Science Redefines the Real

At the opening of his lecture "Psychoanalysis and Cybernetics," Lacan announces, "To understand what cybernetics is all about one must look for its origin in a theme, so crucial for us, of the signification of chance" (296). But rather than take up the signification of chance directly, Lacan turns to a related concern: the relationship of cybernetics to the real. By the real, Lacan initially means what is always to be found in the same place, like the fact that at the same hour and date the same constellation of stars will always appear in the sky. This sense of the real, he asserts, has always been the provenance of science. But cybernetics introduces a new and different kind of science, a conjectural science concerned with the articulation of place as such, regardless of what does or doesn't come to occupy it. (Theorists today would simply call this structuralism.)

By way of explanation, Lacan asks us to consider the opening and closing of a door. As that which opens or closes one space in relation to another, a door marks an obvious threshold and as such serves an inherently symbolic function. But whether it opens onto the real or the imaginary, Lacan says, we don't quite know, because "there is an asymmetry between the opening and closing—if the opening of the door controls access, when closed, it closes the circuit" (302). Lacan thus playfully superimposes onto the familiar function of the door its function as a "logic gate," which is a device—usually part of an electronic circuit—used to implement logical and other computational functions. When a gate is closed, there is flow of an electric current; when open, there is not. In this way the presence or absence of electrical flow (i.e., a voltage potential) digitally encodes a "bit" of information: closed or open, on or off, 1 or 0. There are several types of these gates. In an AND gate both of the two input circuits must be closed for the output circuit to be closed and thus for current to pass; in an OR gate only one of the two input circuits has to be closed; in the more complex XOR gate the circuit is closed if and only if the two inputs are not the same (i.e., one is closed and the other open). Gates are also combined to produce an invert function, in which one circuit is closed if and only if the other is open, and vice versa. By combining logic gates in multiple arrays, Boolean functions can be implemented, mathematical operations like addition and multiplication can be calculated, and information can be stored.[36]

In the lecture, Lacan reproduces the logic tables of OR, AND, and XOR gates, using the binary language of 0s and 1s. (He could just as easily have used the Boolean values true and false.) These tables appear in figure 2.8, with their proper names underneath. (The possible combinations of the two input circuits are indicated in the two columns on the left, the result in the output circuit on the right.)

Curiously, Lacan omits the proper names for these gates and doesn't bother to explain how they function; he merely states that the XOR gate is of "considerable interest" (303), without explaining why. (In fact, all of the Boolean functions can be derived from the XOR function, which is

OR gate	AND gate	XOR gate
0 0 : 0	0 0 : 0	0 0 : 0
0 1 : 1	0 1 : 0	0 1 : 1
1 0 : 1	1 0 : 0	1 0 : 1
1 1 : 1	1 1 : 1	1 1 : 0

Figure 2.8
Logic tables.

not true of the AND and OR functions.) Yet Lacan does provide all he deems necessary to show how circuits of these doors instantiate the tying of the real to a syntax, which then allows a symbolic order to emerge. Since the real is precisely that which admits of no absence—in the real nothing is ever missing from its place—the symbolic order is first of all a succession of presences and absences or, as he later puts it, "of presence on a background of absence, of absence constituted by the fact that a presence can exist" (313). There can be no doubt about what this means: "Once we have the possibility of embodying this 0, this 1, the notation of presence and absence, in the real, embodying it in a rhythm, a fundamental scansion, something moves into the real, and we are left asking ourselves—perhaps for not very long, but after all some substantial minds are doing so—whether we have a machine that thinks" (303–304).

One substantial mind, Alan Turing, had addressed this very question only a few years earlier. In "Computing Machinery and Intelligence," he argued that there is no compelling reason why machines should not be able to think and proposed what became well known as the Turing test for determining whether a particular machine can simulate human-level intelligence.[37] Surprisingly, Lacan takes the question of whether or not the machine really thinks—since humans make the machine, it can only do what they program it to do—as an occasion for play. We know that the machine doesn't think, he adds; but if the machine doesn't think, then neither do we, since when we think we are following the very same procedures as the machine. (Turing argues conversely: when broken down into steps, each part of his own reasoning process can be duplicated by a machine.) In any case, what is truly important about these new machines, Lacan asserts, is that they provide the means by which

> the chain of possible combinations of the encounter can be studied as such, as an order which subsists in its rigor, independently of all subjectivity. Through cybernetics, the symbol is embodied in the apparatus—with which it is not to be confused, the apparatus being just its support. And it is embodied in it in a literally trans-subjective way. (304)

Presciently, here Lacan attributes to cybernetics an idea that will later become a cornerstone of research in artificial intelligence, namely, Newell and Simon's physical symbol system hypothesis.[38] According to this hypothesis, symbol-processing devices do not depend for their operation on the nature of their material substrate; hence they operate equally well in either a digital computer or a human brain. Though symbols are tied to material counters, their operations are not reducible to the physical laws that govern the behavior of the latter. What does govern their behavior is the syntax of symbolic logic, which George Boole called the "language of thought." As Lacan implies in the lecture, this autonomization of the symbol, that is, its being set free from the constraints of nature (the physical laws of matter and energy), is at the heart of the symbolic function and its relationship to the real.

In contrast to phenomena of nature, which are always subject to entropy and therefore tend to equalize levels of difference, cybernetic machines, once set going, autonomously maintain or even increase levels of differentiation. "Everything we call language," Lacan says, is organized around this kind of differentiation, but in order for language to come into being, there must be "insignificant little things such as spelling and syntax" (305). Constituting "a pure syntax," the logic gates and their corresponding truth tables make it evident that cybernetics is a "science of syntax" and thus well positioned to help us perceive that "syntax exists before semantics." Does this mean that it is we who introduce meaning [or "direction," since the French word *sens* can mean either]? Not altogether, Lacan answers; for there to be meaning (*sens*) there must be "a sequence of directed signs" (305). In other words, meaning presupposes syntax, even if it is not reducible to it. What, then, is the meaning of meaning? It is "the fact that the human being isn't master of this primordial, primitive language. He has been thrown into it, committed, caught up in its gears" (307).

While Lacan is careful not to identify the syntax of language with the functioning of the cybernetic circuit, he insists that they are closely related. Considered together, they provide precise evidence of a symbolic order. In the cybernetic circuit it is visible "in its most purified form" (305), whereas in the "impure discourse" in which "the human subject addresses himself to us" (that is, addresses the analyst) it is hidden within and behind this discourse, only visible "in the symbolic function which is manifested through it" (306). Though symbols—images of the body and all kinds of other objects—pervade ordinary discourse, they serve only an imaginary function. This is a necessary function and cannot be eliminated. But in ordinary discourse the imaginary and the symbolic are always and inevitably found to be in-mixed, whereas cybernetics highlights precisely the difference between the two registers. Just as the cybernetic machine is not to be confused with its support, so the symbolic function is not to be confused with the language through which it operates and by which it is made manifest. Thus Lacan makes explicit what was only implicit in the seminar—the irreducible difference between the symbolic order and natural language.

Again we find parallels with Chomsky's theory, which also gives priority to syntax. *Syntactic Structures* contains perhaps his best-known example of this priority: "colorless green ideas sleep furiously" is a recognizably grammatical sentence, whereas "furiously sleep ideas green colorless" is not, even though neither concatenation appears to mean anything (15). What allows a speaker of English to recognize the grammaticality of the first utterance is a certain kind of machine in the head, a linguistic automaton. In this sense Chomsky's theory, like Lacan's theory of the symbolic order, also supposes an in-mixing of machines. In both cases this machine is a device for syntax recognition and production, the exact workings of which appear to be unconscious: Chomsky's machine, or automaton, putatively accounts for the human capacity for language, while Lacan's is identified with a symbolic order that works in and through language but remains distinct from it. Both are "abstract machines," imported from computational theory in order to renovate their respective disciplines.[39]

Lacan closes his lecture on what he unexpectedly admits is the central question: "to know whether the symbolic exists as such, or whether the symbolic is simply the fantasy of the second degree of the imaginary coaptations" (305–306). If cybernetics provides the answer by confirming the existence of the symbolic order, it is because it allows us to demonstrate that "there is something into which [man] integrates himself, which through its combinations already governs" (307). Although there are various ways of saying this—Lacan cites examples from both Freud and Claude Lévi-Strauss—the cybernetic machine provides the most direct and material proof that "man is engaged with all his being in the procession of numbers" (306). Since it is in or through numbers, most simply in the articulation of 1s and 0s, that nonbeing comes to be, Lacan concludes his lecture with a paradox: "The fundamental relation of man to this symbolic order is very precisely what founds the symbolic order itself—the relation of non-being to being" (308).[40]

This sudden shift to metaphysical terms in the last moments of the lecture may seem annoying or mystifying, since up to this point Lacan has spoken of the symbolic order as "tying the real to a syntax" but not in relation to "non-being." Yet a more striking and certainly more significant shift is evinced by Lacan's use of the term *real*, which varies over the course of the seminar. Analyzing Freud's dream of Irma's injection (one of Freud's most famous exercises in self-analysis), Lacan had spoken of the image of the "unnamable" that arises from the abyss of her throat. This anxiety-provoking image "summarizes what we can call the revelation of that which is least penetrable in the real, of the real lacking any possible mediation, of the ultimate real, of the essential object which isn't an object any longer, but this something faced with which all words cease and all categories fail, the object of anxiety par excellence" (164). But later in the seminar Lacan shifts from this "privileged experience, in which the real is apprehended beyond all mediation, be it imaginary or symbolic" (177) to another order of designation in which the real is the domain of chance and the random, and then again to a putatively more scientific conception of the real as that which "is always well and truly in its place" (297).[41] Yet this last conception, Lacan asserts in the lecture, will have to yield in turn to the formulations of cybernetics as "the science of the combination of places as such," for it is in cybernetics that a new relationship of symbols to the real is articulated.

This new relationship follows from the fact that when the science of numbers becomes a combinatory science, we can see that "the more or less confused, accidental traversal of the world of symbols is organized around the correlation of absence and presence" (300). Accordingly, the search for the laws of presence and absence tends toward the establishing of binary orderings, which in turn leads to cybernetics. The whole movement of the theory, Lacan insists, "converges on the binary symbol, and on the fact that anything can be written in terms of 0 and 1" (300). In order for cybernetics to appear, however, this binary symbol must begin to "function in the real":

> It has to function in the real, independently of any subjectivity. This science of empty places, of encounters in and of themselves has to be combined, has to be totalized and has to start functioning all by itself.

> What is required for that? To support this, something must be taken from the real. From the beginning, man has tried to join the real in the play of symbols. He has written things on the wall, he has even imagined that things like *Mene, Mene, Tekel, Upharsin*, get written all by themselves on walls, he has placed figures at the spot where, at each hour of the day, the shadow of the sun comes to rest. But in the end, the symbols always stayed where they were intended to be placed. Stuck in the real, one might think that they were just its landmark. (300)

But cybernetics permits these symbols "to fly with their own wings" (300), that is, to operate simultaneously in the real and in the circuit of pathways Lacan calls "the discourse of the other." The Poe story provides one instance of how the symbol can fly with its own wings: as the letter, or symbol, moves, so does the psychological disposition and distribution of the characters. But it does this only because, as it moves, it extracts something from the real by hollowing out an absence, leaving "something missing from its place." This is not a simple absence but a scansion of presences and absences that appears to self-organize into an autonomous order articulating a structure into which human beings must integrate themselves; indeed, through combinations of presence and absence, the structure "already governs" (307) the place of the human.

From the point of view of traditional semiotics or iconology, what is perhaps most striking about these formulations is Lacan's splitting of the symbol into two distinct functions. According to the first, the symbol is essentially a perceptible image of an object of desire (and of fear and anxiety), entering into discourse through what Lacan calls the imaginary order, or register. According to the second, the symbol is essentially a digital or numeric counter in a written or writeable code that is understood to govern, or regulate, human relationships. Within this framework Lacan appears merely to have reformulated the ancient and traditional opposition between image and word, desire and law, the perceptible and the intellectual. The introduction of cybernetics serves primarily as a new way to distinguish between the two, and to (re)assert the necessary superiority of the second sense of the symbolic. Yet Lacan acknowledges from the outset that these two distinguishable functions are never fully separate but always in-mixed. The cybernetic information machine therefore finds its illustrative value in the fact that the subjective, imaginary aspect of the symbol is absent (i.e., these machines have no imaginary: they cannot see and desire). As Lacan puts it, "The machine is the structure detached from the activity of the subject" (47). However, this doesn't quite work either, since this very same machine is also found to be "inside" the subject, as the mechanism enabling the subject to recognize and follow the syntactic laws that integrate him or her into the human order; at the same time, this mechanism is in-mixed with desires that subvert this very integration. At this stage of his thought, in short, Lacan seems to have no other way to articulate the "impossibility" that defines the human.

Playing against a Machine: Kasparov versus Deep Blue

Over the course of the seminar Lacan wonders several times what it would be like to play against a machine. But how, exactly, would playing against a machine shed light on the Lacanian problematic of in-mixing? In the recent chess matches between the Russian chess master Garry Kasparov and the computer Deep Blue we find a dramatic encounter that suggests several possibilities. In 1996 Kasparov, generally considered to be the greatest chess player in the world, accepted the challenge to play against Deep Blue, which had been built and programmed by a research team at IBM. Until then there had been many chess-playing programs and machines that could easily beat mid-level players, but no machine or program had ever been a match for a grand master. Yet Deep Blue was different. It was faster and more powerful; thanks to new parallel processing hardware, it could conduct multiple searches simultaneously. In fact, it was said to be capable of several million scansions of the board per second and could "think" (i.e., evaluate board positions) as many as seven moves (or "ply") ahead. Confidence among the IBM team was therefore very high.

Two matches were scheduled and played: the first in February 1996, which Kasparov won after losing the first game, and the second in May 1997, which Kasparov lost after winning the first game. Both matches have generated a great deal of commentary and discussion, which in certain ways may be as revealing as the matches themselves. What is immediately noticeable is that most of this discussion is governed by a discourse of rivalry (man versus the machine) and thus, in Lacanian terms, is played out entirely in the domain of the imaginary. But it also seems clear, at least in retrospect, that the outcome of the games could not fail to be determined by the order of the symbolic, or rather, more precisely and in keeping with Lacan, that it would emerge from a continual crossing or in-mixing of the two.

If Lacan is right, the outcome of the first match is meaningless; it is only the sequence that matters. Beyond the utter novelty and nearly unbearable tension generated by the fear or anticipation that Deep Blue might actually win (especially after winning the first game), the lesson of the first match seemed to be that it confirmed what everybody already knew—that chess is not a game of pure calculation but involves strategy or style. Although Kasparov won the first match, many commentators criticized him precisely for changing his style. At the time Kasparov was known as a power player who played very aggressively and usually vanquished his opponent with sheer speed and force. In other words, he played like a machine, although, significantly, no commentator actually says this. Evidently Kasparov realized after losing the first game that if he played to his own strength he would lose, since aggressive play based on rapid calculation is precisely what a computer can do better. So, after the first game Kasparov began to play deliberately odd opening moves, and developed a strategy of underplaying himself. He appears to have adopted the popular wisdom that a computer plays best when it has a situation to respond to, that is, when put into a defensive position, and that the most effective strategy therefore is to make it take the offensive, hoping that it will waste a move or two, which at this level of play is usually fatal. The strategy seems to have worked perfectly and Kasparov won the first match.

In the second match something very different happened. Continuing his prior strategy, Kasparov won the first game, although the quality of the game was very high and it was clear to the experts that Deep Blue was playing at a level never seen before in computer chess. In fact, between the matches the IBM team had doubled Deep Blue's processing power and programmed analyses of Kasparov's previous games and typical strategies into its memory. It was not until the second game, however, that something very surprising, indeed stunning, occurred. At move 36, instead of moving its queen to a strong position deep within Kasparov's territory, which is what everyone expected, Deep Blue hesitated, then spent almost two full minutes calculating other possibilities before finally deciding on a simple pawn exchange. From then on Deep Blue played with what the experts unanimously agreed was a certain style, as if playing with an intuitive feel for the game. "Deep Blue made many moves that were based on understanding chess, on feeling the position. We all thought computers couldn't do that," the women's world champion Susan Polger said afterward.[42] More important, Kasparov himself was so deeply shaken that his strategy immediately unraveled. He would later say that "Suddenly, [Deep Blue] played like a god for one moment," and he dubbed the play "the Hand of God" move.[43] In the postgame analysis another surprise emerged: in playing out the endgame Deep Blue had actually made a mistake, which, had Kasparov caught it, might have led to a draw instead of a loss. During the course of the game Deep Blue had played so brilliantly that Kasparov had come to assume that it couldn't make such mistakes. Consequently, he had missed his opponent's simple error, something normally inconceivable for a grand master like Kasparov.

The next two games were played to a draw. Although Kasparov stated that Deep Blue was back to playing like a computer again, it was apparent to everyone that the human player was slowly being worn down. In game 5, Deep Blue once again displayed strategic brilliance, particularly in the endgame, and Kasparov, known throughout the chess world for his nerves of steel, actually began to exhibit fear. He later acknowledged being disturbed by Deep Blue's evident power: "[Deep Blue] goes beyond any chess computer in the world."[44] Finally, in game 6, Kasparov shocked the chess world by capitulating after only nineteen moves. "I lost my fighting spirit," he confessed. "I was sure I would win because I was sure the computer would make certain kinds of mistakes, and I was correct in game 1. But after that the computer stopped making those mistakes. Game 2 had dramatic consequences, and I never recovered."

Yet this statement only tells half the story. In fact Deep Blue *had* made a critical mistake, while also playing with a widely recognized brilliance. As several observers remarked, it had played strategically: it had begun to play the game instead of responding mechanically after a rapid evaluation of the board's current possibilities. In game 2, in that two-minute pause followed by an unexpected move, Kasparov believes that Deep Blue had actually begun to think and that a spark of true artificial intelligence had been born. But perhaps at that moment Deep Blue simply began to imitate the human whose games and typical strategies it had "learned" to play against. Although imitating a machine was one of Kasparov's most successful strategies, it evidently never occurred to him—why should it?—that the machine might also begin to imitate him. Somewhat eerily, that imitation entailed an almost Medusa-like combination of superior brilliance and near obvious blunder. Apparently it was this imitation of his own strategy in combination with rapid calculation that defeated Kasparov.

Where does this leave us in regard to Lacan? What do we discover when we actually play against a machine? If modern machines like Deep Blue can not only calculate but adopt strategies and imitate us to a certain degree, will the difference between the symbolic and the imaginary orders be impossibly blurred or irrevocably clarified in the harsh light of their dazzling play? In the case of Kasparov, at least, the outcome, if not its meaning, is clear. In facing off against Deep Blue, in imagining a literally acephalic machine as the rival other, Kasparov's normally indomitable will could only dissolve and fade in something like what Lacan calls "a spectral decomposition of the ego function" (165). Speaking of the dream of Irma's injection, Lacan suggests what may also have been true for Kasparov: "What's at issue is an essential alien [*un dissemblable*], who is neither the supplement, nor the complement of [my] fellow being [*mon semblable*], who is the very image of dislocation, of the essential tearing apart of the subject" (177). In these terms Kasparov's visible collapse—particularly evident in his ascription of Deep Blue's winning move to the "hand of God"—follows from a radical disavowal that the game could be anything other than the head-on opposition between a human and a machine. In other words, Kasparov confused a scenario of the imaginary with the real itself.

In this perspective Kasparov's collapse appears to result from a total denegation of everything Lacan is trying to convey in his seminar, namely, that it is in and through the in-mixing of machines that we are integrated into a symbolic order that everywhere exceeds us. Inscribed in a circuit whose "syntax" (if not language) we have not yet learned to recognize, we continue to misrecognize the true nature of the encounter. Thus in ascribing the hand of God to an imitation or appropriation of his own strategy, Kasparov misrecognizes not only what is at stake but what is in play. For what is certain above all is that the game is *not* played out in the opposition between the human and the machine. Just where it is played out—in the crossing levels and in-mixing of the symbolic, the imaginary and the real, where the subject appears only to disappear—is precisely what Lacan's seminar teaches. Nevertheless, if from this perspective Lacan's notion of in-mixing begins to deconstruct the human/machine opposition (since the other is inscribed on both sides of the opposition), it is not a deconstruction in the Derridean sense, inasmuch as the symbolic order—the world of the machine—clearly retains an absolute priority.[45]

Yet if Lacan leaves us with a fairly precise notion of the symbolic order and its relation to the "world of the machine," what makes up or characterizes the human is less well defined. As Lacan himself puts it, "The question of knowing whether the machine is human or not is evidently well decided [*toute tranchée*]: it is not. But it's also a matter of knowing if the human, in the sense in which you understand it, is all so human as that" (319). Lacan no doubt sensed that the introduction of an abstract machine into the nexus of relationships that constitute the human could not fail to have profound consequences for the very definition of the human. In contrast then to Norbert Wiener, who wavered between a conscientious endorsement of the liberal humanist values founded on an assumed "autonomous" ego and a working commitment to the cybernetic viewpoint that eroded it, Lacan confronted the new challenge brought by information machines from a less blinkered perspective.[46] At the very least, Lacan understood that in the era ushered in by these new machines the decentralization of the ego begun by Freud would have to be carried even further and given new theoretical foundations. That he sought to achieve this by means of the very theory that brought these machines into existence redounds all the more to his credit.

There are, however, limits to Lacan's farsightedness. Having introduced the new science of cybernetics into his seminar on psychoanalytic theory in 1954, he also pointed to what he thought was an obvious limit of this new science: "Machines which reproduce themselves have yet to be built, and have yet even to be conceived of—the schema of their symbolic has not even been established" (54). Evidently he was not aware that just such a schema had already been worked out, only a few years before, by John von Neumann.[47]

I conclude with a final historical note on Kasparov and Deep Blue. Although their confrontation occurred only ten years ago, in many respects they are relics—but cultural icons nonetheless—from an earlier historical moment. Whereas Kasparov is a highly egotistical Cold War–style strategist, free to imitate a machine precisely because he believes that he is not one, Deep Blue is a powerful computational machine built on principles of artificial intelligence worked out in the 1950s and '60s. The foundations of both positions—of the computational assemblage that together they articulate—are now obsolete. The limits of classic AI, which attempts to emulate the human by mechanically reproducing certain isolated human cognitive functions, are now widely recognized and have largely given way to other approaches. And similarly for the specious man-versus-machine opposition. To take only one example, J. C. R. Licklider's "Man-Machine Symbiosis" (published in 1960) adumbrates an altogether different kind of cybernetic "machinic subject": "The hope is that, in not too many years, human brains and computing machines will be coupled together very tightly, and that the resulting partnership will think as no human brain has ever thought and process data in a way not approached by the information handling machines we know today."[48] Licklider understands "man-machine symbiosis" as a subclass of human-machine systems that promises to be a distinct advance over the "mechanically extended men" who characterize the industrial era, but without the problems that the artificial intelligence of the distant future will bring. In fact, he predicts a "fairly long interim during which the main intellectual advances will be made by men and computers working together in intimate association" (58).

Deep Blue itself, furthermore, is not so much a "thinking machine" as a chess-playing program implemented on the lightning-fast hardware of a parallel-processing computer. The basis for the program, first worked out by Claude Shannon in 1950, is an evaluative function that rates every possible position on the board in terms of a minimax strategy.[49] But although Deep Blue performs this function faster than any computer ever built and at grand master levels of expertise, it is not, as IBM acknowledges on its Web site, "a learning system ... [and is] therefore not capable of utilizing artificial intelligence to either learn from its opponent or 'think' about the current position of the chessboard." In other words, its ability to "learn" is extremely limited; it can neither "teach itself" nor invent new strategies. More important, it cannot adapt or change its behavior in response to changing environmental conditions.

By today's criteria, this makes it a very limited form of artificial intelligence. As the contemporary roboticist Rodney Brooks explains in his essay, "Elephants Don't Play Chess," whereas traditional AI "tried to demonstrate sophisticated reasoning in rather impoverished domains [i.e., chess]," the new AI "tries to demonstrate less sophisticated tasks operating robustly in noisy complex domains."[50] While the two approaches may seem in some ways complementary, what drives much of current AI research is the excitement generated by using evolution itself, as is evident in evolutionary programming, Artificial Life, and evolutionary robotics. These new sciences suggest that the best route to producing truly creative intelligence may lie in mimicking the processes of natural evolution. Indeed, as culture itself becomes more dependent on advanced technology, it is more likely that strategies of artificial evolution will provide the means by which the culture of the future will be engineered. If cybernetics and information theory can be said to have inaugurated a new discourse network in the 1950s, then the now unforeseeable results of artificial evolution will one day be seen as its full flowering.

3 Machinic Philosophy: Assemblages, Information, Chaotic Flow

Valentine: We can't even predict the next drip from a dripping tap when it gets irregular. Each drip sets the conditions for the next, the smallest variation blows prediction apart, and the weather is unpredictable the same way, will always be unpredictable. When you push the numbers through the computer you can see it on the screen. The future is disorder. A door like this has cracked open five or six times since we got up on our hind legs. It's the best possible time to be alive, when almost everything you thought you knew is wrong.

Hannah: The weather is fairly predictable in the Sahara.

Valentine: The scale is different but the graph goes up and down the same way. Six thousand years in the Sahara looks like six months in Manchester, I bet you.
—Tom Stoppard, *Arcadia*

As a new kind of abstract machine defined solely by its computational functionality, the Turing machine is surely one of the most consequential conceptions of the twentieth century, leading not only to the invention of the modern computer but also to new ways of thinking about human cognition. This new type of machine differs essentially from both the simple machines known in antiquity (wheel, axle, lever, pulley, wedge, and screw) and the motor or power-driven machines that characterize modernity (windmill, turbine, steam and combustion engines). Whereas the two earlier types transmit force or energy, the new type processes information. Whether a calculating machine or logic engine, a finite automaton or universal Turing machine, an information processor does something fundamentally different, touching at the heart of what is often said to be the essence of the human: the capacity to think. Little surprise, then, that with the appearance of the information-processing machines we simply call computers (and the popular press first referred to as thinking machines), symbol systems, cognition, and symbol-dependent behavior became new objects of theoretical scrutiny. As we've seen, in parallel with the early formulations of artificial intelligence, Jacques Lacan was able to isolate what governs the operation and role of the symbolic order. By showing how these new machines digitally encode sequences of symbols articulating presence with absence, Lacan could thus lay bare—at least this was his claim—the specific logic that regulates human existence.

Thanks to cybernetics, Lacan recognized that this symbolic order always exists in tension with another register of human experience, the imaginary order, within which bodies and objects, starting with our own bodies, are internalized in images. Lacan also realized that the new cybernetic machines do not simply exist outside of us. Insofar as many human activities are computational, that is, either perform or depend upon computations, human being assumes an in-mixing of many kinds of information-processing machines. Although not concerned with how these machines would greatly intensify the (for him) imaginary opposition between human and machine—and thus provoke a denial of this very in-mixing, Lacan accepted the challenge that cybernetics posed to traditional human boundaries. The chess matches between Deep Blue and Kasparov subtly display the effects of this boundary disturbance. However, with the flowering of AI and ALife, genetic engineering, cloning, biochip implants and other cyborg technologies, the blurring of heretofore fundamental boundaries becomes more dramatically visible. In order to explore these boundary disturbances and breakdowns in a less negative light—as a new nexus of creative transformation—a wider and more radical framework than Lacan can provide is necessary. What is needed is a framework that enables us to grasp and go beyond the limits of his concept of the symbolic as an abstract machine, which he imported from mathematics and the formal sciences in order to impute universality and deep structure to phenomena that seem neither wholly universal nor historical, neither wholly natural nor cultural, but function at the hinge of these deeply entrenched oppositions.

Such a framework can be found in Gilles Deleuze and Félix Guattari's two-volume *Capitalism and Schizophrenia*.[1] Its original impetus, plainly visible in *Anti-Oedipus* (the first volume), was to explode the limits of the psychoanalytic theory of desire by arguing that desire is produced not from a dialectic of lack embedded within the Oedipal triangle (Daddy-Mommy-Me)—indeed, this societal set-up is seen to be only one of many culturally contingent possibilities—but by particular arrangements of bodies and discourses that define the social field, which can be coded and stratified in a variety of ways. Yet Deleuze and Guattari exceed this initial critical aim by elaborating an alternative theory of "machinic desire." This theory extends Lacan's notion of the in-mixing of machines beyond the boundaries of the individual subject and relocates both subjects and machines on an expansive surface they designate as the *socius*, or "body without organs." Nature itself becomes a machinic process, endlessly producing, inscribing, and consuming flows that are cut into, siphoned off, and connected to other flows. This, in essence, is what Deleuze and Guattari call a desiring machine. In *A Thousand Plateaus* (the second volume), they redefine these machines in a theory of the assemblage (*agencement*), by which they mean a specific functional arrangement of heterogeneous flows and interactions, a concrete set-up of connections between humans and machines that ensures both the coding and decoding of fluxes of matter, energy, and signs (information). Feudalism, psychoanalysis, the war machine, a performance by the composer John Cage—there is no limit to their variety. Assemblages arise, mutate, and disappear at specific conjunctions of material and historical forces; hence they are always dated. However, because they are guided by an abstract machine that reappears in different time periods and unique social formations, they are not simply historical constructions. This aspect of the assemblage leads Deleuze and Guattari to postulate the existence of a special realm they call the machinic phylum, which cuts across the opposition between history and nature, the human and the nonhuman. The term itself suggests a conjunction or interface between the organic and the nonorganic, a form of "life" that combines properties of both. They locate the origins of the machinic phylum in the efforts of the first metallurgists, who fabricated weapons out of the earth's ores and metals.

Following Deleuze and Guattari, Manual DeLanda has extended this realm to include all forms of nonorganic life that occur with the self-organization of matter.[2] Taken as a whole, this entire body of work provides a fresh perspective for thinking about how science and information technology have begun to loosen and transform one of Western culture's most fundamental boundaries, dating back to the Greek opposition between *phusis* and *technē*. In general terms, Deleuze and Guattari's notion of the machinic phylum can be said to arise in the space—physical as well as conceptual—where crossovers and exchanges erode the clear and fixed distinction between the two. In a further extension of this body of work, I suggest that this realm is the site of a "becoming machinic," fully explicable as neither a natural evolutionary process nor a human process of construction. Rather, it is a process of dynamic self-assembly and organization in new types of assemblage that draw both the human and the natural into interactions and relays among a multiplicity of material, historical, and evolutionary forces. In this new space of machinic becomings an increasingly self-determined and self-generating technology thus continues natural evolution by other means.

This chapter focuses on the new window opened onto the machinic phylum by the computer and the computational methods essential to the discoveries of chaos science. Thanks to the computer, new studies of nonlinear dynamical systems have made visible and understandable many unsuspected ways in which common physical forces act unpredictably and even creatively, most notably as new research in physical turbulence stimulated new approaches to modeling the "strange" behavior of such systems.[3] Dynamical systems are systems whose state changes as a function of time; their behavior is usually described by differential equations that can be solved by analytic methods. (The latter break the problem down into parts that are solved separately and then recombined.) Nonlinear dynamical systems, on the other hand, have exponential or other functions that make them intractable, or nearly so; parts of these systems interact in ways that produce disproportionate and strange effects. For example, small changes can produce unpredictably large effects, and these systems usually exhibit chaotic or aperiodic behavior. Because of these difficulties, nonlinearity was mostly ignored, or its effects were approximated. Yet this proved to be a form of blindness, finally rectified with the advent of nonlinear studies and what became widely known in the 1980s as chaos science. Generally speaking, this science is concerned with the emergence and measurement of the new fractal, or self-similar, structures and unpredictable regularities that nonlinear systems exhibit, often revealed by computer modeling and simulation. The result has been a new understanding and redefinition of chaos. Not surprisingly, both the technical and popular literature on chaos theory—a lay term not used by scientists—is now extensive. In fact, two popular and successful books, Ilya Prigogine and Isabelle Stengers's *Order out of Chaos* and James Gleick's *Chaos: Making a New Science*, have created the impression that there are two distinct strands of chaos science.[4]

The first strand centers on the emergence of order out of chaos and is often identified with Prigogine's Nobel Prize–winning research on highly dissipative, open thermodynamic systems, in which large amounts of energy are expended and energy is not conserved. At a state "far from equilibrium," or near chaos, Prigogine argued, many physical systems spontaneously organize themselves and begin to evince new orderly patterns of behavior. The most famous example (though not Prigogine's discovery) is the Belousov-Zhabotinski chemical clock, in which a chemical reaction suddenly produces colorful, pulsating rhythms. In his foreword to Prigogine and Stengers's *Order out of Chaos*, Alvin Toffler likens this jump to a higher level of organization to the visual effect that a million black and white ping-pong balls bouncing around randomly in a tank would make—visually a gray blur—if suddenly they fell into distinct patterns of black on white and white on black.[5] The problem with *Order out of Chaos*, however, is that all too often the authors seamlessly blend science with sometimes unwarranted philosophical speculation. Nevertheless, Prigogine's work with self-organizing chemical systems retains its rightful place alongside other innovative investigations of self-organizing dynamical systems. Hermann Haken's work on laser light theory and Manfred Eigen's work on prebiological forms of organization known as hypercycles are two prominent examples. In providing evidence for the spontaneous emergence of complex self-organizing structures, all three research initiatives bring to fruition the early theoretical formulations of W. Ross Ashby and Heinz von Foerster, in particular the principle of order from noise considered in the first chapter.

The second strand, brought to wide public attention by James Gleick's *Chaos: Making a New Science*, focuses on the application of nonlinear mathematics to diverse phenomena like weather prediction and turbulent flow, population growth, and the rise and fall of market prices.[6] In contrast to Prigogine's interest in a new kind of order that emerges from the self-organization of matter, this second strand is distinguished by the discovery of deeply embedded fractal patterns of order within processes (or data) that formerly appeared to be chaotic or random.[7] Thus whereas the first strand—*pace* both classical and quantum physics—argues for the *ir*reversibility of time, the second demonstrates the limits of predictability, even for completely deterministic systems. Somewhat confusingly, moreover, the first strand is sometimes loosely referred to as a version of complexity theory, whereas the second is simply subsumed into dynamical systems theory. Nevertheless, as we'll see, at least one line of early research in chaos science, Robert Shaw's modeling of a chaotic system, conceived of the two together.

Given Deleuze and Guattari's central concern with the chaotic flux of matter and energy and its capture and stratification by various coding mechanisms, it would be surprising if *A Thousand Plateaus* did not exhibit some kind of influence or relationship with chaos science. Yet when the book was fist published in 1980 much of the groundbreaking research on turbulent phenomena, deterministically chaotic systems, and self-organizing tendencies in a variety of physical systems was still being carried out and not widely known. Nevertheless, in *War in the Age of*

Intelligent Machines Manuel DeLanda boldly claims that Deleuze's notion of singularity anticipates the mathematical modeling in phase space of the attractors and bifurcations that characterize nonlinear dynamical systems and hence that aspects of chaos theory were already inscribed in Deleuze's earlier philosophy.[8] While this claim is perhaps overstated, it is evident that in certain respects *A Thousand Plateaus* does converge and resonate with aspects of chaos science. Of further importance, their central notion of the assemblage clearly owes much to cybernetics. But rather than mapping these convergences, DeLanda conflates Deleuze and Guattari and dynamical systems theory in the making of a revitalized materialist philosophy in which human agency plays almost no part. This poses something of a problem, for although *A Thousand Plateaus* is materialist in intention, it strives not to eliminate the human but to resituate it in a highly distributed field of functions defined by the assemblage. Thus while arguing for a certain kind of naturalism, the book also presents itself as a clear instance of constructivism.

I suggest that Deleuze and Guattari—or the collective authorial agency I shall henceforth refer to as D&G—go beyond such contradictions in their elaboration of what I call machinic philosophy. As I examine aspects of this philosophy in the following sections, it should become evident how an articulation with certain fundamental concepts of chaos science can strengthen D&G's central concept of the assemblage by supplying what is crucially missing, namely, a causal framework that would explain why particular assemblages arise, mutate, and disappear. Conversely, D&G's concepts can also cast light on some of the most radical implications of chaos science research. Thus instead of rewriting D&G in the terms of dynamical systems theory, as DeLanda does, I examine two specific instances of how D&G's concepts and those of chaos science relay and cross-illuminate one another. Specifically, I show how Robert Shaw's modeling of a chaotic system and Jim Crutchfield's ϵ-machine construction function as what I call computational assemblages. In its innovative combination of dynamical systems theory and information theory, their groundbreaking research exemplifies an approach that will become essential to ALife and AI. While D&G's concept of the assemblage involves no explicit computational component, in this chapter I clarify how I have refashioned it into a framework for defining and situating different kinds of information machines and their discourses. Finally, beyond the usefulness of this concept for distinguishing among various computational systems, D&G's discussion of nonorganic life in relation to a new realm they call the machinic phylum will also be of overarching importance to this book as a whole. As later chapters show, with the appearance of new forms of artificial life—digital organisms, mobile autonomous robots, and autonomous software agents—this concept assumes a relevancy far beyond D&G's original formulations.

The Theory of the Assemblage

Deleuze and Guattari's *Capitalism and Schizophrenia* projects an image of thought hardly recognizable to many academic philosophers. Both in its organization and treatment of a vast quantity of material it blurs disciplinary boundaries and enjoins us to rethink themes both ancient and modern through the lens of a novel series of concepts. This originality emerges more fully in the second volume, *A Thousand Plateaus*, which advances well beyond the first volume's unremitting polemic against psychoanalysis and, to a lesser extent, beyond its anthropological and semiotized Marxism, both of which make *Anti-Oedipus* appear less conceptually rigorous than it is. As one sign of this development the new concepts actually shape the second volume's composition. Divided into "plateaus" rather than chapters, sections of *A Thousand Plateaus* are meant to be read and connected up along a variety of paths. Although each plateau is devoted to a central concept and set of related themes, there is "recurrence and communication" among and across the plateaus according to the principles of the rhizome, which provides an image for the way the book is arranged.[9] In contrast to the ordered, forking hierarchy of the tree and other arborescent structures, a rhizome (like crabgrass or a burrow) shoots out in all directions at once, articulating a decentered network of interconnecting paths with multiple entrances and exits. As an alternative to the concept of unity, D&G propose what they call a plane of consistency, where unstructured parts meet and coalesce in a multiplicity. Neither the formation of a dichotomy (e.g., one or many, subject or object) nor the reduction to a single substance with many attributes, the consistency of such a plane can be likened to that of small population or "pack," or a multiagent, highly distributed system. In short, it is a collectivity of diverse forces working together without any centralizing command and control structure or hierarchical organization. In these terms *A Thousand Plateaus* constitutes (and advocates itself as) a system of nonhierarchical relations based upon multiplicity and states of intensity, rather than conventional notions of structure and unity.

Two terms in particular—*machine* and *machinic*—may pose difficulties for some readers. For D&G, the term *machine* designates an ensemble of heterogeneous parts and processes whose connections work together to enable flows of matter, energy, and signs (and consequently desire); a machine, therefore, is not necessarily a mechanical device. The term *machinic* refers precisely to this working together of heterogeneous flows of parts and processes. These terms do not operate as metaphors, as in the phrase "the turning wheels of the bureaucratic machine," but designate arrangements and processes, respectively. Machines, first of all, are assemblages that include both humans and tools, or in modern societies, technical machines. These machines are never simply "out there," awaiting our use; this is a myth fostered by capitalism, like the belief that individualism and the liberal self are *not* constructions of the market and somehow predate it. D&G therefore collapse the boundary between subject and machine. Instead of seeing the machine as a part of objective reality that subjects can take up and use for their own productive or expressive ends, they understand machines and subjects as imbricated in particular distributions across a full body ("the body without organs" or *socius*) defined by the kinds of reciprocal relations that subjects and machines can enter into on this full body's surface. In *Anti-Oedipus* this view gives rise to various typologies and regimes, not only of social organization (primitive, barbarian, capitalist) but also of individual types of psychic structure (perverse, paranoid, schizoid), defined according to how these relations are distributed and coded.

In *A Thousand Plateaus* D&G redefine these typologies and regimes as functions of an assemblage, which is constituted by the conjunction of two subassemblages: a collective assemblage of enunciation (i.e., a specific regime of signs and gestures) and a machinic assemblage of interacting bodies. In this conjunction of the two component assemblages "there are *states of things*, states of bodies (bodies interpenetrate, mix together, transmit affects to one another); but also *utterances*, regimes of utterances where signs are organized in a new way, new formulations appear, a new style for new gestures."[10] To illustrate, D&G consider the feudal assemblage, which on the one hand is composed of a new machinic assemblage of man-horse-stirrup and on the other of a new set of utterances, gestures, and emblems that individualize the knight and give form to his oaths of allegiance and declaration of love, and so on. Thus on one side of the assemblage, human bodies enter into a new relationship with an animal by means of a technology, which in turn makes possible the development of new weapons and a different fighting style, while on the other side human subjects are distinguished and individualized by new expressive possibilities and attendant social formalizations. Yet neither side is viewed as the cause or result of the other; instead, both are functional aspects of the same assemblage, reciprocally determining and coproductive.

Assemblages form, however, only after a series of stratifications produced by a process of double articulation. This double articulation operates by selection (or sorting) and consolidation (or coding), thereby producing a duality of substance and form, territory and code. In the third plateau, the lecturer "Professor Challenger" (a fictional *persona*) explains how flows of unformed, unstable matters are given form and stratified, "imprisoning intensities and locking singularities into systems of resonance and redundancy" (40). This kind of stratification is first seen in the creation of sediment and sedimentary rock in the geological stratum:

> The first articulation chooses or deducts, from unstable particle-flows, molecular units or metastable quasi-molecular units (*substances*) upon which it imposes a statistical order of connections and successions (*forms*). The second articulation establishes functional, compact, stable structures (*forms*), and constructs molar compounds in which these structures are simultaneously actualized (*substances*). Thus ... the first articulation is the process of "sedimentation," which deposits units of cyclic sediment according to a statistical order: flysch, with its succession of sandstone and schist. The second articulation is the "folding" that sets up a stable functional structure and guarantees the passage from sediment to sedimentary rock. (40–41, trans. modified)

Stratification thus involves the formation of substances, which are nothing but formed matters. But forms imply a code, a system of resonance and redundancy, which for D&G always occurs in relation to a specific territory (here a geological stratum). In order for an area of formed matter to be a territory, it must be marked, but coding involves more: in the second articulation it produces "phenomena of centering, unification, totalization, integration, hierarchization, and finalization" (41). Coding thus establishes a form, an order and organization of functions, but always and only in relation to a particular milieu or territory. This understanding of coding (which differs from more familiar semiotic conceptions), becomes clearer when D&G consider two successive forms of stratification, the formation of organic strata and the formation of anthropomorphic strata (which includes language).

The details of these subsequent stratifications are less important here than the fact that the process performs the work of a double articulation. In order to describe these later stratifications D&G mobilize Louis Hjelmslev's theory of glossematics, in which the categories of content and expression are further subdivided on both sides into form and substance.[11]

Hjelmslev's theory thus yields both a form of content and a form of expression, as well as a substance of content and a substance of expression. Since D&G do not want a scheme that projects *only* signifying or meaning-bearing categories onto material processes, this set of distinctions is greatly preferable to purely linguistic relations (as in signifier to signified). With Hjelmslev's categories in place, D&G can say that the first articulation concerns content, the second expression. In the geological or physico-chemical stratum with which they begin, content and expression have no relevance and the primary distinction is between the "molecular" (uncoded) and the "molar" (coded). But with the appearance of organic strata content and expression become distinguishable, and the latter becomes independent: the unit of expression is the linear sequence of nucleotides, and the unit of content is the linear sequence of amino acids that corresponds to the DNA sequence:

> Before, the coding of the stratum was co-extensive with that stratum; on the organic stratum, on the other hand, it takes place on an autonomous and independent line that detaches as much as possible from the second and third dimensions. Expression ceases to be voluminous or superficial, becoming linear, unidimensional (even in its segmentarity). The essential thing is *the linearity of the nucleic sequence*. (59)

This linearity means that the code (here DNA) can be detached from a specific territory and reproduced, thereby constituting *the first* "deterritorialization," which is what D&G call the detaching of a specific code from its associated territory. They also refer to this process as a "decoding."

The advent of the anthropomorphic stratum is defined by yet another new distribution of content and expression, and not by any putative "human essence":

> Form of content becomes "alloplastic" rather than "homoplastic"; in other words, it brings about modifications in the external world. Forms of expression become linguistic rather than genetic; in other words, they operate by means of symbols that are comprehensible, transmittable and modifiable from outside. What some call the properties of human beings—technology and language, tool and symbol, free hand and supple larynx, "gesture and speech"—are in fact properties of this new distribution.... Leroi-Gourhan's analyses [in *Gesture and Speech*] give us an understanding of how contents came to be linked with the hand-tool couple and expressions with the face-language couple. In this context, the hand must not be thought of simply as an organ but instead as a coding (the digital code), a dynamic structuration, a dynamic formation (the manual form, or manual formal traits). The hand as a general form of content is extended in tools, which are themselves active forms implying substances, or formed matters; finally,

products are formed matters, or substances, which in turn serve as tools. (60–61, trans. modified)

The hand "as a formal trait or general form of content" thus marks a new threshold of deterritorialization and accelerates the "shifting interplay" of deterritorialization (decoding) and reterritorialization (recoding). And similarly for speech as a new form of expression. D&G offer a clear example in their earlier book on Franz Kafka.[12] The mouth, tongue, and teeth, they say, "find their primitive territoriality in food," but are then deterritorialized in the articulation of sound; sounds, in turn, are reterritorialized in meaning. Finally, Kafka achieves a deterritorialization of expression itself by pushing further along lines of linguistic impoverishment already evident in the provincial Czech German in which he wrote, arriving at an asignifying, intensive use of language. This trajectory of deterritorialization D&G call a "line of flight," since it marks both a path of escape and a vector along which a code is detached from a territory. Yet according to the same "shifting interplay" or dynamic, a line of flight (a deterritorialization) also presupposes that somewhere in the assemblage something else is being captured, recoded, and stratified, in what amounts to a type of cybernetic feedback effect. For this reason deterritorialization is said to be relative. However, in certain instances deterritorialization becomes absolute, causing a mutation or dissolution of the entire assemblage.

The coding of language, of course, is much more complex than that of DNA. For D&G the main difference is that "vocal signs have temporal linearity, and it is this superlinearity that constitutes their specific deterritorialization and differentiates them from genetic linearity" (62). More specifically, temporal linearity gives rise to a

formal synthesis of succession in which time constitutes a process of linear overcoding and engenders a phenomenon unknown on the other strata: *translation*, translatability, as opposed to the previous inductions and transductions. Translation should not be understood simply as the ability of one language to "represent" in some way the givens of another language, but beyond that as the ability of language, with its own givens on its own stratum, to represent all the other strata and thus achieve a scientific conception of the world. The scientific world (*Welt*, as opposed to the *Umwelt* of the animal) is the translation of all of the flows, particles, codes, and territorialities of the other strata into a sufficiently deterritorialized system of signs, in other words, into an overcoding specific to language. This property of *overcoding* or *superlinearity* explains why, in language, not only is expression independent of content, but form of expression is independent of substance: translation is possible because the same form can pass from one substance to another, which is not the case for the genetic code, for example, between RNA and DNA chains. (62)

This power of translation, and the fact that language achieves a greater degree of independence from its material medium, gives it unequaled expressive power. At the same time, D&G caution against "certain imperialist pretensions on behalf of language" which they see as prevalent in linguistics.

Obviously this brief summary cannot fully explain these complex processes. It is meant only to indicate—in barest outline—how D&G recursively apply the principle of double articulation (selection and consolidation) to material and semiotic processes at ascending strata. In sum, processes of stratification and destratification, coding and decoding, produce formations doubly articulated into form and substance, content and expression. On both molar (coded) and molecular (decoded) levels, these processes are always found at work following the same dynamic, as we move upward from geological to organic to anthropomorphic strata. Yet the principle of double articulation alone can only describe the dynamic of stratification and deterritorialization, coding and decoding, evident throughout the processes that D&G analyze. As we'll see below, to fully account for the dynamic working of force and material—the "shifting interplay" of doing and undoing—they postulate a virtual force, or abstract machine, that, while not fully present, actualizes these doubly articulated processes.

At the last, anthropomorphic series of strata, the process of double articulation yields a more complex formation, which D&G call an assemblage. It is here, within the assemblage, that human activities can finally be considered. Nevertheless, what is human and not human within the assemblage is never given in advance; in fact, for D&G the human never transcends the assemblages within which it is always to be found. To map an assemblage in human terms, one must determine both what captures desire and sets it circulating, by tracing flows and blockages, the codings and decodings that make it meaningful, and the intensities and "black holes" where it is sustained or disappears. Desire is always a machinic setup, a particular arranging and assembling of connections that bring about a specific flow of bodies and signs. It is not, therefore, a "lack in being" that would define the subject as a set of impossible desires (as in Lacan). From D&G's perspective, moreover, the encoding of the real effectuated by the symbolic is merely one type of machine, which has been singled out and elevated to the status of a formal principle defining and regulating human existence. Whereas individual agency for Lacan is mostly an illusion of the ego, for D&G it operates within the larger set of distributed relations that define the assemblage. Assemblage, arrangement, setup: these are possible translations for D&G's unusual term, *agencement*, which suggests a locus of agency within which specific functionalities are distributed and the human subject deploys a specific regime of signs as the basis on which to form an identity.

An assemblage, however, is not always or necessarily the intentional construction of an individual or a collectivity; more often it arises when converging forces enable a particular amalgam of material and semiotic flows to self-assemble and work together, and within which forms of subjectivity and subject positions can emerge. This contingency is reflected in the fact that D&G define an assemblage only pragmatically, in terms of its components and actions, rather than in relation to meanings that can be attributed to it. Composed of relations, liaisons, and affiliations among and across heterogeneous materials and functions that work together, an assemblage possesses no inherent structural unity or totality, only a functional consistency that emerges from the "jelling together" of its parts and processes. It is, in short, nothing more than a functional arrangement of material and semiotic flows, with no other meaning than the fact that it "works." Specifically, an assemblage works by conjoining two reciprocally interacting realms—a machinic assemblage of bodies and a collective assemblage of enunciation—that hold it in a state of dynamic equilibrium. It is essentially a cybernetic conception, and in a loose (nonscientific) sense, self-organizing, though D&G do not use the term. Within the assemblage agency is completely distributed; no centrally located seat of power, control, or intentionality allocates and guides its functioning. Nevertheless, assemblages are always subject to capture and overcoding, most egregiously by the State apparatuses that form despotic and imperialist regimes of power.

While resembling a cybernetic machine, the assemblage differs from it in this respect: instead of positive and negative feedback, there is both a coding and decoding, or territorialization and deterritorialization, in each of the two realms, or sides, of the assemblage. Diagrammatically, this yields a "tetravalent" set of relations, meaning that on *both* sides of the assemblage—the one of material bodies, the other of semiotic expressions—forces pull in opposed directions, both toward stabilization of the assemblage through the creation of molar forms (stratification) and the articulation of a code with a territory, or milieu, and, simultaneously, toward its dissolution or mutation through a particle-izing (molecularization) and decoding, that is, the detaching or unhinging of a code from a territory. This deterritorialization usually remains relative, offset or counterbalanced by reterritorializations and recodings within the assemblage.

In instances of absolute deterritorialization, however, the assemblage itself loses coherence and dissolves, and the subject positions it makes possible disappear. This eventuality remains undertheorized by D&G and suggests one place where chaos science might provide a complementary perspective on what happens between the dissolution of one assemblage and the appearance of another. In short, the moment of crisis and dissolution may well be described by dynamical systems theory as a bifurcation in which constituent processes become chaotically unstable and vacillate among a set of possible solutions. The appearance in the same milieu of another assemblage that successfully self-organizes into a working functional arrangement would then be the solution to this crisis.

By way of summary, D&G delineate what a full description of the feudal assemblage would entail:

> Taking the feudal assemblage as an example, we would have to consider the intermingling of bodies defining feudalism: the body of the earth and the social body; the body of the overlord, vassal, serf; the body of the knight and the horse and their new relation to the stirrup; the weapons and tools assuring a symbiosis of bodies—a whole machinic assemblage. We would also have to consider statements, expressions, the juridical regime of heraldry, all of the incorporeal transformations, in particular, oaths and their variables (the oath of obedience, but also the oath of love, etc.): the collective assemblage of enunciation. On the other axis, we would have to consider the feudal territorialities and reterritorializations, and at the same time the line of deterritorialization that carries away both the knight and his mount, statements and acts. We would have to consider how all this combines in the Crusades. (89)

To map the assemblage's tetravalence, therefore, means tracing two kinds of lines: the lines of stratification in which forces acting on both sides of the assemblage produce stability and redundancy, both of substance and form, territory and code; and, at the same time, the "lines of flight," or vectors, along which the assemblage is destratified and particle-ized until it dissolves and is swept away. In the example of the Crusades, codings and reterritorializations like oaths of allegiance and commitments of faith reinforce the assemblage's stability and counter the pull of deterritorializing forces that sweep toward the "outside" and limits of the assemblage. Although D&G do not discuss the dissolution of the feudal assemblage, a similar pairing of terms in Thomas Carlyle's famous declaration that feudalism was destroyed by gunpowder and the printing press accords perfectly with their theory: on the one hand, gunpowder brings about a new war technology that produces a different machinic assemblage of bodies; on the other, the printing press as a new technology for producing and disseminating statements brings about a different type of collective assemblage of enunciation. Their integration in a new assemblage leads to the formation of the modern European state.

We arrive finally at the assemblage's piloting function, which D&G call the abstract machine. Defined "diagrammatically" (141–145), as what "constitutes and conjugates all of the assemblage's cutting edges of deterritorialization" (141), the abstract machine is responsible for what defines and dissolves the assemblage. As such, it is one of D&G's most complex and elusive concepts:

> The diagrammatic or abstract machine does not function to represent, even something real, but rather constructs a real that is yet to come, a new type of reality. Thus when it constitutes points of creation or potentiality it does not stand outside history but is instead always "prior" to history. Everything escapes, everything creates—never alone, but through an abstract machine that produces continuums of intensity, effects conjunctions of deterritorialization, and extracts expressions and contents.... Abstract machines thus have proper names (as well as dates), which of course designate not persons or subjects but matters and functions. (142)

Much of the difficulty in this passage will drop away if we assume, as Manual DeLanda asserts (see the section below on the machinic phylum), that D&G's abstract machine is conceptually equivalent to attractors and bifurcations in dynamical systems theory. No doubt the similarity is strong. Rather than take up the issue here, however, I want to develop the point that the abstract machine is not formed by and does not act in opposition to specific or concrete machines; rather, it involves a relation of the virtual to the actual.

This usage follows from Henri Bergson's distinction between the possible and the real on the one hand, and the virtual and the actual on the other—two oppositions that philosophy often tends to blur. Given a situation or specific state of affairs, the possible is what might happen or could be, as in the simple example: if only the driver of the car had stopped for the red light, the accident would not have occurred. As such, the possible is always formed on the image of the real, hovering like ghostly images of other possible realizations. The relation of the possible to the real is thus one of resemblance and limitation, since only one of these possible images can be realized, to the exclusion of all others. Conceptually, the only difference between the possible and the real is a logical difference, the predicate of existence. The relation of the virtual to the actual, on the other hand, is completely different, and accounts for the creative process of biological evolution in contrast to the mechanical repetition of physical law. Creative evolution, Bergson says, is a virtuality in the process of actualization. This actualization occurs through ceaseless differentiation, dissociation, and division; in short, from the splittings, forkings, and branchings of an always burgeoning ramification. The virtual is like a seed in time, but unlike a plant seed, its actualization can take any number of forms or directions. (In this sense, a car accident is not a contingent possibility but one inherent actualization of driving a car.) Thus the virtual is what enables and ensures the creative aspect of time—instead of cyclic repetition, the creation of new lines of actualization in positive acts. What is actualized does not resemble the virtuality it embodies; rather, it is differentiated as it ramifies into divergent series in a becoming other, or hetero-genesis. This creative force, which Bergson calls the *élan vital*, explains how the genuinely new can come into existence.[13]

In Bergson's sense, the virtual is abstract but not transcendent in relation to the concrete, as would be the case for a Platonic form or essence. Instead, it is immanent to the concrete, "real without being concrete, actual although not effectuated," as D&G put it. It is precisely this sense of the abstract—as the relation of the virtual to the actual—that informs D&G's use of the concept of the abstract machine. The abstract machine, accordingly, points to the virtual dimension of an assemblage, its power to actualize a virtual process, power, or force that while not physically present or actually expressed immanently draws or steers the assemblage's "cutting edges of deterritorialization" along a vector or gradient leading to the creation of a new reality.

Biological Assemblages

It is striking that one of its most important aspects of the assemblage—its doubly articulated structure—resonates with similar theoretical formulations found within science itself, particularly in biophysics. Howard Pattee's theory of the physics of symbols is a clear instance. It is based on the fact that while all physical processes are determined by the laws of physics, once an element is introduced that has the capacity to direct, alter, or constrain these processes, the laws of physics alone are no longer adequate to account for the outcome.[14] This is most evidently true for living systems, where a complementary or double model is called for. In an early essay, "How Does a Molecule Become a Message?" Pattee focuses on how "molecules can function symbolically, that is, as records, codes, and signals."[15] *Communication* between molecules must first be distinguished from the "normal physical *interactions* or forces between molecules that account for all their motions" (1). As Pattee argues, even for a molecule to function as a simple on-off switch it must be situated "in a more global set of geophysical and geochemical constraints, which we could call the primeval 'ecosystem language'" (3). In other words, only within the context of a "primitive communication network" understood as a "larger system of physical constraints" (8) can a molecule become a message. This primitive ecosystem must exist first, as a more or less random network of switching catalysts, before real functionality can occur. Pattee points to Stuart Kauffman's experiments using random Boolean networks to simulate genetic networks (described in chapter 5, below) as a fruitful confirmation of his theoretical perspective.

In later work Pattee focuses more specifically on symbol systems capable of generating and holding in memory the self-descriptions necessary for cell replication and the reproduction of a living organism, while asserting the necessity of a dual and complementary model. Again, this necessity stems from the fact that physical processes are governed by material laws, whereas structures of control are determined by a semiotic, or symbolic, "syntax." Since neither is reducible to the other, both are deemed necessary for a complete description of a living organism. Pattee calls the principle underlying this reciprocal complementarity "semantic closure" and points to its implications for both a theory of the origin of symbols and the possibility of open-ended evolution.[16] At the same time, Pattee's terms make it easy to think of a living system like a cell as a kind of molecular assemblage. On the one hand, what is unique about every specific form of biological organization is symbolically encoded in its genetic information. On the other hand, while this genetic information "spells out" the organism's identity, it is not fully given in the organism's DNA. In fact, the expression of the genetic information that results in the phenotype results from a complex of self-organizing dynamic processes that are very sensitive to and dependent on the material environment. For example, although DNA specifies the linear sequence of amino acids for building proteins, the folding and unfolding of the proteins that result are governed by complex dynamic processes dependent in turn on chemical properties, temperature, the presence of specific enzymes, and so forth.[17] As in D&G's assemblage, the functioning of each side of the cell assemblage presupposes the functioning of the other.

Another assemblage of considerable interest to artificial life is Richard Dawkins's conception of biological organisms as replicator-survival machines.[18] Basically Dawkins reinvigorates evolutionary theory by reversing the customary perspective: instead of viewing our genes as the means by which we reproduce, he argues that the fundamental and incontrovertible lesson of Darwin is that we are "robot vehicles [built and] programmed to preserve the selfish molecules known as genes" (v). In other words, the function of our genes is not to reproduce *us*; rather, we are the means by which the genes themselves replicate. Our bodies are only instruments, highly crafted vehicles or "survival machines" that house and protect the genes, so that they can replicate and propagate more effectively and for longer periods of time. In short, we exist so that genes can replicate. Dawkins casts this stark perspective in narrative terms. Some four billion years ago a molecule that could copy itself emerged by accident from a primeval chemical soup. This was the first replicator, and soon the soup would have been filled with identical replicating molecules, each producing more copies of itself. Copying errors would inevitably have led to variations and then competition among the variant replicators for the necessary building blocks. Some may have "discovered" how to break up the molecules of rivals and use their materials, thus becoming proto-predators. Others may have discovered how to shield and protect themselves with protein coatings. (Perhaps this was the origin of the first living cells.) In any case, "the replicators that survived were the ones that built survival machines for themselves to live in" (19), and over time these survival machines only got bigger and more elaborate. Giving up their freedom, the replicators no longer float in primordial seas:

> Now they swarm in huge colonies, safe inside giant lumbering robots, sealed off from the outside world, communicating with it by tortuous indirect routes, manipulating it by remote control. They are in you and in me; they created us, body and mind; and their preservation is the ultimate rationale for our existence. They have come a long way, those replicators. Now they go by the name of genes, and we are their survival machines. (19–20)

Dawkins's replicator-survival machine is really a stripped down, bare-bones assemblage, its dual and reciprocal functionality easy to specify: the genetic code is the collective assemblage of enunciation; the physical body, its machinic assemblage. Viewed in these terms, a biological species appears not as an organic whole or embodiment of a particular essence but as a contingent amalgam of parts and functions wholly selected by the rigors of the evolutionary process. Citing Dawkins, Manual DeLanda reflects this contingency when he writes: "Much as our bodies are temporary coagulations in the flow of biomass, they are also passing constructions in the flow of genetic materials" (*A Thousand Years of Nonlinear History*, 111). Furthermore, the entire process of natural selection that defines and shapes the life of the replicator-survival machine is algorithmic—the mechanical execution of a set of steps that invariably produces a specific result, in this case the replication and propagation of the most viable and efficient genes. Thus evolution is an ongoing, highly distributed and parallel computation on a vast scale. Not only is natural selection a computation, but so is the construction of the individual survival machine by the replicator code, the building process itself being only the execution of a specific set of instructions, a morphogenetic computation both constrained and enabled by the laws of physics and chemistry.

This example brings to light something that was not fully evident in Pattee's cell assemblage: how over the course of evolution a single functional entity (the replicator) becomes a double entity or assemblage in the sense that the replicator code, on the one hand, and the physical body that houses it, on the other, become quasi-independent; it is their conjunction and reciprocal functionality that makes the whole an assemblage. In the replicator-survival machine the role of computation also becomes more prominent. Indeed, this machine is at bottom not very different from what I call a computational assemblage, where the survival machine is simply the material computational device and the replicator the instructions, or code, of its associated discourse. In basically the same sense, a desktop PC, cellular automaton, neural net, and even the immune system are all computational assemblages, distinguished by both their physical computational mechanisms and the codes they instantiate and execute. Yet in these instances the discursive side of the assemblage (D&G's "collective assemblage of enunciation") includes not only the code but the entire discourse that explains its usage, justification, or purpose. A simple way to think of the distinctions this concept introduces is to consider the evolution of the computer, from simple calculators to the stored-program machines capable of executing more complex instructions like conditional jumps, to the multiplicity of PCs, mainframes, work stations, and supercomputers, which run complex simulations and instantiate vast communicational networks. All these computational assemblages compete and evolve, get copied and improved at both hardware and software levels as new computational spaces, or "niches," open up and gradually become saturated. Overall, these computational assemblages may even appear to constitute a new ecology.

The Machinic Phylum

Like the psychoanalytic theories of Freud and Lacan, D&G's theory of the assemblage decenters the human subject in relation to the social field in which it arises and by which it is circumscribed. But as we have seen, their theory goes much further in relating the human subject to a multiplicity of material and semiotic processes, not all of which are signifying in the linguistic sense. This opening of the assemblage to various nonsignifying (i.e., nonanthropomorphic) flows of matter and energy leads D&G to the notion of what they call nonorganic life. Specifically, they consider how metals and metallurgy bring to light "a life proper to matter, a vital state of matter as such ... ordinarily hidden or covered, rendered unrecognizable, dissociated by the hylomorphic model" (411). The hylomorphic model assumes that matter passively receives the imposition of an extrinsic form and therefore ignores a number of features and behaviors in the flow of matter as it is brought out of the ground, variously transformed (usually by heat), and combined with diverse elements. Seeking a model that can account for these other properties, D&G postulate a realm where, contrary to the hylomorphic model, matter appears to be active and exhibits this hidden kind of life. They call this realm the machinic phylum, which they define as "materiality, natural or artificial, and both simultaneously; it is matter in movement, in flux, in variation, matter as a conveyor of singularities and traits of expression" (409). First rendered visible by metallurgists working with and on different metals and ores, this new phylum eventually comes to constitute a technological lineage comprehending different "lines of variation" (405). In short, what D&G propose is a different way of conceptualizing the development of technology.

They begin by asking how nomads first found or invented their weapons and note that nomadism and metallurgy formed a confluent flow. While it is generally assumed that metallurgists were always controlled by a State apparatus, it is more likely that they enjoyed a certain autonomy. Our lack of a sufficiently developed concept of a technological lineage, considered as a line or continuum with variables of extension, poses another obstacle to understanding the early nomad metallurgists. But in fact, metallurgy is inseparable from "several lines of variation" (405) that arise from differences in kinds of metal (meteorites and indigenous metals, ores, and proportions of metal), kinds of alloys (natural and artificial), and kinds of operations that can be performed on the metals given their inherent material qualities. D&G group these variables under two rubrics, singularities and traits of expression, and illustrate the differences by considering the saber and sword:

> [The saber] implies the actualization of a first singularity, namely, the melting of the iron at high temperatures; then a second singularity, the successive decarbon-

ations; corresponding to these singularities are traits of expression—not only the hardness, sharpness, and finish, but also the undulations or designs traced by the crystallization and resulting from the internal structure of the cast steel. The iron sword is associated with entirely different singularities because it is forged and not cast or molded, quenched and not air cooled, produced by the piece and not in number; its traits of expression are necessarily very different because it pierces rather than hews, attacks from the front rather than from the side; even the expressive designs are obtained in an entirely different way, by inlay. (406)

Since saber and sword have distinguishable lines of variation with respect to both their singularities and traits of expression, they can be said to derive from different metal phyla. But the analysis can also be situated on the level of singularities that extend from one phylum to another, thus yielding "a single phylogenetic lineage, a single machinic phylum, ideally continuous: the flow of matter-movement, the flow of matter in continuous variation, conveying singularities and traits of expression" (406).

Before describing how the machinic phylum enters into different assemblages, D&G make a more general point: "This operative and expressive flow is as much artificial as natural," as if it were the "unity of human beings and Nature" (406). Yet it is *not* the unity, precisely because "it is not realized in the here and now without dividing [and] differentiating." In other words, every assemblage is at once a unity of human beings and nature and a different differentiation of both. Since every assemblage is a "constellation of singularities and traits deducted from the flow—selected, organized, stratified—in such a way as to converge ([providing] consistency) artificially and naturally," the limits to how this unity and differentiation can be accomplished and to the corresponding variety of subjects called forth are determined only by the physical laws of nature. While assemblages can be grouped into large constellations constituting whole "cultures" or "ages," specific assemblages still "differentiate the phyla or the flow, dividing it into so many different phyla, of a given order, on a given level, and introducing selective discontinuities in the ideal continuity of matter-movement. Yet at the same time that assemblages cut the phylum up into distinct, differentiated lineages, the machinic phylum cuts across them all, taking leave of one to pick up again in another, or making them coexist" (406). To summarize, as matter-in-variation flows into and out of various assemblages, these flows are in turn selected, differentiated, and distributed into phyla. Yet cutting across them all, appearing and disappearing from one assemblage to another, is the machinic phylum itself.

As the passages cited above indicate, within the assemblage the opposition between the artificial and the natural disappears, and a kind of unity of human beings with nature is glimpsed. D&G develop this idea specifically in relation to the artisan and the apparent fact, which they insist on, that the matter-flow or machinic phylum "can only be followed." Thus the artisan is an "itinerant," one who follows the matter-flow. The first specialized artisan is the metallurgist, who forms a collective body distinct from, while always having relations with, other collectivities—hunters, farmers, nomads, the State apparatus, and so on. In a sense that D&G only suggest, metallurgists not only follow the matter-flow but join with and become part of it, extracting ore and metal and using their knowledge of its singularities to extend and transform it as it is brought forth from the earth, worked, and then passed along into other domains. In this role as itinerant artisan the metallurgist is thus also part prospector. However, once the social organization separates the prospector from the artisan and the artisan from the merchant, the artisan is "mutilated" into a mere "worker." Finally, in this double function of extracting and transforming, which often occurs at the same time, the artisan operates as a probe head, or searching device (*une tête chercheuse*). In this capacity the artisan functions as the metallic head of the machinic phylum and as such the bearer of a certain kind of thought, the "prodigious idea of *Non-organic Life*" (411).

Following Deleuze and Guattari, Manual DeLanda has reformulated these singularities and traits of expression in the language of dynamical systems or "chaos theory," as bifurcations and instances of self-organization. The basic idea behind these terms is that the behavior of a dynamical system can be modeled by constructing its "phase portrait," which maps how the system's global behavior changes as its variables change.[19] The phase portrait thus provides a visual snapshot of the configuration of forces that determine the system's behavior over successive moments of time. Most dynamical systems exhibit a tendency to move toward one or more stable or limit states called "attractors," which are recognized by the pattern of the system's trajectory through phase space. These attractor states vary in complexity, from single point to periodic to "strange" (chaotic), and unstable systems often undergo continuous deformation from one state to another. The point of change or transformation from one attractor to another is called a bifurcation. This forking in the phase portrait occurs because the system has begun to oscillate between two possible states. Finally, the term self-organization (as we saw in chapter 1) is generally applied to highly dissipative systems "far from equilibrium" that suddenly and unpredictably jump to a more organized state or structure. What chaotic attractors and self-organization have in common is simply that they only occur in *non*linear dynamical systems, whose exponential variables interact in complex ways and produce unexpected and disproportionate effects.[20]

In *War in the Age of Intelligent Machines* DeLanda cites a number of self-organizing phenomena in the domains of chemistry, physics, biology, and human social history (migrations, crusades, invasions) which can all be described by the dynamical systems model. He then extends D&G's concept of the machinic phylum to include the "overall set of self-organizing processes in the universe" (6). Thus, in DeLanda's formulation, at certain critical moments all material processes, whether organic or nonorganic, are traversed or subtended by a few abstract mechanisms that can be said to constitute the machinic phylum. More specifically, under conditions of instability the nonlinear flows of matter and energy spontaneously evolve toward a limited set of possible states that can be mapped as a "reservoir" of abstract mathematical mechanisms (i.e., attractors). What Deleuze calls a virtual or abstract machine, therefore, is really a "particular set of attractors" (236):

> We have then two layers of "abstract machine": attractors and bifurcations. Attractors are virtual machines that, when incarnating [*sic*], result in a concrete physical system. Bifurcations, on the other hand, incarnate by affecting the attractors themselves, and therefore result in a mutation in the physical system defined by those attractors. While the world of attractors defines the more or less stable and permanent features of reality (its long-term tendencies), the world of bifurcations represents the source of creativity and variability in nature. For this reason the process of incarnating bifurcations into attractors and these, in turn, into concrete physical systems, has been given the name of "stratification": the creation of the stable geological, chemical and organic strata that make up reality. Deleuze's theory attempts to discover one and the same principle behind the formation of all strata. (237)

Regrettably, however, this important discussion is relegated to a long footnote, for here we glimpse how dynamical systems theory might account more fully for the processes of stratification and destratification that D&G discuss only in terms of a double articulation and implicit dynamic. And while their dynamic is surely *not* the dynamic of dynamical systems theory (it is not expressed mathematically in terms of variables that change over time), it could easily be strengthened and completed by it. Indeed, one could easily extend the application of dynamical systems theory in order to explain how assemblages arise, mutate, and dissolve, as I suggested earlier. Yet perhaps it is because D&G lacked a more precise understanding of nonlinear dynamical systems (the theory of which was still being developed when they wrote *A Thousand Plateaus*) that they show little interest in the mutations and instabilities that lead to transformations of specific assemblages.[21]

In any event, whereas D&G emphasize how specific material properties are brought to consciousness by itinerant artisans and metallurgists who follow and develop the metallic matter-flow, DeLanda focuses on the complex history of changing military strategy and weapons development, particularly of "smart weapons." Basically it is a story about the military's attempt to get human beings "out of the loop"—that is, removed from the production and use of new weapons systems, and, as a consequence, the increasing necessity of decentralized command and control structures. DeLanda begins *War in the Age of Intelligent Machines* by asking us to imagine a book similar to his own but written by some future "robot historian." Such a historian would not look at things from a human point of view. Rather, since it would be concerned with its own history, it would trace the gradually increasing transfer of intelligence from humans to machinic processes and engineered systems. It would see the emergence of fully autonomous artificial intelligence as the inevitable outcome of an evolutionary process in which humans had become increasingly dependent on machinic arrangements and technical assemblages.

Yet DeLanda's purpose is not to offer techno-scientific speculation about futuristic possibilities but to trace the actual "history of several military applications of AI as much as possible from the point of view of our robot historian." He understands the military as "a coherent 'higher level' machine: a machine that actually integrates men, tools and weapons as if they were no more than the machine's components" (4). From this nonanthropomorphic perspective, human social assemblages appear to be immanent to nature, different in scale but not in essence, and the history of war only incidentally concerned with success or failure of policy. For DeLanda, looking through the eyes of the robot historian, what matter most are moments of singularity, transitions, or thresholds, when physical processes unpredictably change state or achieve new levels of self-organization. These changes are recognized when an aggregate of previously disconnected elements suddenly jells and begins to "cooperate," resulting in the formation of a higher-level entity with new and unforeseeable properties. The emergence of the canoidal bullet, for example, which gave the rifle unprecedented accuracy and allowed the foot soldier a new independence, in turn demanding a restructuring of military tactics and organization, is one such moment. (This new formation is precisely what D&G mean by an assemblage.) However, these thresholds of emergence in self-organizing processes could hardly escape the notice of the robot historian, since its own "intelligence" or "consciousness" would be the result of this very same ongoing process. In the robot historian's narrative, "humans would have served only as machines' surrogate reproductive organs until robots acquired their own self-replication capabilities"; consequently, "both human and robot bodies would ultimately be related to a common phylogenetic line: the machinic phylum" (7).

DeLanda's next book, *A Thousand Years of Nonlinear History*, takes up D&G's discussion of the three major strata. Again applying nonlinear dynamical systems theory, he shows how the self-organization of matter directs three layered histories—in geology, biology, and linguistics—from about 1000 AD to 2000. Much could be said about this important work, which greatly elucidates (and fills out) D&G's overly schematic discussion of the formation of the three strata, but here I will confine myself to one methodological point. In his account of living structures DeLanda distinguishes between two abstract machines ("one generating hierarchies, the other meshworks"), but goes on to assert the necessity of a third, which has no counterpart in the geological world and is closely related to the mechanism of Darwinian evolution (variation plus natural selection). Generalizing from the latter, DeLanda states that scientists have come to realize "that *any variable replicator* (not just genetic replicators) coupled to *any sorting device* (not just ecological selection pressures) would generate a capacity for evolution" (138–139). This leads to a central idea: "The coupling of variable replicators with a selection pressure results in a kind of 'searching device' (or 'probe head') that explores a space of possible forms (the space of possible organic shapes, or birdsongs, or solutions to computer programs)" (139). Like the mechanisms of Darwinian evolution, this searching device is blind (or shortsighted), though often highly effective. It is, DeLanda asserts, "the abstract machine we were looking for, the one that differentiates the process of sedimentary-rock formation from the process that yields biological species" (139). And later, discussing how linguistic norms can evolve within language, he advances an analogous claim, that "language may be considered to embody a probe head or searching device" (320).

However, in explaining these mechanisms DeLanda uses the term *probe head* in a more general sense than D&G, and the distinctive meaning they give to it is lost. Whereas for them it is the artisan who functions as a probe head in the process of the machinic phylum's emergence and extension along diverse lines of variation (hence the term doesn't appear in their text before the discussion of anthropomorphic strata), for DeLanda the term designates a blind and random exhaustion of a search space and thus a form of agency without a human agent. The computer, for example, is often used as a searching device; but for D&G that would not make it a probe head. Yet DeLanda uses *probe head* and *search device* synonymously—in one instance (the discussion of biology) to indicate nature's effort to produce an effective combination of genes but in another (explaining the evolution of language) to designate "the abstract machine we are looking for." While making the two terms synonymous may emphasize and compensate rhetorically for an absence of human agency, it also blurs D&G's usage, according to which the concept of the probe head is reserved for the human agent's function in a process that, while not altogether anthropomorphic, nevertheless requires human intervention and agency. By assimilating *probe head* to the general meaning of a search device DeLanda can rhetorically eliminate "the human" and "naturalize" the processes described. But this seems plainly redundant, given that the processes he describes in *A Thousand Years of Nonlinear History* are already natural and nonanthropomorphic. On the other hand, if we should want to consider the role of those who have used the computer to develop new theoretical sciences, the term *probe head* taken in D&G's original sense seems not only pertinent but highly suggestive.

Silicon Probe Heads

With the invention of the digital computer, physical processes characterized by nonlinear relationships and deterministically chaotic effects could be modeled and made widely visible for the first time. Because the computations could be rendered graphically, the computer made both quantitative and qualitative methods available for the study of chaotic phenomena. For this reason nonlinear systems science—that is, chaos theory—owes its widespread and rapid acceptance, if not its actual existence, to the extensive use of computers, especially after the late 1970s, when desktop computers became available. By opening a "window" onto the machinic phylum, the computer enabled many scientists, engineers, and hackers to begin tracking and directing the emergence and flows of the machinic phylum in the medium of silicon, much as the artisans and metallurgists had done with various ores and metals taken from the earth. These modern silicon probe heads constructed and worked with a number of distinct computational assemblages, several of which we examine below.

This is not, of course, the customary perspective. It is surprising, nevertheless, that there haven't been detailed narrative accounts of how nonlinear dynamical systems theory has been relayed by the computer's development and particularly by its capacity to simulate physical systems. To be sure, historical accounts of the emergence of chaos theory are sprinkled with acknowledgments of the importance of nonlinear mathematics and the computer's crucial role in making both old and new data "visible" in a new light. What remains inadequately discussed is the extent to which the computer, rather than being simply a new tool in a constellation of research programs, has both made new sciences possible and changed the ways in which normal science is done. The real issue lies not in the computer's unparalleled capacity for calculation, although that has certainly been essential to the discovery of the new phenomena studied by chaos science. But beyond its capacity for calculation, the computer can automate search strategies and simulate physical processes with extraordinary precision and speed. These new capabilities have inaugurated what many claim is a paradigm shift in contemporary science.

The computer has also brought about a new kind of blurring of human and machine functions. In contrast to the telescope and microscope, which make visible what is only contingently invisible to the naked human eye, the computer renders visible what is inherently invisible, as in the case of the configuration of forces or relationships that define a "strange attractor." In and through these new functions the computer exceeds its prosthetic and calculating instrumentality and becomes a new type of abstract machine. Functioning in a variety of ways within science and contemporary society at large, new computational assemblages have brought into existence not only a new sense and understanding of physical reality but a new reality tout court. Although this larger story cannot be told here, some of the consequences of the computer's productive and performative capacities will be discussed. In general, the work of three distinct (but sometimes overlapping) groups of silicon probe heads were essential to the evolution of both computer hardware and software: the scientists and mathematicians who developed nonlinear systems science, the computer scientists who developed new computational techniques (particularly simulations, evolutionary programming techniques, and automated search strategies), and the hackers who emerged as a new social and functional type in the 1980s. In each instance the computer operates as part of a larger and very specific computational assemblage, and as such it allows a new extension and development of the machinic phylum. Though I focus here on the activities of a few individuals within the first group, in later chapters I discuss significant instances from within the second and third groups as well.

In *Chaos: Making a New Science*, James Gleick offers a vivid account of what would mark the beginnings of chaos theory in the 1960s. Using a Royal McBee computer that barely fit into his office, Edward Lorenz discovered that very small differences in the initial conditions of weather systems lead rapidly to wide divergences in their behavior and hence predictability.[22] Taking into account obvious variables, such as temperature, barometric pressure, and wind speed, the relationships among which he calculated by means of a group of twelve equations, Lorenz produced a skeletal but reasonably realistic model of a weather system. Although the physics of weather prediction were thought to be well understood, the variables could not be accurately measured nor the calculations made rapidly enough to make long-range weather prediction feasible. With the advent of the digital computer and the assumption that small influences could be ignored or approximated, hopes grew for more accurate and longer-range predictions. These hopes were dashed when Lorenz began to simulate weather systems by feeding real values into the variables of his bare-bones model and computing the results. These were graphically displayed in a simple line drawing, which showed how a specific weather system was changing over time. As Gleick describes it, one day Lorenz decided to rerun one of the sequences, but instead of starting the whole run over again he began midway by punching in the initial conditions from the first printout. Rather than entering the numbers to the sixth decimal point (0.506127), he rounded them off to three (0.506). When he later returned after grabbing a cup of coffee, he was startled by the results: after duplicating the first pattern for several wavy humps, the graphic display began to diverge rapidly, until it was clearly no longer the same pattern.

What did this mean? Obviously a weather system is nonperiodic (it never repeats itself exactly), but here the same values were plugged into the equations and therefore the results should have been the same. However, because Lorenz had not started at the exact same point as in the prior run, he had produced a second weather system, not a replication of the first. Because the equations were nonlinear, the calculated outputs changed disproportionately as the variables changed; specifically, the tiny difference in decimal points was greatly magnified as the equations were iterated again and again to calculate the changes for each successive time step. This peculiar sensitivity to initial conditions accounted for the growing divergence between the two (at first) very similar systems. It later became known as the butterfly effect, meaning that small differences in initial conditions ramify into disproportionately large effects.[23]

An explanation is to be found in Lorenz's essay "Deterministic Nonperiodic Flow," one of the most often cited and by all accounts deeply beautiful of the early articles on chaos theory.[24] The paper explores just what Lorenz had seen in his simulated weather system: a lack of periodicity coupled with the growth or amplification of small differences. (The latter effect, a hallmark of nonlinear systems, is now referred to as "sensitive dependence upon initial conditions.") Instead of using the twelve-equation model, he had pared it down to three, having started with his colleague Barry Saltzman's seven equations for describing convection currents (*The Essence of Chaos*, 137). These three differential equations describe the behavior of a general type of nonlinear deterministic system, although in the article the weather remains the ostensible reference:

$$dx/dt = -\sigma x + \sigma y$$

$$dy/dt = -xz + rz - y$$

$$dz/dt = xy - bz$$

The equations look disarmingly simple; but as Lorenz notes (188), they have been intensively studied and have even become the subject of a book-length study by Colin Sparrow at Cambridge University. The three constants b, σ, and r determine the system's behavior. After suitable values are chosen, solutions can be plotted to graph points in a three-dimensional coordinate system such that each point indicates the system's state at that particular moment and the sequence of points traces a path of the system's changes over successive moments $(t_1, t_2, t_3 \ldots t_n)$.

The resulting "phase portrait" sketches a visual image of how the system behaves over a period of time. Such phase portraits provide qualitative information about the attractors acting on, or "pulling," a dynamical system. If the traced path arrives at a single point and then remains there, it means that the system's behavior is governed by a single point attractor; if the path forms a circle or loop, it means that the system has settled into a pattern of periodic behavior, its behavior determined by a periodic attractor. Lorenz's system did neither; rather, it looped endlessly in two quadrants, without the paths ever crossing or repeating (see fig. 3.1). As Gleick describes the figure, it "traced a strange distinctive shape, a kind of double spiral in three dimensions, like a butterfly with its two wings" (*Chaos*, 30). This double spiral became known as the Lorenz attractor, one of several types of strange attractor that signaled a deterministically chaotic system. Although its phase portrait resulted from deterministic equations, its exact path was unpredictable. Even so, the path always

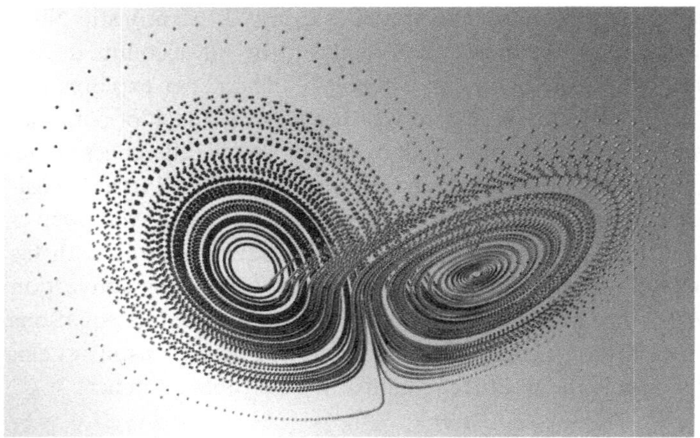

Figure 3.1
Lorenz strange attractor.

stayed within certain bounds, while tracing a clearly recognizable shape. Deterministic chaos was thus a new kind of disorderly order, chaotic but not random, unpredictable but visually coherent.

Soon other scientists would get a glimpse into this strange new realm, and each time the computer played a central role. Nevertheless, in a review of Gleick's book the mathematician John Franks takes Gleick to task for not making the computer's role central enough. Indeed, according to Franks, what Gleick chronicles is not the "chaos revolution" but the "computer revolution, and the current popularity of chaos is only a corollary."[25] Stephen H. Kellert, discussing the claim that chaos theory had to wait for the appearance of computers as its technological precondition, stresses the extent to which the institutionalized study of physics had trained scientists to ignore phenomena not amenable to linear analysis, thereby blinding them to the importance of nonlinear systems and the kinds of effects that made chaos visible.[26] Yet this is a theme treated by Gleick as well, notably in the chapter he devotes to the Dynamical Systems Collective at the University of California at Santa Cruz.

The Santa Cruz Collective is crucially important, not only because of the fundamental contributions to chaos science made by the four graduate students who comprised it but also because of the special intensity with which they were captured by a vision of how the computer had opened a new realm of study, as if chaos theory and the computer (or new ideas about computation) were simply two sides of the same assemblage. It started when Robert Shaw began to spend more time with the old analog computer at the lab than with his doctoral thesis on superconductivity. (I note in passing the connection of superconductivity with metallurgy.) Then one of his professors handed him Lorenz's equations to tinker with, Shaw having already learned to do phase-space portraits of simple systems on the old analog machine. The Lorenz equations didn't look more complicated than those with which he was already familiar, and it was a simple matter to plug in the right cords and adjust the knobs. Gleick describes the result:

> [Shaw] spent several nights in that basement, watching the green dot of the oscilloscope flying around the screen, tracing over and over the characteristic owl's mask of the Lorenz attractor. The flow of the shape stayed on the retina, a flickering, fluttering thing, unlike any object Shaw's research had shown him. It seemed to have a life of its own. It held the mind just as a flame does, by running in patterns that never repeat. The impression and not-quite-repeatability of the analog computer worked to Shaw's advantage. He quickly saw the sensitive dependence on initial conditions that persuaded Edward Lorenz of the futility of long-term weather forecasting. He would set the initial conditions, push the go button, and off the attractor would go. Then he would set the same initial conditions again—as close as physically possible—and the orbit would sail merrily away from its previous course, and yet end up on the same attractor. (*Chaos*, 246–247)

This fascination was soon imparted to the other members of the collective—Doyne Farmer, Norman Packard, and James Crutchfield—who were all drawn irresistibly to this newly visible realm of nonlinear behavior, which their training in physics had largely taught them to ignore. Eventually the four would author an article on chaotic attractors that remains the best short introduction to chaos theory.[27]

Following that first glimpse of the Lorenz attractor, the Collective quickly absorbed the research of other early pioneers of chaos science (notably Lorenz, Hénon, Rössler, May, Ruelle and Takens, and Gollub and Swinney) and began to explore for themselves the behavior of chaotic attractors. One of their most significant early achievements was to demonstrate how the state space of an attractor could be constructed from a time series of experimental measurements.[28] Focusing specifically on how the unpredictability of chaotic systems could be measured, the Collective further developed the work of the Russian mathematician A. R. Lyapunov, whose "Lyapunov exponents ... [there was one for each dimension] provided a way of measuring the conflicting effects of stretching, contracting, and folding in the phase space of an attractor" (*Chaos*, 253). The group also made a number of animations that visually display the dynamics of different attractors, both singly and in combination, using the analog computer as well as film and video equipment.[29] This rendering visible of what had never been seen before was extremely valuable and merits attention in its own right. But more important here, the effort to determine the unpredictability of chaotic systems led two members of the group to pursue the study of these systems in information-theoretical terms.

Strange Attractors and Information Flow

The combining of information theory with dynamical systems theory in the study of chaotic or complex systems marks a conceptual accomplishment of great import. As noted in chapter 1, the first quantitative theory of information was proposed by Claude Shannon in a technical paper published in 1948 and in book format in 1949. Based on his work at Bell Labs to reduce noise in telephone lines, Shannon's mathematical theory of communication became a core text of the cybernetic movement, and his association with Wiener, von Neumann, and Turing probably ensured that its influence would extend far beyond its practical value in the technology of signal processing. But in fact the theory was groundbreaking and soon attracted the attention of many interested in dynamical systems.[30] As Shannon's schematic diagram of a general communication system indicates (see fig. 3.2), the theory takes into account the entire process of encoding a message in a signal or medium, transmitting it through a channel impeded by noise, and decoding the message at the receiver's end.[31]

Because noise in a communication system causes uncertainty—uncertainty as to how closely the message received compares with the

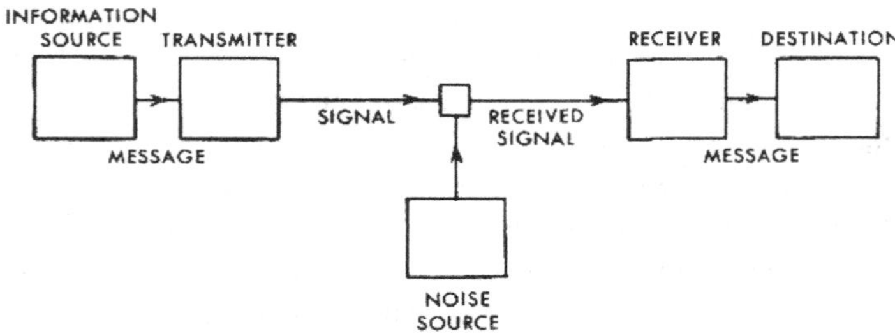

Figure 3.2
Shannon's diagram of a general communication system. Claude E. Shannon and Warren Weaver, *The Mathematical Theory of Communication* (Urbana: University of Illinois Press, 1963), 34.

message sent—the theory requires a precise measure of this uncertainty. A concept that makes this uncertainty quantitative and measurable lies therefore at the heart of Shannon's theory. This concept is *information*, defined as a function of probabilities and not in any way to be confused with meaning. Simply put, information expresses the probability of any single message in relation to all messages possible in a given communication system. Mathematically, it takes the "p log p" form of Ludwig Boltzmann's well-known logarithmic equation for computing entropy in a closed system, where p designates the probability that a single molecule will be in a specified state at a moment in time. Shannon consciously adopted Boltzmann's formula,

$$H = -k \sum p_i \log p_i$$

stating that it would "play a central role in information theory as measures of information, choice and uncertainty" (50). Whereas in Boltzmann's statistical mechanics "p_i is the probability of a system being in cell i of its phase space," Shannon explains, in his own theory p_i is the probability of a single symbol's occurrence. The total probability of the message is found by adding up all the probabilities of the individual symbols that compose it. The constant k refers to the choice of unit of measure. The minus sign is necessary to ensure that the information value is positive, since probabilities are given as fractions or decimals (like 1/2 or .5) and the log of a number less than 1 is negative. Simple calculation shows that if the probability is 1 (i.e., there is no uncertainty or choice in the message), then the information value is zero, and that information is greatest when there is an equal probability of all possible messages. However, in most messages the probabilities of the symbols are not equal. In English, for example, the probability of the letter *e* is much higher than that of a *z*; furthermore, the probability in any given message that an *e* will follow *th* is much greater than that it will follow a *z*. The statistical inequality of such groupings makes the probabilities more difficult to compute. Fortunately, many such sequences are ergodic, meaning that they exhibit statistical properties that make the job of calculating their probabilities much more feasible (see *Mathematical Theory of Communication*, 45–48).

In Boltzmann's theory entropy is the measure of a system's disorder, or of our uncertainty about individual molecular states. We can therefore think of entropy as missing or invisible information. In physical terms, entropy measures the loss of energy available for doing work. (In a closed system, heat loss from the random collisions of molecules inevitably causes entropy to increase over time.) Given the impossibility of keeping track of the individual positions and velocities of these molecules, Boltzmann defined entropy as a property of statistical distribution. As a statistical measure, the entropy of a system tells us how far it has evolved in the drift from a less probable to a more probable state, with uniform distribution or thermal equilibrium being the most probable. The drift seems inevitable because there are infinitely more pathways to increasing disorder than to increasing order, making increasing order far less likely. However, in Shannon's concept of information there is no inevitable increase or decrease of information. The probability of the message (or of the symbols that compose the message) alone determines the amount of information it contains. Hence a surprising fact or unexpected phrase transmits more information than a cliché. However, this example is misleading because it concerns meanings rather than probabilities of symbols. Since information is measured in bits (short for binary digits)—the outcome of a coin toss or the answer to a yes or no question yields *one* bit—the game of twenty questions provides a better analogy for determining how much information a message contains.

Shannon's theory of information entropy and use of Boltzmann's formula generated a number of intriguing theoretical questions. For example, since an increase in uncertainty (i.e., more symbols to choose among, with an equiprobability of choosing any one) increases the measure of information of any single message, would not utter uncertainty about the message (as in the case of a completely random string of symbols) mean maximum information? And does computation or the manipulation of information itself expend energy?[32] In fact, the first question blurs the distinction between two sources of uncertainty: the uncertainty of the message and the uncertainty generated by noise in the communication channel. In contrast to information defined as a purely relative, statistical concept divorced from questions of meaning or significance, noise is something we all think we understand. In Shannon's theory, however, noise acquires a paradoxical complexity. In a communication channel, noise appears to be both what impedes the transmission of information and what is not (yet?) coded as information. Because of noise, in fact, the amount of information received can be *greater* than the amount transmitted. (This is because the received message is selected out of a greater set of possibilities than the transmitted message.) This apparent paradox notwithstanding, Shannon's theory was basically directed toward solving a very practical problem: how to encode a message in such a way that its transmission rate through a noisy channel (in bits per second) could be maximized and the error rate minimized. Using Boltzmann's formula to define both the entropy of an information source or input to a communication channel and the entropy of the received signal or output, Shannon was able to give a very precise definition of channel capacity: It is the maximum of the mutual information between source and receiver computed over the whole range of possible messages. Mutual information thus measures the reduction in the uncertainty that results from receiving a message from a sender.[33] On this basis, Shannon was able to establish a fundamental theorem for the maximum rate of symbol transmission for a particular channel. Previously it had been assumed that noise could be overcome simply by repeating the message or adding redundancies to the code, at an obvious cost in transmission capacity. Shannon's theorem demonstrated contrarily that there were inherent, insurmountable limits. At the same time, it suggested unexpected ways to encode the message to take advantage of the statistical properties of a particular channel, resulting in a more efficient signal-to-noise ratio and pointing to more efficient methods of error detection and correction. Indeed, these discoveries are actually what made modern digital communications possible.

Turning to the study of dynamical systems, we find that the most striking historical development is how claims for the complete predictability of all physical systems—stated most famously by Pierre Simon de Laplace—have given way to the realization that even simple, completely deterministic systems can become chaotic over time.[34] In other words, the universe is not as orderly and stable as Newton's deterministic equations for classical mechanics would suggest. The first to grasp that nonlinearity in a dynamical system is what gradually erodes stability (and hence predictability) was Henri Poincaré. In the 1890s, while working on the differential equations used to compute the positions and orbits of multiple bodies in celestial mechanics (the *n*-body problem), Poincaré was able to prove that even for three bodies the problem could not be solved quantitatively. Forced to take a different approach, he adopted qualitative and geometrical methods. It was while attempting to resolve these difficulties that Poincaré saw evidence of sensitivity to initial conditions and the growing divergence of nearby trajectories.[35] Thus he wrote presciently in 1903, "It may happen that small differences in the initial conditions produce very great ones in the final phenomena. A small error on the former will produce an enormous error in the latter. Prediction becomes impossible, and we have the fortuitous phenomenon."[36] However, given the mathematics of his time, Poincaré had no way to integrate these observations into a more complete theory. Nevertheless, in working out a geometric method for representing the possible states of a deterministic system in phase space, he laid the foundations for the new field of dynamical systems.

When we consider that nonlinearity in the study of physics in the 1970s was a peripheral problem to be avoided through approximation, that the use of computers was generally disdained, and that information theory was thought to be the provenance of electrical engineering, we begin to see the originality and importance of the work of the Dynamical Systems Collective. Though several early articles published by members of the Collective were concerned with how to measure the influence of chaotic attractors in a physical system, it was Shaw's article, "Strange Attractors, Chaotic Behavior, and Information Flow," published in 1981, that first makes the case for information theory.[37] Indeed, Shaw shows how chaotic attractors can be seen in a new light, as *generators of information*. In developing this powerful idea, Shaw synthesizes several sets of related ideas: first, equations like Lorenz's that produce a deterministically chaotic stream of numbers are articulated with the qualitative transitions marking turbulent flow; second, the problematic gap between energy states of a system at the microscales and those at the macroscale is bridged; and finally, the difference between dynamical systems that are time reversible and those that are not is understood in terms of the production and destruction of information.

Shaw distinguishes at the outset between conservative and dissipative dynamical systems. When mapped into phase space, the former turn out to be "well behaved," meaning predictable and describable in terms of the laws of physics and the extensive mathematical formalisms developed to quantify their behavior. As Shaw notes, "Implicit in the use of this formalism for the description of dynamical systems is the *assumption of reversibility*, i.e., no information is lost as time passes" (82). Since each dimension of phase space represents a variable and hence a degree of freedom for the system, laws of nature, such as the conservation of energy, act as a constraint, reducing the dimensionality of the state space available to a given system. Whether governed by a single-point or periodic attractor, the state space remains low dimensional, and its phase portrait shows how over time the system continually revisits each of its possible states. Put another way, if the system is abstractly represented by a flow in phase space, this flow remains "compact" and predictable. Furthermore, if the information it generates is computed by the change in predictability of its movement through phase space per unit of time (in bits per second, for example), then this system generates no information at all, since its behavior remains predictable for an indefinite time into the future. In the terms of classical mechanics this modeling of a conservative system corresponds to temperature measurements made at the macroscale, which ignores the fact that the total "free energy" of a system at the microscale may be far greater. Classical mechanics thus keeps the two scales artificially separated, with no interaction or communication between them.

Everything changes, however, when the phase portrait for a highly dissipative, nonconservative system is constructed. As shown above, when Lorenz's system of equations is mapped in phase space, it yields the portrait of a chaotic attractor. As the system flows through its possible states, it twists and turns from one quadrant to another in what first appears to be a three-dimensional space, without ever exactly retracing its path. In fact, the space is fractal.[38] Apparently composed of many ultrathin layers, it is a space articulated in fractional dimensions. Topologically, the trajectories stretch and then fold over onto themselves, increasing the dimensionality or volume of the phase space. As Shaw argues, the flows of such strange, or chaotic, attractors through phase space actually create new information:

> In three and higher dimensions it is possible to have flows which in a compact region continuously expand volumes of phase space in some dimension or dimensions while contracting it in others. . . . The effect of these flows is to systematically create new information which was not implicit in the initial conditions of the flow. Given any finite error in the observation of the initial state of some orbit, and the Uncertainty Principle guarantees such an error, the position of an orbit will be causally disconnected from its initial conditions in finite time, thus any prediction as to its position after that time is in principle impossible. (85–86)

Not only is new information created as the system moves away from its initial state into an expansion of phase space, but information also becomes irretrievable as other regions of phase space are contracted as a result of this same flow. With this loss of information, the system becomes time *ir*reversible.

But where does this new information come from? In one sense it merely reflects the fact that the system is becoming chaotic and unpredictable; that is, applying Shannon's formula, this information would simply be the quantitative measure of the increasing uncertainty of a particular state. But in terms of the system's physical alteration, the increase in information means that the uncertainties of the microscale, "the bath of microscopic randomness in which anything physical is immersed," as Shaw vividly puts it, are being systematically pulled up to *macro*scopic expression. In physical systems driven by a chaotic attractor, in other words, events at the macroscale and events at the microscale can no longer be artificially separated. In the case of turbulent flow, for example, as increasing energy flows into the system downward to microscales, it produces eddies, whorls, and vortices, and then eddies, whorls, and vortices within them at ever smaller scales, in a telltale sign of the system's increasing turbulence. As the turbulence increases to chaotic flow, information in the form of increasing unpredictability over a unit of time can be said to flow upward from the microscale to the macroscale, where it can then be measured. Using a Systron Donner analog computer Shaw was able to verify in quantitative terms that indeed, in the case of turbulent flow, information is continuously generated by the flow itself.

In his article Shaw points out that equations like Lorenz's and Robert May's logistic equation belong to a special class of theoretical object, since they define an iteration procedure that produces a stream of numbers that eventually becomes unpredictable. He doesn't deny their relevance to the study of deterministically chaotic physical systems—in fact, to the contrary—but their applicability is not his primary interest. Rather, he is searching for "a common thread" running through a diversity of research fields that all yield this very peculiar object, with these specific mathematical properties. Hence his central question: "What is responsible for the same qualitative behavior in these systems?" (81). His answer, hardly surprising now that chaos science is a respectable field of study, is that "the many examples of systems exhibiting chaotic behavior which are appearing and will no doubt continue to appear, are different guises of what are actually a limited number of basic forms. Furthermore, these forms arise naturally in a geometry where irreversibility is taken as a *postulate*, in contrast to the reversibility implicit in the usual description of systems in phase space" (81).[39] As Shaw demonstrates, this irreversibility can be quantified in terms of the loss and gain of information that a chaotic attractor produces in a nonlinear dynamical system. Ultimately this means, as Shaw concludes, that "new information is continuously being injected into the *macroscopic* degrees of freedom of the world by every puff of wind and swirl of water," an effect that may put severe limits on our predictive ability but also "insures the constant variety and richness of our experience" (103).

The Dripping Faucet as Model Chaotic System

In 1984 Shaw published a short book, *The Dripping Faucet as a Model Chaotic System*, which continued his investigation into the relationship between a chaotic dynamical system and its production of information.[40] As the title indicates, Shaw takes a specific physical system—a simple dripping faucet—as his model of study. The simplicity of the model is belied, however, by both the complexity of its behavior and the degree of philosophical questioning to which Shaw submits "the idea" of the model. Indeed, there is a subversive, even revolutionary idea lurking in chaos science that Shaw's little book carefully (and rigorously) brings to the foreground. What is at stake is the difference between two correlations: the correlation of observed experimental data and the explanatory model in classical experiments versus the correlation in experiments when a chaotic attractor is shaping the behavior of a physical system. Shaw primes us with an initial sense of this difference with a quotation from Steven Hawking's lecture "Is the End in Sight for Theoretical Physics?": "It is a tribute to how far we have come already in theoretical physics that it now takes enormous machines and a great deal of money to perform an experiment whose results we cannot predict" (*The Dripping Faucet*, 1). The irony is obvious: Shaw will demonstrate that even an ordinary dripping faucet can generate data that is quantifiably unpredictable. Far from being the exception to the rule, chaotic behavior pervades the physical world.[41]

Shaw notes in the first few pages that the dripping faucet exhibits behavior typical of a deterministically chaotic system. Although many variables determine how and when individual drops form and fall, Shaw assumed that measuring the intervals *between* drops would reveal something essential about the system's behavior. The experimental apparatus he constructed therefore measured the drop intervals in addition to the flow rate of the falling water. (Shaw's drawing of the apparatus is reproduced in figure 3.11.)

Taking successive intervals as the x and y coordinates, Shaw plotted a two-dimensional "return map" of the data. If, for example, the time interval between the first and second drops was 0.168 seconds and the interval between the second and third was also 0.168 seconds, that would yield a single point on the map (with x and y coordinates 0.168 and 0.168). If the second interval measured 0.168 but the third was 0.180, there would now be two dots; and if these intervals continued to repeat it would suggest a

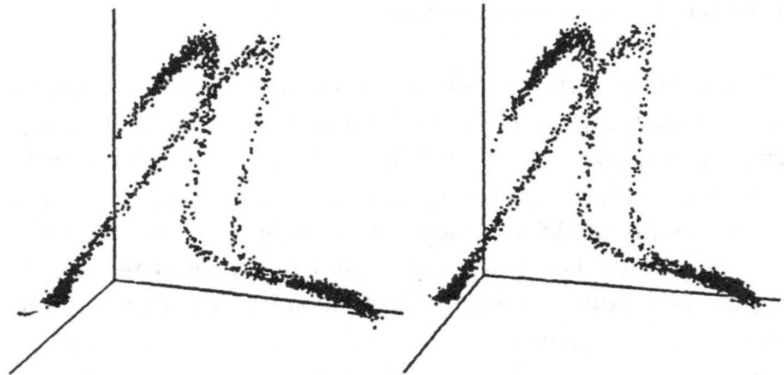

Figure 3.3
Chaotic attractors with increasing flow rates, 1. Robert Shaw, *The Dripping Faucet as a Model Chaotic System* (Santa Cruz: Aerial Press, 1984), 11.

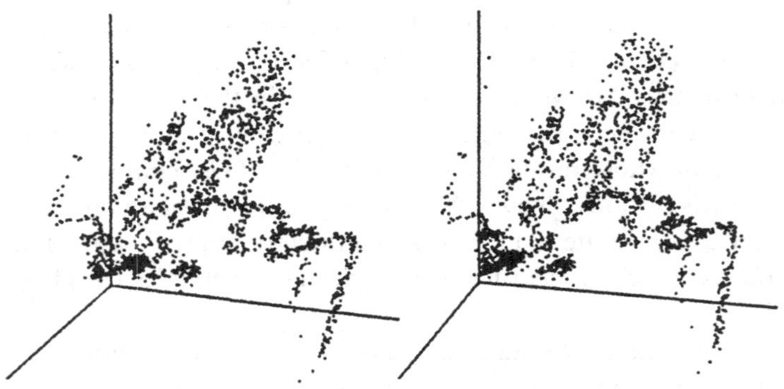

Figure 3.4
Chaotic attractors with increasing flow rates, 2. Robert Shaw, *The Dripping Faucet as a Model Chaotic System* (Santa Cruz: Aerial Press, 1984), 12.

periodic pattern. But if successive intervals varied randomly, the result would be a stochastic scattering of dots. Finally, if the dots formed fuzzy curves or combinations of fuzziness with a visible geometric pattern, that would indicate the presence of one or more chaotic attractors. Three such sets of mappings—which result from increasing the flow rates—are indicated by figures 3.3, 3.4, and 3.5 (reproduced from *The Dripping Faucet*).

As expected, when the tap was opened slightly, drops of water began to fall in regular intervals. As the tap was opened further, the drops began to fall in periodic intervals. When the water pressure was further increased, a period-doubling bifurcation appeared, in what is now the

Figure 3.5
Chaotic attractors with increasing flow rates, 3. Robert Shaw, *The Dripping Faucet as a Model Chaotic System* (Santa Cruz: Aerial Press, 1984), 106.

telltale sign of increasing turbulence and the route to chaotic flow. While noting this behavior, Shaw cautiously observes that this increasingly chaotic behavior is not a simple function of flow rate, since it depends upon a variety of different initial conditions. In fact, although the physical forces producing individual drops can be modeled as a problem of falling bodies in classical mechanics (see *The Dripping Faucet*, 15, as well as *Chaos*, 266), in the end what mattered to Shaw was only the *variation* in drop rate over time.

In Shaw's experimental apparatus this variation is registered as a data stream, or flow of numbers. According to classical experimental procedure, the experimenter would then construct a discourse that tells us what these numbers mean—interpreting the data by showing how it confirms or invalidates predictions based upon a specific model. Although Shaw certainly doesn't ignore this objective, he is primarily interested in how this data might be used to compute other quantities, specifically "the amount of information a system is capable of storing, or transmitting from one instant of time to the next, and . . . the rate of loss of this information" (*The Dripping Faucet*, 3). A loss of information about a system, of course, means an increased uncertainty about its future state. This is because, as Shaw had demonstrated in his article, as a system moves into a chaotic regime its present state is causally disconnected from its initial conditions and past states. This information, in effect, is lost. But if a system becomes unpredictable, *how* unpredictable is it? And can the difference between randomness arising from the interaction of many degrees of freedom and randomness due to the chaotic dynamics of only a few

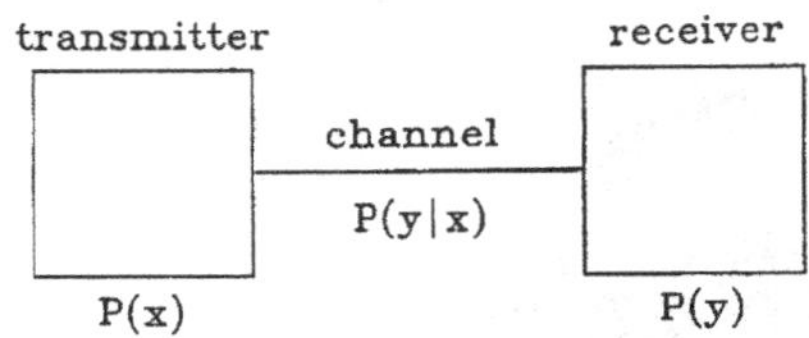

Figure 3.6
Distribution through a communication channel. Robert Shaw, *The Dripping Faucet as a Model Chaotic System* (Santa Cruz: Aerial Press, 1984), 21.

degrees of freedom be discerned? More specifically, how could the amount of information *stored* in a system be computed and its *rate of loss* be measured?

Again information theory provided the framework. Every physical system is subject to noise, which of course limits the total amount of information it can store and transmit. Moreover, in order to be useful the information stored in a system must be a property of the system and not of the particular type of measurement used to measure it. This implies, Shaw says, "an *invariance under coordinate transformations*, a property which appropriately defined measures of information possess" (6). It is hoped, he adds, that "*any* dynamic variable [in this case, the drop interval] will serve to characterize the predictability" (7). His objective will be to use this varying predictability as a way to measure how the system's capacity to store and transmit information changes over time.

Shaw takes as his point of departure Shannon's theory of the transmission of information from a "transmitter" to a "receiver" through a specific "channel," designating their known statistical properties as shown in figure 3.6. Here $P(x)$ denotes the distribution of possible transmitted messages x, $P(y)$ the distribution of possible received messages y, and $P(y/x)$ the probability of receiving message y given that message x was transmitted.

Assuming that a dynamical system "communicates" some (but not necessarily all) information about its past state into the future yields the diagram shown in figure 3.7. Here x and x' denote observable past and future system variables, the other time independent variables being "lumped into the conditional distribution $P(x'/x)$ which describes the causal connection between past and future given by the system dynamic" (22). This means that $P(x)$ and $P(x')$ represent (as probability distributions) knowledge of the system's past and future states.

These probability distributions can then be rendered by a series of iterated mappings (F), from t_1 to t_2 and so on into the future. Figure 3.8 dis-

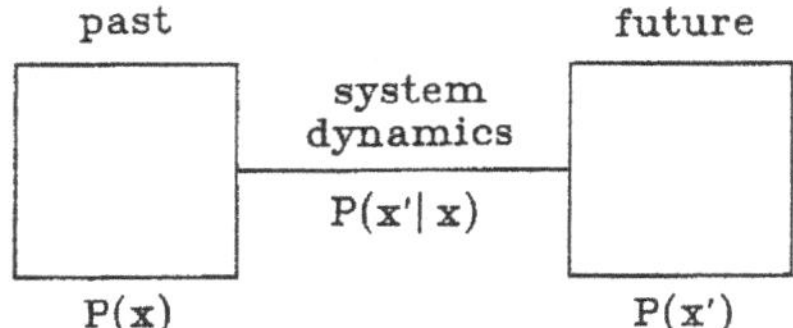

Figure 3.7
Distribution from past to future. Robert Shaw, *The Dripping Faucet as a Model Chaotic System* (Santa Cruz: Aerial Press, 1984), 22.

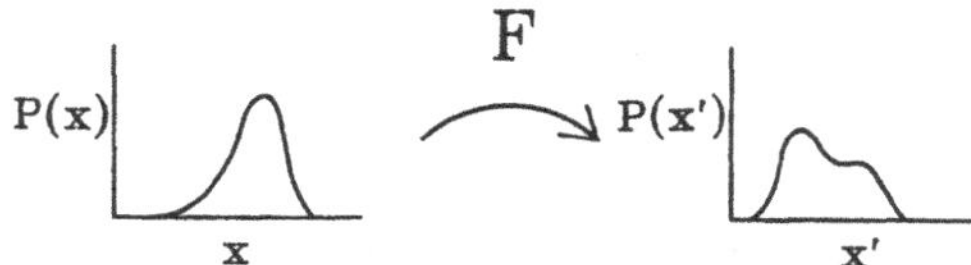

Figure 3.8
Mapping distributions. Robert Shaw, *The Dripping Faucet as a Model Chaotic System* (Santa Cruz: Aerial Press, 1984), 23.

plays the change in the distribution of x over one time step (i.e., from x to x'). Keeping in mind that a highly peaked probability distribution (as on the left) represents a fairly definite knowledge about the variables and that a broader distribution (as on the right) indicates an increase of uncertainty, the broadening of the distribution of x' indicates that there has been a loss of knowledge (i.e., information) and thus of predictive ability.

Shaw's task is to quantify this procedure, thus providing a measure of the system's predictability. Not surprisingly, it requires considerable mathematical ingenuity and we need not follow Shaw's inventive path in detail. Part of the difficulty has to do with the nature of information itself: as a purely relative concept, the information in a distribution $P(x)$ is always measured relative to an a priori expectation, which is the system's minimum information distribution. We must keep in mind therefore that anything coming out of the channel that could be predicted from the input or system's past state is not information, and there is no surprise associated with it. Another problem is that the iterated mappings from one distribution to the next have to be much more fine-grained than indicated above, since for some dynamical systems the "points" can spread apart exponentially fast and errors grow uncontrollably. Indeed, in certain cases, Shaw states, the concept of the point is but a "convenient name for an arbitrarily sharp probability distribution" (31). Following the work of Kolmogorov and Sinai,[42] Shaw resolves this difficulty by applying "a

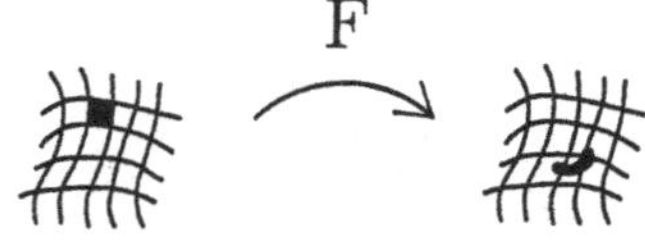

Figure 3.9
Mapping with partition. Robert Shaw, *The Dripping Faucet as a Model Chaotic System* (Santa Cruz: Aerial Press, 1984), 31.

'partition' to the domain of the continuous variables x and x', breaking it up into small elements. The deterministic dynamics can now map an image of each little block, conserving the underlying probability" (31). Shaw represents this more refined mapping of distributions as shown in figure 3.9.

With this more refined mapping, he can begin to define the amount of information stored in a system:

A "system" might be considered a body of information propagating through time. If there is any causal connection between past and future system states, then information can be said to be "stored" in the state variables, or *communicated* to the immediate future. This stored information places constraints on future states, and enables a degree of predictability. Again, *if* the system dynamics can be represented by a set of transition probabilities $P(x'/x)$, this amount of predictability is directly quantifiable, and corresponds, in Shannon's vocabulary, to a particular *rate of transmission of information* through the "channel" defined by $P(x'/x)$. (37)

There are two possibilities, as figures 3.9 and 3.10 indicate.

Suppose, as the system moves into the future, its state is restricted to a narrow interval in x, described by $P(x)$ (indicated by the shaded area in fig. 10). If the distribution relaxes back to the minimum knowledge state (as in the top half of the diagram) regardless of the input distribution over the time interval, then no information is stored in the system. If, on the other hand, the output distribution varies with the input (as in the lower portion of the diagram), then some information is passing into the future.

An unchanging and completely known "variable" communicates no information into the future. Although a static structure is certainly a predictable element, establishing continuity through time, it becomes a part of the fixed time-independent system description. The "information" measures described here are a property of the system *dynamics*. In mechanics, as in the human sphere, the transmission of information requires the possibility of change. (38)

In short, only those parts of a system that are changing over time can convey information.

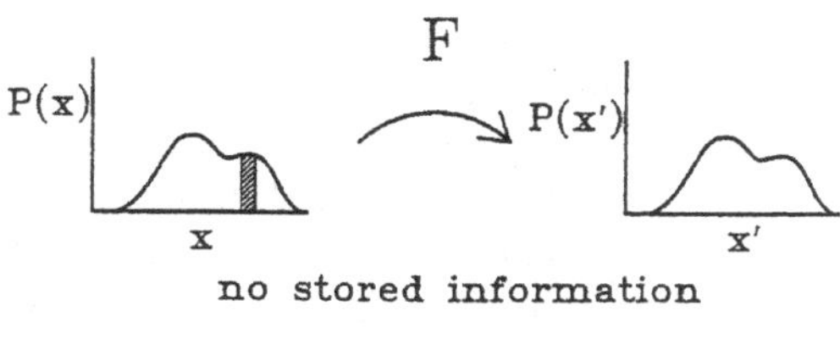

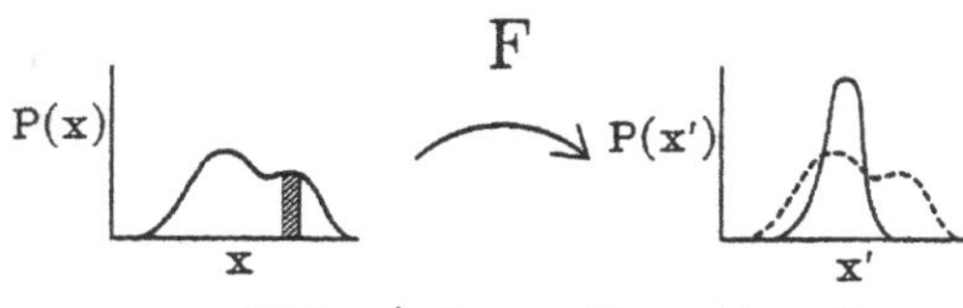

Figure 3.10
Measure of stored information. Robert Shaw, *The Dripping Faucet as a Model Chaotic System* (Santa Cruz: Aerial Press, 1984), 38.

We now arrive at the crucial step of defining quantitatively the information stored in a system: it is simply "the Shannon channel rate for an input (and output) distribution given by the equilibrium distribution for the system" (39). This formula, Shaw explains, "quantifies the average increase in our ability to predict the future of a system when we learn its past" (40). In other words, it is the quantity derived by Shannon's formula for the mutual information shared by sender and receiver. Thus Shaw's exact formula provides a measure of "the *difference* in the expected randomness of a system with and without knowledge of its past" (40). With this quantity in hand, to compute the rate of loss of the stored information or the system's entropy becomes a relatively simple matter. This computation in turn enables Shaw to measure quantitatively the changes in the system's predictability over time, which is his primary objective.

Omitted from this bare-bones exposition are not only the mathematical details but also several essential considerations, most importantly the problem of noise and the system's entropy. In order for the system's stored information to be a meaningful quantity, it has to be distinguished from both the noise in the physical system and the noise in the system of measurement. These two kinds of noise reflect our inability to measure the system's behavior accurately and thus add to our overall uncertainty. However, what Shaw seeks is a way to measure a third kind of noise—the uncertainty that the system itself *produces* over time as it moves from a single-point or periodic regime to a chaotic one. In this movement the system's total entropy (which combines all these forms of noise) would rapidly increase. Since Shaw wants to quantify this increase in unpredictability precisely in informational terms, he first has to establish, following Shannon on channel capacity, that the information measurement channel has a higher capacity than the system's entropy; otherwise, he would simply be measuring fluctuations in the heat bath. Then, subtracting out the entropy caused by the first two kinds of noise—subtracting a smaller entropy from a larger one (40)—he arrives at the precise amount of the system's stored information.[43]

Shaw's experiment thus performs a necessary integration of two concepts: the system's physical entropy as given by Ludwig Boltzmann's famous formula and the system's stored quantity of information as derived from Shannon's formula. As he establishes in his article and restates here, "Entropy describes the *rate* at which information flows from the microscopic variables to the macroscopic" (34). In the dripping faucet model, however, he is able to demonstrate that "the maximum entropy, or loss of predictability, will thus occur when the stored information is at a maximum" (49). This makes intuitive sense, because "sharper distributions representing greater stored information will spread faster than broad distributions" (49). This discovery, of course, provides further empirical evidence for what Shaw had proposed in his article. But what distinguishes *The Dripping Faucet as a Model Chaotic System*—indeed, makes it a methodological tour de force—is how it throws light on the problem of predictability in the precise modeling of a chaotic system. Not only is Shaw's central intuition concerning the loss of information and predictability borne out by precise experimental results, but a contradiction in classical experimental methodology is resolved: specifically, a discrepancy that grows over time between the number stream generated by empirical observation of the physical system and the number stream generated by the explanatory model does not automatically or necessarily mean that the experiment is a failure, since this discrepancy itself can be a positive source of *new* information if the growing rate of that difference can be measured. Which is exactly what Shaw demonstrates.

While it is not my purpose to assess Shaw's unique contribution to the then fledgling field of chaos science, there is no denying the ingenuity of his experiment—based simply on counting the intervals between drops of water—and the sheer amount of intellectual work he accomplishes in elaborating its significance for a theory of modeling chaotic systems. But above all, what gives this work special value and importance is the con-

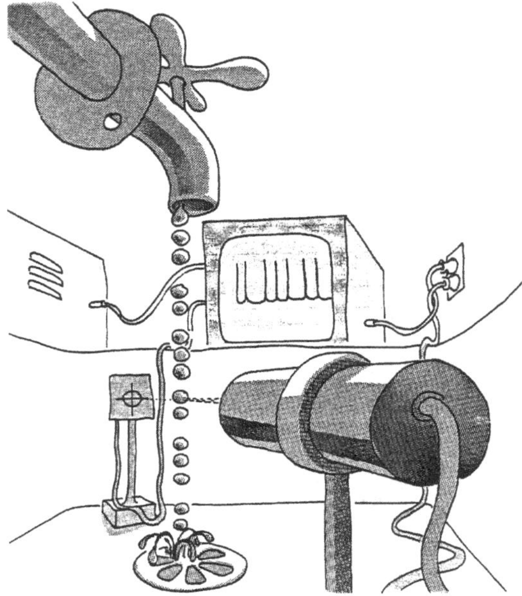

Figure 3.11
The apparatus (drawn by Chris Shaw). Robert Shaw, *The Dripping Faucet as a Model Chaotic System* (Santa Cruz: Aerial Press, 1984), 3.

ceptual synthesis Shaw achieves by introducing information theory into a dynamical systems framework. At the same time, this work also marks a pioneering exploration of the machinic phylum. First, it can easily be shown that Shaw's experimental apparatus and published discourse together constitute a concrete assemblage in Deleuze and Guattari's specific sense. On one side, the experimental apparatus and observer define a machinic assemblage of bodies; on the other side, the data stream and discourse that accompanies it define a collective assemblage of enunciation, woven from many textual sources but primarily from the technical languages of physics and mathematics (see figs. 3.11 and 3.12).[44]

At first glimpse, what stands out is the obvious contrast between the everyday banality of the dripping faucet and the abstruse fecundity of this mathematical discourse. D&G, in fact, characterize mathematics as a "fantastic and nomadic argot" (*A Thousand Plateaus*, 24). More striking is what draws the two sides of the assemblage together onto a "plane of consistency" while also defining its "cutting edges of deterritorialization": it is clearly the force of the strange, or chaotic, attractor, which on both

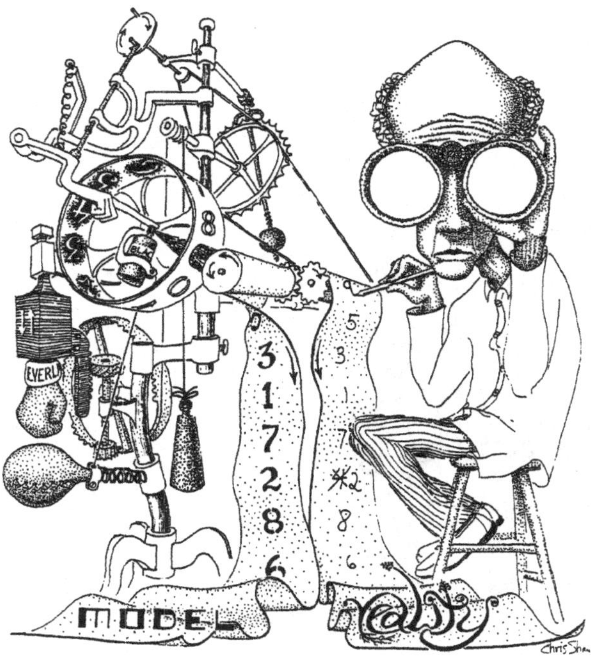

Figure 3.12
The two data streams (drawn by Chris Shaw). Robert Shaw, *The Dripping Faucet as a Model Chaotic System* (Santa Cruz: Aerial Press, 1984), 87.

sides is pulling the assemblage into the realm of the unpredictable. On one side, the behavior of falling water becomes unpredictable, even though mapping a fuzzy but recognizable visual pattern; while on the other side, a certain amount of the system's stored information becomes irretrievable, but at a quantifiable rate. Yet these paradoxes derive from the chaotic attractor itself, which is at once a deterritorializing set of physical forces and a very complex mathematical object with singular properties.

As modeled by Shaw, moreover, the chaotic attractor exhibits the precise attributes of what Deleuze and Guattari call an abstract machine, which inhabits equally the realms of matter-energy and abstract mathematical function:

The abstract machine in itself is destratified, deterritorialized; it has no form of its own (much less substance) and makes no distinction within itself between content and expression, even though outside itself it presides over that distinction and dis-

> tributes it in strata, domains, territories. An abstract machine in itself is not physical or corporeal, any more than it is semiotic; it is *diagrammatic* (it knows nothing of the distinction between the artificial and the natural either). It operates by *matter*, not by substance; by *function*, not by form. Substances and forms are of expression "or" content. But functions are not yet "semiotically" formed, and matters are not yet "physically" formed. The abstract machine is pure Matter-Function—a diagram independent of the forms and substances, expressions and contents it will distribute. (141)

The primordial and dynamic quality of an abstract machine, which is what D&G are trying to elucidate, cannot be conveyed in the language of a "thing" and its "representation." If both thing and representation could change in a dynamic relationship of reciprocal determination, then perhaps the "diagrammatic" quality of the abstract machine could be conveyed. Viewed as a type of abstract machine, however, the peculiar qualities of the chaotic attractor begin to make a unique kind of sense. As both an array of forces and a mapping of their vectors, the chaotic attractor is what deterritorializes the assemblage, both pulling it into a state of chaotic unpredictability and pointing to a new mathematical coding that allows this process to be measured. Rather than view the attractor as a kind of Platonic form that exists independently of its instantiation in a particular nonlinear dynamical system—which is how attractors are sometimes viewed—we should say, as D&G say of the abstract machine, that it "plays a piloting role" (142): it neither preexists nor represents the real, but constructs it and holds it in place.

Although D&G's concepts provide an unusual perspective on Shaw's work and chaos science more generally, this perspective is not incompatible with that of chaos science itself. This is not to downplay or ignore their many obvious differences, which reflect fundamental differences between physics and philosophy. While Shaw's research stands firmly on its own as part of an already validated and ongoing scientific revolution—and similarly for D&G's accomplishment in relationship to philosophy—when brought together these two lines of inquiry create a conceptual space for considering how the new realm of phenomena made visible by nonlinear systems science actually opens a window onto the machinic phylum. In one sense this new perspective simply re-asserts what Shaw and other nonlinear systems physicists have made clear—that chaotic attractors pervade the natural world and that recognition of nonlinear phenomena restores complexity to the classical deterministic universe, which is the world we actually inhabit. At the same time, D&G's concepts give special emphasis to the chaotic attractor's most radical and deterritorializing effects.

Strikingly original research like Shaw's often generates questions about how it is to be interpreted: Is this research primarily empirical or theoretical, physical or mathematical, creation or discovery? However, by considering his modeling of a chaotic attractor in relation to the machinic phylum and the absolute deterritorialization of an abstract machine we can easily see that it falls on neither side of these false oppositions but represents an innovative conjunction made possible by physical measurement and a creative extension of information theory. As noted earlier, the whole point of Shaw's experiment is *not* to use the observed data merely to confirm the model as a correct understanding of a physical process, as would be the case for a traditional experiment. Rather, it is to use it as the basis for further computations that may account for the data's increasing *difference* from the data that the model was supposedly constructed to explain. In fact, the novel use of computational theory is a theme that characterizes much innovative contemporary science. In using experimental data in a manner not to confirm what is already known but to measure the rate of its destruction, Shaw becomes a kind of probe head, the human part of a machinic assemblage that functions like a dynamic feedforward device, relentlessly pushing into the unknown (or at least the unpredictable future) all while insistently measuring the rate of that advance. What is affirmed and confirmed—for both physics and philosophy, and against their prior and respective idealizations—is that processes of dynamic change follow the irreversible arrow of time.

Further consequences follow. First, as suggested earlier, Shaw's research possesses an epistemological complexity different in kind from the methodologies made by means of the telescope or microscope, whose prosthetic powers made visible phenomena previously imperceptible to the naked eye. Although the physical evidence required a conceptualization in order to become part of scientific theory—in Newton's theory of gravity or the germ theory of disease, for example, it also retained an autonomy within the empirical order. But in chaos science the behavior of numbers points to kinds of changes that can't be seen directly, except in a simulation. Lorenz's research, for example, did not so much explain the phenomena of weather as provide numerical evidence—and a new understanding—of its unpredictability. In Shaw's experiment the primary data are measured intervals between drops of water, but what matters are how these intervals are correlated as indices of information storage and loss in a simple physical system. Although we can see turbulence in water flowing from a faucet, the irretrievable loss of information it instantiates as a process can only grasped in a complex conceptualization. In Shaw's experimental apparatus what counts as empirical data and as theoretical construct are mediated not only by the computer but by a new understanding of computation, which brings the two together in a new kind of dynamic relationship. This mediation, which is both a material and discursive conjunction, can best be understood diagrammatically, in relation to a computational assemblage that opens out onto the machinic phylum.

A Complexity Metric: ϵ-machine Reconstruction

This reading of Shaw's work is confirmed and can be extended by considering some very original research carried out by Shaw's younger colleague in the Santa Cruz Collective, James P. Crutchfield. Like other members of the Collective, in the late 1970s and early '80s Crutchfield published a number of pathbreaking articles that greatly advanced our understanding of chaotic attractors, specifically in terms of the modeling and measurement of their behavior. Study of the coupling of dynamical systems with external information sources and the influence of noise soon led Crutchfield to focus on the computational capacity of physical processes. Given a data stream obtained by a measuring instrument (like Shaw's waterdrop intervals), how could the physical processes that produced it be understood as instantiating a computation and therefore be modeled as a specific type of computational machine? Thus, whereas Shaw had sought to measure how the information stored and transmitted in a system varied as it gradually entered a chaotic regime, Crutchfield became interested in how a dynamical system's intrinsic computational capacity might be measured in relation to its changing state. His objective, in short, was to (re)construct a computationally equivalent machine, or what he would call an ϵ-machine. Given that there are many different computational models, classes, and languages, the complexity of the particular "reconstructed" machine or automaton would necessarily reveal a great deal about the complexity of the physical process that produced the data stream. The point of reconstructing such a machine, however, was not simply to reproduce this stream (or a future time series of data) but to measure quantitatively the complexity of the computational process required to produce it. In essence, the ϵ-machine reconstruction would provide a complexity metric for nonlinear or chaotic systems.

For Crutchfield, complexity means a productive tension between order and chaos. More specifically, he assumes that the data stream will be neither orderly and predictable, like the motion of a clock, nor utterly random and unpredictable, like a series of coin flips. Research on nonlinear systems suggests that a rich spectrum of unpredictability lies between these two extremes, and that complexity would appear to be an amalgam of both order and randomness, the unpredictable result of their interplay. "Natural systems," Crutchfield writes, "that evolve with and learn from interaction with their immediate environment exhibit both structural order and dynamical chaos."[45] The interface between structure and uncertainty often results in increased complexity, which can appear "as a change in a system's computation capability." Indeed, the present state in the course of evolutionary progress suggests that we need "to postulate a force that drives in time toward successively more sophisticated and qualitatively different computation." We ourselves, and our ongoing attempts to construct more sophisticated models, both of ourselves and of the processes around us, are the evidence for this, in addition to the fact that we can look back to earlier times when there were "no systems that attempted to model themselves" (46). Answering the question of how "lifeless and disorganized matter [can] exhibit such a drive" is therefore essential to the understanding of our own complexity.

As we might expect, the task of measuring complexity involves both new conceptual and technical tools. As Crutchfield conceptualizes it, two distinct kinds of processes must be taken into account. On the one hand, reconstructing an ϵ-machine is fundamentally an inductive problem (detecting a pattern in the data stream) and involves formal learning theory in its inference of a regularity. Treated as a sequence of symbols, the data stream implies a specific language with its own type of grammar, in the sense that formal language theory defines these terms. On the other hand, to reconstruct an ϵ-machine requires the application of computational ideas to dynamical systems and therefore a statistical mechanical description of the machines themselves. Critically, the reconstruction must be of "minimal computational power yielding a finite description of the data stream."[46] This minimality is essential, since it restricts "the scope of properties detected in the ϵ-machine to be no larger than those possessed by the underlying physical system" (230). In other words, the ϵ-machine must be able to reproduce exactly—nothing greater and nothing less than—the physical system's intrinsic capacity for computation. The very idea of an ϵ-machine thus implies a hierarchy of computational models and types, differentiated by the order of complexity each type of computation is capable of. As we saw in the previous chapter, Noam Chomsky produced the first and best-known of these computational hierarchies, in which the complexity of the type of language is ranked according to the complexity of the machine or automaton that recognizes or can generate it. Specifically, this hierarchy provides the following correlation between formal languages (grammars) and automata (machines), in an order of increasing complexity:

1. Regular languages are recognized by finite-state automata.
2. Context-free languages are recognized by pushdown automata.
3. Context-sensitive languages are recognized by linear-bounded automata.
4. Phrase-structure or recursively enumerable languages are recognized by Turing machines.

Epsilon-machine reconstruction assumes and builds on this hierarchical correlation. To be sure, computation theory cannot simply be applied to the construction of computational metrics for natural physical processes, as Crutchfield is well aware. Many differences have to be taken into account, and the theory itself has to be extended in several ways. For example, whereas research in computation theory is primarily occupied with scaling the difficulty of different computational and information-processing tasks, any attempt to use computation theory for scientific modeling must be able to measure structure in stochastic processes.[47] In fact, in order to construct a useful computational hierarchy, a wide range of variables must be considered, such as how input-output configurations correspond to internal states of the machine, the amount and kind of memory required to perform the computation, and so on. In "Reconstructing Language Hierarchies," Crutchfield includes a table of eight different types, ranked according to specific formal criteria such as grammar, state, memory, and so on; however, in "The Calculi of Emergence" the computational hierarchy contains a much greater diversity of computational mechanisms, as we see from his diagram (fig. 3.13). Indeed, looking at the array of twenty-four interrelated but distinguishable types, one cannot help but wonder if this mapping of abstract machines might represent a new kind of computational phylum.

Like Shaw's experimental assemblage, Crutchfield's ϵ-machine reconstruction can be reconceptualized as a two-part computational assemblage: on the one hand it is a machine for producing a data stream (i.e., measuring physical quanta); on the other, a machine for (re)producing the structure of this data in a correlative computational discourse. If the data stream is conceived of as a symbol string that obeys the "grammar" of a formal language, then it should correspond to a specific type of machine or automaton.[48] The difficulty arises when the data stream exhibits

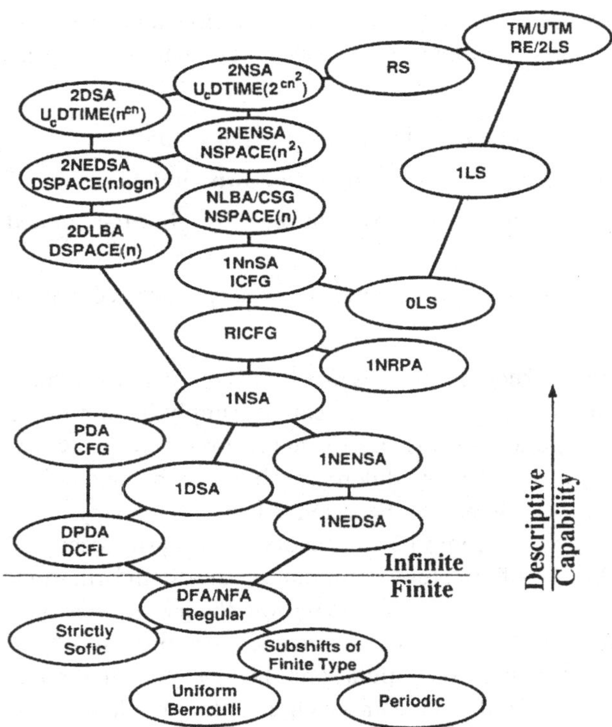

Fig. 2. The discrete computation hierarchy. *Adjective legend*: 1 = one way input tape, 2 = two way input tape, D = deterministic, N = nondeterministic, I = indexed, RI = restricted I, n = nested, NE = nonerasing, CF = context free, CS = context sensitive, R = recursive, RE = R enumerable, and U = universal. *Object legend*: G = grammar, A = automata, FA = finite A, PDA = pushdown A, SA = stack A, LBA = linear bounded A, RPA = Reading PDA, TM = Turing machine, LS = Lindenmayer system, 0L = CF LS, 1L = CS LS, and RS = R set. (After [31,35–39].)

Figure 3.13
The discrete computation hierarchy. James P. Crutchfield, "The Calculi of Emergence: Computation, Dynamics and Induction," *Physica D* 75 (1994): 21.

the effect of a nonlinear parameter and thus the presence of a chaotic attractor in the dynamical system producing it. Given that Crutchfield's ϵ-machine reconstruction is explicitly designed to model complexity in the form of noticeably irregular or chaotic data and is therefore predicated on the conjunction of a physical machine and discursive machine, it is not surprising that the result resembles what I have been calling a computational assemblage. As with Shaw's computational assemblage, the relationship between the physical machine and the discursive machine is one that reflects dynamical differences rather than simple numerical correlations.[49] Only this more complex kind of correlation, it would seem, makes the quantification of complex behavior possible.

In several articles, but most notably in "Computation at the Onset of Chaos," Crutchfield focuses on a complex behavior well known in chaos science: the period-doubling cascade observed when a system first advances along the route to chaos. The most familiar of such nonlinear systems is the logistic map plotted by the difference equation

$$x_{n+1} = rx_n(1 - x_n)$$

which exhibits steady states, periodic cycles of different lengths, and chaos for different values of r. Specifically, with increasing values of r above 3.0, the system enters a period-doubling regime. In effect, the system cycles twice through phase space before exactly repeating its periodic orbit, then cycles four times, then eight times and so forth, in what was recognized early in the development of chaos science as a telltale sign of the system's entry into a chaotic regime.[50] In the terms of dynamical systems theory, with each cycle the attractor bifurcates, or splits into two attractors. These bifurcations indicate a dramatically increasing instability in the system and correspond to the system's phase transition from a solid and orderly structure to a fluid or gaseous one. Typical bifurcation diagrams of this transition (see fig. 3.14) show a slowly increasing series of pitchfork splittings that eventually give way to a series of blurry bands unpredictably punctuated by clear stripes, indicating brief reappearances of periodic (i.e., orderly) behavior. Reproducing a similar bifurcation map, Gleick notes that "the structure is infinitely deep. When portions are magnified . . . they turn out to resemble the whole diagram" (74–75). In other words, the structure is fractal.

How might this behavior be emulated by a set of computational machines or automata whose output would provide data sets that (re)produce such a mapping? To produce the equivalence of periodic behavior would be easy: it would only require a stored pattern that repeated itself,

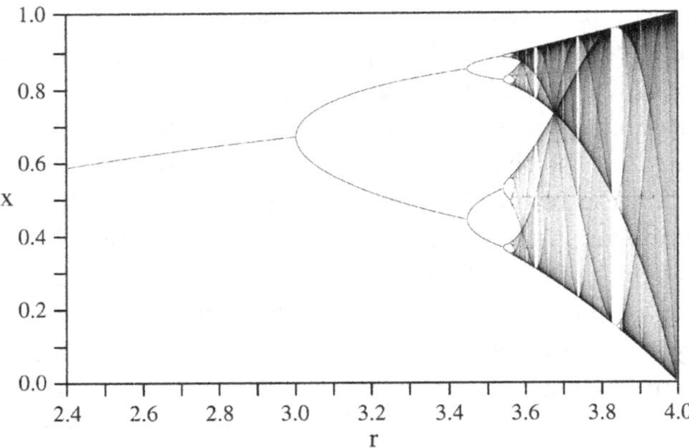

Figure 3.14
Bifurcation map.

like 001100110011.... Likewise, a random behavior would be easy to generate using a Bernoulli-Turing machine, which could feed stochastic input to a Turing machine through contact with a heat bath information source, such as a boiling pot of water.[51] As we have seen in previous sections, the first type of behavior would be low in information content, while the second would be high. Furthermore, according to measures of complexity derived from Shannon's information entropy and ergodic theory, most notably Kolmogorov's "algorithmic complexity,"[52] the first would also be low in complexity and the second high. But this seems counterintuitive, given the just noted ease with which both behaviors can be reproduced. Crutchfield thus argues for a measure of complexity that would reflect the relative simplicity of both regularity and randomness, compared with the behavior observed in phase transitions. As noted, he believes that a measure of the computational effort needed to model and reproduce the behavior of nonlinear dynamical systems would reflect such differences. This other kind of complexity he calls statistical complexity.[53] Indeed, the whole point of ϵ-machine reconstruction is to provide a suitable way to measure it.

Returning to the problem at hand, we note that a system in a period-doubling cascade en route to chaos seems to partake of both types of behavior in an impure mix, so to speak, with a marked transition from one toward the other. Since this complexity of behavior may not be captured with measures of entropy and mutual information, the classification of system behavior based on an application of the entropy metric to the storage and transmission of information (as in Shaw) must be seen as limited. That approach, Crutchfield demonstrates, can be complemented with one based on the combination of computation theory with nonlinear dynamical systems theory in such a way that the behavior under study would be graded according to the intrinsic amount of computation that is performed by the dynamical system, "quantified in terms of machine size (memory) and the number of machine steps [required] to reproduce behavior" (226). Put simply, since a dynamical system consists of a series of states and their transitions, its computational machine equivalent would consist of a directed graph of nodes and links (like the examples in chapter 2) that would accept or generate only those sequences of symbols allowed by its computational grammar or formal language. Crutchfield goes on to show that the behavior of the period-doubling cascade at its initial stage has as its machine equivalent a series of deterministic finite-state automata, which grow in size as the periods double. As the system enters more deeply into the chaotic regime, its states, though still visited deterministically, exhibit nondeterministic branching, meaning that at certain points in the system's orbit through phase space—the points of bifurcation—it must "choose" between two equally viable states. To compute this effect requires a higher-order machine in the Chomsky hierarchy. Specifically, the critical limits of the phase transition require a computational machine with an "indexed context-free grammar," in other words, an automaton with a nondeterministic nested stack (that is, a memory with a "last in first out" structure). Crutchfield then demonstrates that this jump in level of machine complexity is explained by the fact that "nontrivial computation, beyond information storage and transmission, arises at a phase transition" (257).

"Computation at the Onset of Chaos" thus makes fully evident that Crutchfield's project goes well beyond a categorization of levels of complexity such that, for example, thanks to his research we might know that a particular physical process P requires a computational machine of complexity M to fully account for it. His underlying aim, as he plainly avows, is to establish the relationships between levels and to know precisely which changes in a physical process require or instantiate a jump in computational level. And indeed, these are precisely the terms of his actual achievement. Having devised a workable "complexity metric," he has been able to show that the emergence of complexity in certain physical processes is associated with the innovation of new computational model classes. In simpler terms, as certain physical systems tend toward the onset of chaos, they exhibit a marked increase in computational capacity and thus in their information-processing capacity. But there is more. Crutchfield's complexity metric reveals a very precise relationship between the increasing complexity and the increasing randomness of a dynamical system: the two increase together only up to a certain point—the point of a phase transition and the onset of chaos—beyond which complexity begins to decrease as the randomness (or Shannon information entropy) further increases ("Computation at the Onset of Chaos," 258–262).

These unexpected discoveries are absolutely essential for future work in complex systems and theories of complexity. Yet this work has also been of special interest to those studying evolutionary dynamical systems, where the complex interplay between dynamical behavior and information processing is clearly essential to their full understanding.[54] It should come as no surprise, therefore, that the combination of chaos science (or nonlinear dynamical systems theory) and computational theory has been a sine qua non for the development of the new science of ALife and complex adaptive systems, as we'll see in part II.

II MACHINIC LIFE

4 Vital Cells: Cellular Automata, Artificial Life, Autopoiesis

Durham tried again, but the Autoverse was holding fast to its laws. Red atoms could not spontaneously appear from nowhere—it would have violated the cellular automaton rules. And if those rules had once been nothing but a few lines of a computer program—a program which could always be halted and rewritten, interrupted and countermanded, subjugated by higher laws—that was no longer true. Zemansky was right: there was no rigid hierarchy of reality and simulation anymore. The chain of cause and effect was a loop now—or a knot of unknown topology.
—Greg Egan, *Permutation City*

The very idea of self-reproducing machines disrupts our conceptual boundaries and sets us wondering how such self-reproduction could be initiated and sustained. Specifying the formal conditions that would have to be met in order for such a process to occur, John von Neumann first broached the question from a logical point of view. In "The General and Logical Theory of Automata" (1948) and then in his five-part lecture "The Theory and Organization of Complicated Automata" (1949), von Neumann considered differences between natural and artificial automata in computation, control, and complexity. Then in "The Theory of Automata: Construction, Reproduction, Homogeneity," a manuscript begun in 1952 but not completed during his lifetime, he worked out the design criteria and specifications for a self-reproducing automaton.[1] However, it was not until the late 1970s that von Neumann's schema came to fruition, when Christopher Langton embedded a self-reproducing machine in a cellular automaton running on an actual computer. Then, in order to gather together scientists interested in "the synthesis and simulation of living systems," Langton organized a conference (which took place in 1987) that led to the founding of the new science of ALife.[2] Drawing on both theoretical biology and computer science—but reducible to neither—artificial life fomented new questions about life and its relation to information, mainly by creating digital organisms that could replicate and evolve within computer-generated virtual machines. Moreover, as ALife dovetailed with and relayed theories of self-organization and emergence in nonlinear dynamical systems, it took a leading role in the formation of a new conceptual framework in the interdisciplinary study of complex adaptive systems.[3]

From the outset ALife constituted a new kind of computational assemblage, one that produced both new forms of lifelike behavior and a scientific discourse that was performative and synthetic rather than simply descriptive and analytic. In challenging conceptual boundaries, it also raised novel philosophical questions. Daniel Dennett, for example, has suggested that ALife is "a new way of doing philosophy" and not simply a new object for philosophy.[4] Like its putative older cousin artificial intelligence—although the filiation is misleading in key respects—ALife involves the creation and testing of thought experiments, with the difference that the experimenter is kept honest by the fact that the experiments are, as Dennett puts it, "prosthetically controlled" by the "simulational virtuosity of computers." Although Dennett does not discuss specific experiments, he understands that ALife is no ordinary science, since it creates both a new kind of object and a new methodology. Initially opinion differed on whether this object was only the simulation of natural life processes and behaviors or the actual realization of life in a nonorganic medium—ALife's "weak" and "strong" forms, respectively. About its methodological significance, however, there could be no doubt. Proceeding by what Dennett calls "bottom-up reverse engineering,"[5] ALife opened up a new space of exploration, a search or design space predicated on an innovative synthesis of computational theory, dynamical systems theory, and the Darwinian theory of evolution.

Even more than AI, ALife experiments loosen hard-and-fast oppositions between the natural and the artificial—*phusis* and *technē*—oppositions that are increasingly being challenged and transformed.[6] To be sure, the opposition between organism and machine has never been a simple empirical given, even while functioning, at least since Aristotle, as a mainstay of Western metaphysics. What has enabled the boundary line between machines and biological organisms ("life") to be drawn and maintained is the basic fact that machines have never been able to reproduce themselves. In Kant's classical formulation, a machine possesses only "*motive* force," not "*formative* force." In contrast, a product of nature is an "organized being that has within it *formative* force, and a formative force that this being imparts to the kinds of matter that lack it (thereby organizing them). This force is therefore a formative force that propagates itself [*fortpflanzende bildende Kraft*]—a force that a mere ability [of one thing] to move [another] (i.e., a mechanism) cannot explain."[7] What all machines thus lack (Kant's specific example is a watch) is "natural purpose," which is exhibited by all "organized beings." Natural purpose, in turn, follows from two conditions, both of which must be met: the parts of organized being are produced and exist for one another, and they are all part of a self-organizing unity of the whole. Again, machines lack this purposiveness, or the capacity to be self-organizing and self-directed, and receive not only their formal cause or purpose but also their efficient cause from outside themselves. As Kant summarily states, machines exist only "for the sake of the other."

Not surprisingly, then, when machines begin to self-organize and reproduce, they attain a dramatic kind of "life" never before imagined, except under the literary or mythic aegis of the demonic and the infernal. Yet it is not from these categories and the thinking they imply that machinic life will liberate itself. What we now see taking place, rather, is a complex process of involution and rearticulation: not only have new kinds of abstract machines been constructed that can reproduce themselves, but biological life itself has been reconceived in terms of "living systems," that is, as self-organizing biochemical machines that produce and maintain their own boundaries and internal regulation. The theory of autopoiesis proposed by the Chilean biologists Humberto Maturana and Francisco Varela is the earliest and most influential version of this reconceptualization. Inspiring what is generally considered to be cybernetic theory's "second wave," Maturana and Varela defined a living system as a machine that constantly produces, regenerates, and maintains itself in all of its parts and processes as a function of its dynamic and self-directed organization. Arguing that autopoiesis is *the* essential defining feature of all living systems, they decried the information-processing bias of molecular biology as misleading, precisely because it centralizes DNA and cellular reproduction, which they regard as a subsidiary or secondary process.

When ALife and the theory of autopoiesis are considered together, as they will be in this chapter, they provide striking evidence of a double inversion in which each side of the opposition between machines and biological organisms gives way to the other: nonorganic machines become self-reproducing quasi organisms, and organisms become autopoietic machines. This double inversion participates in a contemporary tendency of technological development that Kevin Kelly has vividly described as neobiological.[8] Marshaling an impressive array of examples from systems theory, ecology, computer simulations, robotics, e-commerce and evolutionary theory, Kelly argues that we have entered a new era defined by a "biology of machines," in which "the realm of the *born*—all that is nature—and the realm of the *made*—all that is humanly constructed—are becoming one" (1). But whether this is a becoming "one" or a new and dramatic exfoliation of the machinic phylum—the position I take, it is clear that the opposition between machine and organism no longer marks the site of a simple conceptual breach or collapse; rather, it has become a nexus from which new conceptual possibilities and technologies are rapidly emerging. Computer viruses and recent attempts to construct artificial immune systems, both discussed below, are only two of many compelling examples.

Life on the Grid

When John von Neumann first began to think about how self-reproducing machines could be constructed, he imagined a Rube Goldberg–like apparatus floating in a lake. This apparatus, or automaton, consisted of various "organs" that would enable it to build a copy of itself out of parts freely available in the lake. The most important of these organs were a Turing machine with a tape control that could both store and execute instructions and a "universal constructor" with a constructing arm and various "elements" that could sense, cut out, and connect other elements. In the design of the automaton certain elements called "girders" provided both structural rigidity and encoded information (i.e., their presence or absence at certain joints denoted 1s or 0s). The "lake" was simply a reservoir of elements that differed from the universal constructor only in that it was unbounded and lacked internal organization.[9]

This rather crude and unwieldy model could nonetheless serve its intended purpose, for what von Neumann sought was not so much a material device that could be constructed as a visualizable instantiation of a set of abstract logical functions. Intrinsic to the automaton's design was von Neumann's insight that in order for an entity to reproduce itself—whether that entity be natural or artificial—two separate but imbricated functions were necessary. First, the entity would have to provide a blueprint with instructions that when activated could direct the production of a copy of itself. Second, the blueprint and instructions would have to be passively copied into the offspring, which otherwise would not be able to reproduce itself in turn. Summarily, then, the reproductive mechanism would have to contain information that could function as both executable instructions and passive, or uninterpreted, data to be copied. Watson and Crick's discovery of the structure of DNA in 1953 confirmed that nature itself used the same dual functionality for the reproduction of life.

Von Neumann's student Arthur Burks called this first attempt to instantiate the principles of self-reproduction in an abstract logical structure the kinematic model. Its obvious limitation was its material complexity. Consequently, in order to avoid unwieldy problems like the sensing, cutting, and connecting of the girders randomly dispersed in the lake, von Neumann decided, following the suggestion of the mathematician Stanley M. Ulam, to try a different medium—a cellular automaton. In its most common form, a CA is a two-dimensional lattice, or checkerboard array, divided into square cells, each of which is a finite-state automaton. Recall that at each time step every cell takes its present state as well as the states of its surrounding neighbor cells as its input. Its output will then be its own state in the next time step, as defined by a specific set of rules called a state-transition table. For example, a transition rule might be that if three or more of the adjacent cells are "alive" ("on"), then the central cell will also be "alive" for the next time step; if not, the cell will "die" or go quiescent. Such simple rules can generate very complex behavior.[10] In von Neumann's cellular automaton there were twenty-nine possible states for each cell: one quiescent state and twenty-eight ways to be on. Von Neumann thought that such a large number of states would be necessary in order to work out the state-transition table that would yield a configuration of cells with the specific property that it could reproduce itself.

As with the kinematic model, the automaton consisted of two parts: a control unit (basically, a universal Turing machine) and a constructor arm. Given the right programming (i.e., the right set of rules), the constructor control would "grow" a constructor arm out into an empty, or off, region of the cellular space. As the "head" of the constructor arm scanned back and forth, it would produce new on cells. Eventually a copy of the original configuration would be built from the surrounding cells. Finally, when complete, it would separate from the original configuration and begin to reproduce itself in turn. In contrast to the kinematic model, self-reproduction in the cellular automaton takes place simply by extending into and directly organizing the adjacent space according to the logic of a predefined set of rules. Despite the fact that von Neumann never fully worked out the state-transition tables—an essential but arduous task—the entire conception possesses an eerie brilliance. Yet von

Neumann also realized that by "axiomatizing" natural automata in this way, "one has thrown half the problem out the window and it may be the more important half," since "one has resigned oneself not to explain how these parts are made up of real things.... One does not ask the most intriguing, exciting and important question of why the molecules or aggregates which in nature really occur in these parts are the sort of things they are" (*Theory of Self-Reproducing Automata*, 77). In other words, there could be other forces at work in the physical process of self-reproduction that this formalization fails to capture, a possibility that would come to haunt ALife research.

Following Arthur Burks's editing and completion of von Neumann's manuscripts, both Edgar Codd and John Conway devised cellular automata that could self-reproduce, but with far fewer states per cell.[11] Conway also initiated an eventually successful effort to prove that some cellular automata are computationally universal (that is, equivalent to Turing machines). Then, in the late 1970s and early 1980s, just as the study of cellular automata was about to tail off into complete obscurity, a number of physics-oriented researchers—among them J. Doyne Farmer, Norman Packard, Edward Fredkin, Tommaso Toffoli, Norman Margolus, and Stephen Wolfram—began to explore CA as dynamical systems. In 1983 an invigorating interdisciplinary workshop at Los Alamos National Laboratory brought these scientists together and consolidated this new perspective. The key idea was that a change of state in a discrete dynamical system was also a computation. Since in a CA each cell in the whole array of cells is changing state simultaneously, the computation is actually parallel, rather than serial, as Stephen Wolfram explains in the preface to the workshop *Proceedings*:

> The discrete nature of cellular automata allows a direct and powerful analogy between cellular automata and digital computers to be drawn. The initial configuration for a cellular automaton corresponds to the "program" and "initial data" for a computation. "Processing" occurs through the time evolution of a cellular automaton, and the "results" of the computation are given by the configurations obtained. Whereas typical digital electronic computers process data serially, a few bits at a time, cellular automata process a large (or infinite) number of bits in parallel. Such parallel processing, expected to be crucial in the architecture of new generations of computers, is found in many natural systems.[12]

One of the main themes of the conference, in fact, was that cellular automata, in their capacity to perform parallel and highly distributed computations, could provide a new and better tool for studying the behavior of complex dynamical systems. In short, CA could simulate many physical processes that when modeled using differential equations were computationally intractable.

Overall, then, CA research following von Neumann underwent two distinct developments: the mechanism of self-reproduction was simplified and the parallel processing capacities of CA were actualized in computer simulations. But yet another possibility was waiting in the wings, which would eventually give birth to ALife. This was the realization that cellular automata, by virtue of their capacity to generate self-reproducing structures and, like many natural systems, to process information in parallel streams, might actually constitute another form of life. In other words, computers could not only simulate natural forms but synthesize and actualize life in a computational medium.

Christopher Langton, who had pursued the trajectory of CA research from von Neumann to Burks, Cobb, and Conway more or less on his own, made this idea a concrete reality. He would later argue that biological life as we know it may be only one possible form—it is carbon-based—of a more general process whose logic and essential properties remain to be discovered. This would mean that some of the properties of life as biologists currently study it may be only contingent properties, due solely to local conditions on Earth. It is easy to see how research with CA, in which "cells" interacting according to simple rules produce highly complex patterns of behavior, would lead Langton to the idea that life is "a property of *form*, not *matter*, a result of the organization of matter rather than something that inheres in the matter itself."[13] Neither nucleotides nor amino acids are alive, he observes, yet when combined in a certain way "the dynamic behavior that emerges out of their interactions is what we call life." Life is thus a kind of behavior—not a kind of stuff—that emerges from simpler behaviors. The complex behavior that emerges from the nonlinear interactions among many physical parts are "virtual parts" of living systems, in the sense that they cease to exist when the physical parts are isolated.[14] However, when these virtual parts—in Langton's metaphor, "the fundamental atoms and molecules of behavior" (41)—are simulated in a computer, the dynamic behavior resulting from their interaction is no less real and can lay claim to reality as a new form of "life."

Langton shaped these ideas into the guiding principles of a new research program. First, he organized and launched the conference on the "Synthesis and Simulation of Living Systems" mentioned above. It took place in Los Alamos in 1987 and drew a diverse assortment of 160 scientists working on "a wide variety of models of living systems, including mathematical models for the origin of life, self-reproducing automata, computer programs using the mechanisms of Darwinian evolution to produce co-adapted ecosystems, simulations of flocking birds and schooling fish, the growth and development of artificial plants, and much, much more."[15] Second, in the essay that would become the new research program's manifesto and the title of the conference proceedings, Langton gave this diversity both a conceptual unity and a name. Appropriately, the new discipline was called Artificial Life, and Langton would become known as its founder.[16]

Langton's path to conceptual innovation began as an attempt to pursue a mysterious feeling he had experienced while working on the mainframe computer at Massachusetts General Hospital, which employed him as a programmer in the early 1970s. Debugging code one night while running a long configuration of Conway's Game of Life on the computer, he was suddenly overwhelmed by a sense that what was on the screen was somehow alive, and that something akin to life was evolving before his very eyes. He soon sought out books on cellular automata theory and retraced von Neumann's work. A telephone call to Arthur Burks brought the unexpected news that a self-reproducing cellular automaton had yet to be programmed on a computer. Fortunately for Langton, desktop computers were beginning to appear on the market, and in the summer of 1979 he bought an Apple II. Within months he was able to duplicate Codd's work with eight-state CA, and by October he had created little digital loops with short tails—were they organisms or machines?—that could replicate and form colonies.

The centrality of the computer in Langton's research cannot be overstated. Like von Neumann before him, he experienced a shift of interest that can be viewed retrospectively as a two-stage process. Von Neumann's work with cellular automata grew out of his work developing the new computing machines; in fact, his design for EDVAC, the first electronic general purpose, stored-program computer, was one of his most far-reaching achievements. Like the cyberneticists, von Neumann thought about these new machines in relation to both natural and artificial automata. His objective, nonetheless, was not to simulate or synthesize biological life but to understand how its organizational logic might be deployed to achieve greater reliability in computational devices. When his interest turned to constructing a self-reproducing automaton, he reasoned roughly as follows. Since self-reproduction in nature occurs, there must be an algorithm that describes it. If there is such an algorithm, then it can be executed on a universal Turing machine. The task, therefore, is to embed a UTM that can both execute and copy its own instructions, along with a universal constructor, in a cellular automaton. Langton continued this line of thinking, which he also experienced in two stages. First, he realized that with CA a self-reproducing structure could be achieved without the necessity of either universal computation or a universal constructor. Having demonstrated this insight with his self-reproducing digital loops, he widened his objective to consider how new ideas about computation might provide the conditions of possibility for new forms of artificial life.

Langton defines this larger agenda in his introduction to *Artificial Life*. In a section entitled "The Role of Computers in Studying Life and Other Complex Systems," he distinguishes between AI and Artificial Life.[17] Whereas the former takes the technology of computers as a model of intelligence, the latter attempts "to develop a new computational paradigm based on the natural processes that support living organisms" (50). As Langton points out, this shift was already implicit in the development of "connectionist [neural] net" models that had revived an early strand of AI research based on Warren McCulloch and Walter Pitts's classic work on neural nets.[18] Langton therefore explicitly aligns ALife with connectionism and its rejection of the digital computer's architecture as a model of intelligence. This alignment signals both an underlying conceptual and a methodological affinity, inasmuch as both ALife and connectionism espouse an understanding of intelligence and life based on the concepts of dynamic self-organization and parallel, highly distributed information processing.

This new approach to computation, later called emergent computation, inverts the rigidly hierarchical, centralized command-and-control structure that dominated early AI approaches to modeling intelligence and that is still deployed in most computer architectures today. The inversion is necessary for the simple reason that, in contrast to AI's top-down approach, ALife "starts at the bottom, viewing an organism as a large population of *simple* machines, and works upward *synthetically* from there—constructing large aggregates of simple, rule-governed objects which interact with one another nonlinearly in the support of life-like, global dynamics" (*Artificial Life* [1989], 2). The result Langton calls emergent behavior:

> Natural life emerges out of the organized interactions of a great number of nonliving molecules, with no global controller responsible for the behavior of every part. Rather, every part is a behavior itself, and life is the behavior that emerges from out of all of the local interactions among individual behaviors. It is this

bottom-up, distributed, local determination of behavior that AL employs in its primary methodological approach to the generation of life-like behaviors. (*Artificial Life* [1989], 2)

Clearly it would not be appropriate, therefore, to instantiate this bottom-up approach using a traditional computer program—"a centralized control structure with global access to a large set of predefined data-structures"—since such programs are specifically designed to halt after producing a final result, not to allow "ongoing dynamic behavior" (*Artificial Life* [1989], 3).[19] Accordingly, Langton proposes the following as essential features of computer-based ALife models:

- They consist of populations of simple programs or specifications.
- There is no single program that directs all of the other programs.
- Each program details the way in which a simple entity reacts to local situations in its environment, including encounters with other entities.
- There are no rules in the system that dictate global behavior.
- Any behavior at levels higher than the individual programs is therefore emergent. (*Artificial Life* [1989], 3–4)

As discussed below, the new computational paradigm defined by emergence extends well beyond its application to ALife and has been generalized in what is loosely called complexity theory. Langton contributed further to its development by creating Swarm, a software platform that deploys a bottom-up, highly distributed architecture designed to simulate a range of complex systems comprised of many interacting agents (Swarm even allows for the hierarchical nesting of such complex systems).[20] It should be emphasized—particularly since Langton himself does not always receive proper credit—that it was within the context of ALife research that the basic principles of emergence in the contemporary sense were first fully elaborated.[21]

Langton points out in "Artificial Life" that the computer is both "an alternative medium within which [one can] attempt to synthesize life" and a laboratory tool that replaces the "wetware paraphernalia" that would normally stock a typical biology lab. This is because the computer itself can incubate informational structures: "Computers themselves will not be alive, rather they will support informational structures within which dynamic populations of informational 'molecules' engage in informational 'biochemistry'" (50–51).[22] As both medium and tool, computers can simulate complex processes such as turbulent flow as well as show that "complex behavior need not have complex roots" (51). Indeed, one of Langton's major points is that complexity often arises from the interactions of many simple elements. Summarily, then, ALife involves two fundamental themes: first, life is an emergent, bottom-up form of behavior arising from decentralized, nonliving elements in interaction; second, as a form of behavior, it can be synthesized in media other than that of organic chemistry, specifically in a computational medium in which informational structures can replicate and propagate. While these themes owe their condition of possibility to the computer, ALife itself participates in the computer's use within and further development of new kinds of computational assemblages, which vary in material organization, means of functioning, and purpose.[23]

Artificial Life: The First Formulations

In his programmatic essay "Artificial Life," Langton recontextualizes his earlier research with cellular automata, for the most part presented in articles published in *Physica D*,[24] within a broad overview of this new field of scientific inquiry. In addition, he discusses the work of those he deems necessary or exemplary for ALife research. Beyond conveying "the 'spirit' of the Artificial Life enterprise" (92), Langton opens a new beginning that necessarily concerns a moment of origination. What begins with the conference (at least officially or professionally) is a new kind of science grounded in a reconceptualized understanding of life based on simulation and synthetic (re)production and not limited by what has been observed in the natural realm. But as discussed below, it also leads to a new theorization of the origins of life. Indeed, the question, what is life? is perhaps more central to ALife than it is to biology, which restricts itself to documenting and explaining the empirical order of organic life. In contrast, ALife gives itself a double objective: to advance scientific understanding of the mechanisms and logic of life regardless of medium and to bring into existence new forms of nonorganic life. This double objective necessarily means that ALife has an ambiguous status in relation to traditional biology. For in taking what Deleuze and Guattari call "the prodigious idea of *Nonorganic Life*" as a sanctioned objective of scientific research, Langton and fellow ALife scientists in effect join the ranks of other silicon probe heads who follow and develop the flow of a new phylum. Yet ALife is less a window onto the "machinic phylum" than its burgeoning and increasingly rich extension into new forms of behavior and even a new mode of being. With this double perspective in mind, let us review the intellectual framework that Langton proposes.

In contrast to biology's analytic approach, which breaks down living systems into their component parts in an attempt to understand how they work, ALife seeks to synthesize not simply life as we know it but, more importantly, life as it could be. Theoretically this means that ALife is no longer simply a mimetic undertaking but is a performative and productive one. Langton wisely confines himself to what this means in practice—producing diverse and complex lifelike behavior through evolution in an artificial or wholly constructed medium. Specifically, in its attempts to generate lifelike behavior, ALife will first identify the mechanisms by which behavior is generated in natural systems and then recreate these mechanisms in artificial systems. Unlike AI, which seeks to produce the effects of intelligence without concerning itself with the methods by which it occurs naturally, ALife endeavors to follow nature in at least one fundamental aspect, which Langton succinctly emphasizes: "Living systems are highly distributed and quite massively parallel" (41). Furthermore, if life results from a particular organization *of* matter, rather than from something inherent *in* matter, then nature suggests that this organization emerges from dynamic, nonlinear interactions among many small parts. In other words, life does not result from the infusion of some universal law or life-principle into lower, more localized levels of activity but emerges spontaneously from the bottom-up. Having rejected vitalism, modern biologists generally believe that life can be explained by biochemistry. In principle, this means that they believe that "living organisms are nothing more than complex biochemical machines" (5). In Langton's view, however, a living organism is not a single, complicated biochemical machine but a large population of relatively simple machines. The complexity of its behavior results from the highly nonlinear nature of the interactions among all the members of this polymorphic population. "To animate machines," he states, "is not to 'bring' life to a machine; rather it is to organize a population of machines in such a way that their interactive dynamics is 'alive'" (*Artificial Life* [1989], 5). In practice this means creating a population of machines that evolve and self-organize.

To model the complex behavior characteristic of life as a multiplicity of machines, Langton appropriates the biological terms *genotype* and *phenotype*, applying the first to the set of local rules that define the local agents, or elements, of the system, and the second to the behavior that results from their interactions. Thus the genotype is a "bag of instructions," and the phenotype what happens when those instructions are executed. Applied to nonbiological situations, the terms also yield a distinction of levels: at the level of the genotype (GTYPE), local rules produce nonlinear interactions among simple elements, or agents, while at the level of the phenotype (PTYPE) global behaviors and structures emerge. Thus defined, the model exhibits the essential key features of rich behavior and unpredictability: nonlinear interactions among the constituent objects specified by the GTYPE provide the basis for an extremely rich variety of possible PTYPES, which draw on the full combinatorial potential implicit in the set of possible interactions among the low-level rules. This means, however, that we cannot predict the PTYPES that will emerge from specific GTYPES, due to the general unpredictability of nonlinear systems (57–58). In other words, one cannot look at a GTYPE and determine what kind of behavior or properties it will generate in the PTYPE. Inversely, one cannot begin with a desired behavior or property (evident in a given PTYPE) and work backward to the specific GTYPE that produced it. Again, this is because any specific PTYPE is the outcome of many nonlinear interactions among local elements.[25] For example, what changes would have to be made in the human genome in order to produce six fingers on each hand instead of five? No answer can be calculated; there is only trial and error. Or rather, there is nature's way—trial and error guided by the process of evolution by natural selection. Langton concludes that "it is quite likely that this is the only efficient, general procedure that could find GTYPES with specific PTYPE traits when nonlinear functions are involved" (58). Thus evolution enters the picture not as an external theme that ALife will have to address but as the inevitable and necessary solution to a dilemma internally generated by the very nature of the object or process considered.

Turning to the building of actual GTYPE/PTYPE systems, Langton considers only examples based on the methodology of "recursively generated objects"—objects generated by repeatedly applying the same rule or procedure within a larger procedure.[26] (See the Lindenmayer system below for an example.) The appeal of this approach "arises naturally from the GTYPE/PTYPE distinction: the local developmental rules—the recursive description itself—constitute the GTYPE, and the developing structure—the recursively generated object or behavior itself—constitutes the PTYPE" (59). As a consequence, the resultant behavior occurs in the same medium as the rules written to generate it. Specific examples are taken from three areas: Lindenmayer systems (L-systems), cellular automata, and computer animations.

L-systems are produced by simply iterating over a set of substitution rules, as in: for A, substitute CB, for B substitute A, for C substitute DA, and for D substitute C. Taking A as the initial "seed," one can quickly generate a linear growth in a relatively short number of steps. Thus A → CB → DAA → CCBCB, and so forth. This kind of growth can be correlated with the behavior of a specific type of finite-state machine. To achieve a branching growth one only has to introduce context-sensitive rules to make the substitution rule change depending on what lies to the left or right of the element in question (see Langton, 60–61, for further discussion). With context sensitive rules one can generate grammars equivalent to Noam Chomsky's "context-sensitive or Turing languages" (63) as well as propagate a signal in the sense of moving a symbol from one position to another in a symbol string—from the far left to the far right positions, for example. Context-sensitive rules thus enable one to embed a computational process that will directly effect the structure's development.

This type of embedding is even more evident in the example of cellular automata, where Langton introduces his own self-reproducing loops as the "simplest known structure that can reproduce itself" (64). Every cell in the CA lattice is a finite automaton whose state-transition table is defined by a single set of rules applied homogeneously across the lattice. In this sense the rules constitute the "physics" of a discrete space/time universe. However, although the same set of rules is iteratively applied to update the cell states, the resulting individual cell states constantly differ because of changing local configurations. (In order to compute its next state, each cell takes as input its own present state as well as the states of its neighboring cells.) This context sensitivity of the rules is what allows the embedding of general-purpose computers and structures that compute and construct other structures (64). Moreover, since these computers are "simply particular configurations of states within the lattice of automata, *they can compute over the very set of symbols out of which they are constructed*" (64, author's emphasis). To illustrate, Langton explains how his self-reproducing loops are embedded in a recursively generated structure. Each loop consists of an inert sheath of cells within which there is a data path. Along the data path, signals (carried by contiguous cells) propagate counter-clockwise until they reach a T-junction between loop and tail, where another loop begins to form. Through four cycles of instructions an offspring-loop is gradually constructed, at which point a collision of signals disconnects the two loops. These two loops in turn replicate, then those replicate, and so on, eventually producing a colony of loops.[27] In this way, Langton states, a "double level of recursively applied rules . . . makes use of the signal propagation capacity to embed a structure that itself *computes* the resulting structure, rather than having the 'physics' directly responsible for developing the final structure from a passive seed" (65). For Langton, this kind of embedding captures the flavor of what happens in natural biological development, where the genotype codes for the constituents of a dynamic process in the cell and this dynamic process then "computes" the expression of the genotype.

For his third example Langton cites Craig Reynolds's computer simulation of flocking behavior. Reynolds discovered that only three easily programmable rules were necessary to make his artificial birds, or "boids," exhibit realistic flocking behavior:

• to maintain a minimum distance from other objects in the environment, including other boids,

• to match velocities with boids in the neighborhood, and

• to move toward the perceived center of mass of the boids in its neighborhood. (Langton, 66)

The vivid computer simulations that instantiate Reynolds's flocking algorithm allow Langton to make an essential point about the "ontological status of the various levels of behavior in such systems," namely, that even though boids are obviously not real birds, "*flocking Boids and flocking birds are two instances of the same phenomenon: flocking*" (68, author's emphasis). In other words, the flocking behavior of boids is not simply a "lifelike" imitation but emerges within an artificial system in the same way that it emerges in nature—through the interaction of independent agents following a simple set of rules. Here the complex behavior (the PTYPE) that emerges from the interaction of simple agents whose local behavior is determined by simple rules (the GTYPE) is just as real or genuine as its naturally occurring counterpart. This means that terms like *model* or *simulation* do not quite convey what is going on here: Reynolds's boids exhibit flocking behavior tout court and do not simply imitate or simulate flocking behavior.[28]

Langton states this explicitly when he refers back to the L-systems and self-reproducing loops in a summary statement: "The constituent parts of the artificial systems are different kinds of things from their natural counterparts, but the emergent behaviors that they support are the same kinds of thing as their natural counterparts: genuine morphogenesis and differentiation for L-systems, and genuine self-reproduction in the case of the loops" (68). From here it is a short step to ALife's "strong claim," with which Langton concludes this section: "A properly organized set of artificial primitives carrying out the same functional roles as the biomolecules in natural living systems will support a process that will be 'alive' in the same way that natural organisms are alive. Artificial Life will therefore be *genuine* life—it will simply be made of different stuff than the life that has evolved here on Earth" (69, author's emphasis).

Having established the cogency of this fundamental idea, Langton turns to evolution and how the principles of natural selection might be embedded in a population of machines. In the original (conference proceedings) version of the essay, Langton considers only John Holland's genetic algorithms, which apply the process of natural selection to the problem of machine learning.[29] Genetic algorithms, basically, are the result of methods Holland invented to "breed" algorithms that are more efficient at performing specific tasks. Starting with a population of algorithms (each typically encoded as a symbol string), Holland selected the most successful and applied "genetic operators" to generate a population of offspring. The basic genetic operators are: (1) random mutation, (2) inversion or bit-flipping, and (3) "crossover," where the algorithm is split in half and the two halves "mated" with the two halves of another algorithm. The entire process is then repeated several times. The method turns out to be a very effective way to search the "schema space" of possible algorithms and to produce algorithms meeting ever higher fitness criteria. Holland also formulated theorems explaining this remarkable efficiency.

In the expanded version of his essay Langton discusses several other research projects that apply evolutionary principles to computation. In connection with Holland's work, John Koza extends genetic algorithms to parse trees in the programming language Lisp. In what Koza calls genetic programming, genetic operators are embedded in the programming language itself.[30] Another example of "computational artificial selection" is the "biomorph breeder" program that Richard Dawkins wrote to illustrate the ideas in his book *The Blind Watchmaker*. Evolutionary programming techniques, or "algorithmic breeders," are also used by Danny Hillis to design more optimized sorting programs and by Kristian Lindgren to evolve better strategies for playing Prisoner's Dilemma. In general, these various projects are preliminary stages along the way toward eliminating the human hand from the selection/breeding process, which would thus allow something like genuine, autonomous evolution to occur. Langton's crowning example in these terms is Thomas Ray's computer-created environment Tierra, one of the most successful instantiations of the evolutionary dynamic in ALife research. In Tierra, digital organisms (actually blocks of code) reproduce, mutate, and compete for the computer's resources, with the result that a complex ecology rapidly evolves. Since this work will be discussed at length in the next chapter, I turn instead to Langton's subsequent research on computation and the origin of life, as set forth in his essay "Life at the Edge of Chaos."

Life and Information at the Edge of Chaos

Whereas in "Artificial Life" Langton presents a framework for programmatic research on the "biology of possible life," that is, life defined without the historical and possibly contingent restrictions of carbon-based chemistry as it has developed on Earth, in "Life at the Edge of Chaos" he attempts to answer a specific question: "Under what conditions can we expect a dynamics of information to emerge spontaneously and come to dominate the behavior of a physical system?"[31] Langton's answer is emblazoned in his title. Taking it as a given that living systems are precisely instances in which "information processing has gained control over the dynamics of energy, which determines the behavior of most non-living systems" (41), Langton argues that it is "at the edge of chaos" that information processing is most likely to gain the upper hand over the dynamics of energy exchange. To be sure, we may wonder if it is always so easy to draw a clear line between information processing and the dynamics of energy exchange, especially in biological processes. In these general terms, in fact, many systems, such as the stock market, could perhaps be said to exhibit the features of a living system.[32] But Langton does not concern himself with such possibilities and proceeds directly to his argument.

The evidence, once again, is culled from his research with cellular automata. He offers the following list of features to justify their appropriateness:

- CA are spatially extended, nonlinear dynamical systems.
- As nonlinear dynamical systems, CA exhibit the entire spectrum of dynamical behaviors, from fixed-points, through limit cycles, to fully developed chaos.
- CA are capable of supporting universal computation. Thus they are capable of supporting the most complex known class of information dynamics.

• There is a very general and universal representation scheme for all possible CA: a look-up table. This form of representation allows us to parameterize the space of possible CA, and to search this space in a canonical fashion.

• CA are very physical, a kind of "programmable matter." ... Thus what we learn about information dynamics in CA is likely to tell us something about information dynamics in the physical world. (42)[33]

The idea of CA as a kind of programmable matter is especially significant here. Toffoli and Margolus explain that conventional models of computation like the Turing machine are structurally divided into fixed parts (hardware) and variable parts (data and instructions). But a computer built on this principle, they assert, "cannot operate on its own 'matter,' so to speak; it cannot extend or modify itself, or build other computers" (9). However, a special cellular automata machine, or CAM, can be specifically designed to do so. It would constitute a computationally amorphous machine that can be programmed to act as a numerical wind tunnel one moment and a sea of fermions the next. In a later article, "Programmable Matter: Concepts and Realization," Toffoli and Margolus acknowledge that all computers "can realize programmable matter to some extent," but a CAM would do it "on a sufficiently large scale (in terms of spatial resolution and updating rate) and with sufficient flexibility (in terms of the underlying fine-grained model) for the concept of 'programmable matter' to come to life."[34] In these terms Langton's underlying question becomes, To what extent can the idea of programmable matter be extended to an understanding and reproduction of life processes?

For Langton the answer requires a specific way to correlate the information-processing capacity of CA with their dynamical state or regime. Accordingly, he devises a tuning knob, which he calls the lambda parameter, that reflects how changes in the CA rules correspond to the changing behavior of the CA themselves. This behavior is mapped in phase space in exactly the same way as the behavior of dynamical systems is mapped in dynamical systems theory. The lambda parameter λ varies from 0, which corresponds to a lifeless, or frozen, state of the automata, to 1, which corresponds to chaotic, completely aperiodic behavior. At low values the system is rigid and exhibits little life. As the value of λ slowly increases, some activity appears, but then dies out, as if determined by a limit point attractor. Around $\lambda = 0.2$ local structures begin to persist and may propagate in a single direction. Around $\lambda = 0.3$ another dramatic

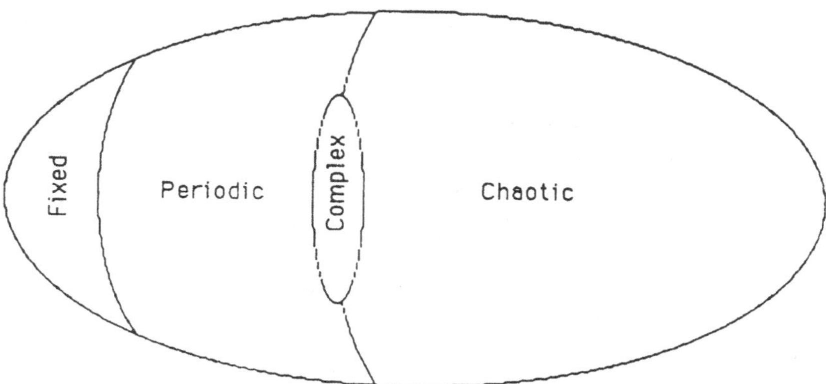

Figure 4.1
Langton's CA rule space. Christopher G. Langton, "Life at the Edge of Chaos," in *Artificial Life II: Proceedings of the Workshop on Artifical Life Held February, 1990 in Santa Fe, New Mexico*, Santa Fe Institute Studies in the Sciences of Complexity 10 (Redwood City, Calif.: Addison-Wesley, 1992), 76.

change occurs, as self-sustaining structures propagate in every direction. Behavior in this range seems to be determined by a periodic attractor, the difference being that around 0.2 interactions are only local, while around 0.3 they propagate across the entire array. At values greater than 0.4, on the other hand, the system begins to go chaotic, as if pulled by a strange attractor.[35] The area of greatest interest therefore lies somewhere in between, in "the sweet spot" of the CA rule space that Langton indicates with the diagram in figure 4.1 (76).

As the behavior of the CA moves from fixed or periodic to chaotic, it doesn't alter continuously from one regime to another but passes through the space of a critical phase transition. Far from being "a smooth surface between the other two domains," Langton argues, this transition regime is a complex domain with its own "complicated structure" (75) that can support "complex interactions between propagating and static structures" (69). But if this is the case, then these structures can in turn "be pressed into useful service as logical building blocks in the construction of a universal computing device" (69). For example, glider guns (a particular configuration of cells in Conway's Game of Life) can be constructed to emit on-off pulses that can be arrayed together to function as AND, OR, and NOT logic gates, which can in turn function as the basic building blocks of a computational device. Such an array of glider guns instantiates what Langton calls "constructability," which refers to whether or not a particular set of CA rules can be used to construct logic gates and thus be said to support computation. Some rule sets, for example, do not yield glider guns.

Langton acknowledges that there is no sure determination or absolute guarantee that constructability will always be the case: "Similar constructions can be made for other complex rules, but not all complex rules tried have yielded to such simple constructions" (75). Yet this is not a stumbling block, since what matters is the remarkable similarity observed between "the surface-level phenomenology of CA systems" and "the surface-level phenomenology of computational systems," specifically with respect to "the existence of complexity classes, the capacity for universal computation, undecidability, etc." (75). Langton thinks that this fundamental similarity can only be explained by concluding that "the structure of the space of computations is dominated by a second-order [or critical] phase transition" (75). In other words, the complex behavior of CA and the conditions of universal computation exhibit the same underlying structure because both are instances of a more general structure of change, which can best be characterized by the phase transitions of matter—from solid to liquid to gas.[36]

To define this underlying structure Langton draws on Stephen Wolfram's classification of the CA's dynamical behavior. Wolfram found that CA behavior falls into four general classes, with "analogs" (46) in dynamical systems theory. In class 1, the cells go quiescent or "die" within several time steps (they are drawn by a single-point attractor). In class 2, the cells are livelier but settle down to static or periodically oscillating configurations, sometimes with fractal, self-similar structures (they are drawn by a periodic attractor). In class 3, the cells behave chaotically, with no patterns forming or lasting for more than a few time steps (they are drawn by a strange attractor). In class 4, where the behavior is the most interesting, the cells endlessly form localized structures, which often move across the grid, break apart, and reform. Neither periodic nor chaotic and having very long transients, the behavior of the class 4 cells does not correspond directly to any analog "identified among continuous dynamical systems" (46), although this kind of complex behavior is sometimes exhibited in Conway's Game of Life. Langton notes, however, that there is an exact correspondence between the behavior of class 4 CA and the λ values that mapped the space of a phase transition in the CA rule space diagram. While low λ values correspond to classes 1 and 2, and high values to class 3, the complexity of class 4 behavior corresponds to a peculiar CA rule space that seems to define a specific regime with its

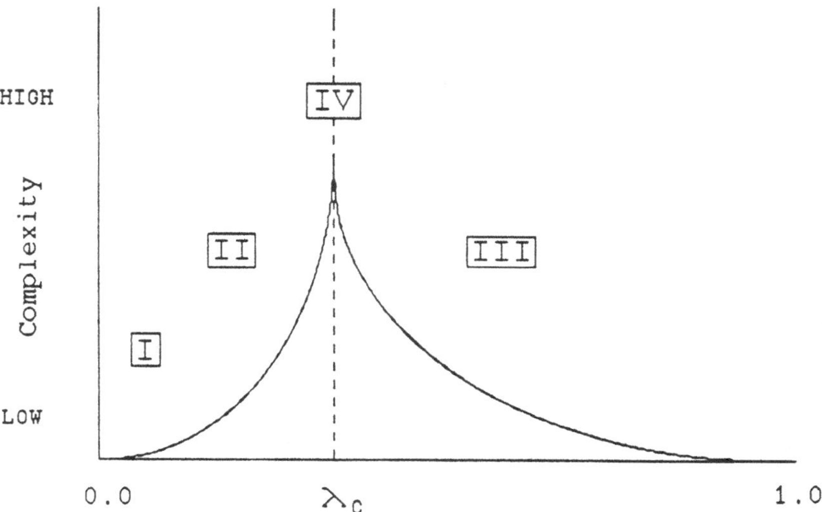

Figure 4.2
Complexity versus lambda in Langton's CA rule space. Christopher G. Langton, "Life at the Edge of Chaos," in *Artificial Life II: Proceedings of the Workshop on Artifical Life Held February, 1990 in Santa Fe, New Mexico*, Santa Fe Institute Studies in the Sciences of Complexity 10 (Redwood City, Calif.: Addison-Wesley, 1992), 77.

own intrinsic features and phenomenology (see Langton's diagram, fig. 4.2). Approaching this special regime from low λ values, the CA exhibit longer and longer periods of transient time as well as increasing sensitivity to array size. When moving away from this regime toward high values, the inverse occurs: at values that fall exactly within the domain of complexity the transient time becomes immeasurably long or undecidable. From these correlations Langton concludes that the CA phenomenology is structured by an underlying phase transition.

Furthermore, since the same or a very similar phenomenology holds true for the space of computation, Langton concludes that "the structure of the space of computation is dominated by a second-order phase transition" (75). Although mathematicians often develop a practical sense of whether or not particular computations can be carried out in finite time, Turing proved that in many instances computability cannot be determined in advance. Computational problems can therefore be classified accordingly: the first class is comprised of computations that halt (i.e., those that can be solved in finite computational time); the second, of those that do not halt (and thus that cannot be computed); and the third, of those that are undecidable. Not surprisingly, the phenomenology of these three classes suggested to Langton the same structure of the CA dynamics, with its critical transition regime: the halting computations correspond to class 1 and class 2 CA, non-halting computations to class 3 CA, and undecidable computations to class 4. Once again this would mean that as one approached the complex domain from either side, the transient time of computability would increase exponentially. In short, the most complex computations are to be found at a phase transition.[37]

Langton thus believed that he had found a clear correlation between the complex behavior of nonlinear dynamical systems—in this instance, that of cellular automata—and the structure or conditions of computability. Hence the question, What kind of structure is necessary if information is to be computed? The phase transition regime turns out to provide both the underlying structure and the key to understanding its phenomenology. To clinch his argument Langton presents a third instantiation—the phase transitions of matter itself, from solid to liquid to gas and the reverse. As ice melts, for example, it moves through a phase transition from solid to liquid in which each molecule must "decide" whether to be in either of two qualitatively different states. As the temperature approaches the limits of the phase transition, the molecules seem to require more and more time to reach a decision. (Their transient time is said to increase.) Accordingly, Langton understands the phase transition as a moment of critical "slowing down," when "the system is engaged in 'solving' an intractable problem" (82). Though this sounds anthropomorphic, he thinks that the results justify our thinking that such systems are "effectively *computing* their way to a minimum energy state" (82, author's emphasis).

For Langton, these correlations provide "evidence for a *natural domain of information* in the physical world" (81, author's emphasis). To be sure, the question of whether information exists in the natural world or is only a human measure of structure is a thorny one, for it raises the semantic problem of how information is to be defined as well as the epistemological issue of its status.[38] For Langton, nevertheless, the evidence that information processing—the storage, transmission, and modification of information—takes place in the physical world is not in question.[39] Nor is there any question about where we should expect this process to emerge and come to dominate the dynamics of a physical system: it is in a space between orderly and chaotic regimes, in the vicinity of a critical phase transition that can be defined precisely using the lambda parameter.

In the essay's final section Langton summarizes what this work implies about the relationship of life to matter, computation (or intelligence), and evolution. It seems very likely, he suggests, that "the origin of life occurred when some physico-chemical process underwent a critical phase transition in the early history of the earth" (81). His work lends credence to this claim by offering compelling evidence for a series of subsidiary claims. First, there is "a fundamental equivalence between the dynamics of phase transitions and the dynamics of information processing" (82). Not only does the phenomenology of phase transitions explain the phenomenology of computation, but also the reverse; thus these are not two distinct phenomenologies, but one: "We are observing one and the same phenomenon reflected in two very different classes of systems and their associated bodies of theory" (82).

Second, solids and liquids are dynamical, rather than material, categories, and these two universality classes of dynamical behavior are separated by a phase transition.[40] Furthermore, since the dynamical behavior of systems operating near this phase transition provide a basis for embedded computation, "a third category of dynamical behaviors exists in which materials—or more broadly, material systems in general—can make use of an intrinsic computational capacity to avoid either of the two primary categories of dynamical behaviors by maintaining themselves on indefinitely extended transients" (83). More simply, a dynamical system operating near a phase transition can use its acquired computational capacity to maintain itself near this enabling regime.

Third, "living systems can perhaps be characterized as systems that dynamically *avoid* attractors," and "have learned to steer a delicate course between too much order and too much chaos—the *Scylla* and *Charybdis* of dynamical systems" (85). In these terms evolution can be viewed "as the process of gaining control over more and more 'parameters' affecting a system's relationship to the vital phase transition" (85). In sum, after a living system has emerged near a critical phase transition, "evolution seems to have discovered the natural information-processing capacity inherent in near-critical dynamics, and to have taken advantage of it to further the ability of such systems to maintain themselves on essentially open-ended transients" (85).[41]

Whereas his earlier theoretical objective was to synthesize artificial life by abstracting and simulating its complex dynamical behavior, here Langton arrives at a scenario for understanding not only the basic conditions of life but also a specific mechanism by which life might have arisen from nonlife and then maintained and perpetuated itself. He has argued along the way that "hardness, wetness, or gaseousness are properties of the organization of matter, rather than properties of the matter itself"

(84), implying that it is only a matter of organization to turn "hardware" into "wetware" and that hardware should eventually be able to achieve everything achieved by wetware. Fittingly, Langton ends with a paean to water the writer James Joyce would no doubt have appreciated. While theorists of the origin of life have looked to "the dynamics of molecules *embedded* in liquid water," Langton states, it may well be that life "originated in the dynamics of water itself" (85).

Autopoiesis: The Self-Organization of Living Systems

Beyond its paramount importance in the founding of ALife, Langton's research possesses a methodological and intellectual interest comparable in one respect to Robert Shaw's and James Crutchfield's work on chaos: it combines (and to a certain extent synthesizes) an information-processing, computational approach with a dynamical systems perspective, making them complementary aspects of the same computational assemblage. In the next chapter I consider the critique of Langton's work on computation at "the edge of chaos" proposed by Melanie Mitchell and James Crutchfield—one problem is that Langton doesn't define computation in relation to specific tasks—as well as further developments in ALife that similarly contribute to the fields of theoretical biology and nonlinear dynamical systems theory. Here I want to turn to another nonstandard theoretical approach to the definition of life, most readily recognized by its central concept of autopoiesis. This approach was developed in the 1970s by Humberto Maturana and Francisco Varela. As they define it, an autopoietic system—for them all living systems are autopoietic—is organized in such a way that its only goal is to produce and maintain itself. Directly opposed to the computational/informational approach to the study of life, they object to the centrality of DNA and genetic coding in molecular biology and seek to put the autonomy and individuality of all living systems on a firm theoretical footing.

Maturana and Varela first presented their theory in *Autopoiesis: The Organization of the Living*, a collaborative work first published in Chile in 1972.[42] Later, in *Principles of Biological Autonomy*, Varela re-presented the theory in a more biologically detailed version, with attention to several organs as well as the nervous system.[43] This later volume is of special interest because in it Varela discusses his early work with cellular automata, which he used to model a simple autopoietic system that forms self-enclosing boundaries and repairs them when they break down.[44] In the early 1990s, many European scientists found in these formulations the basis for an alternative approach to ALife, no doubt spurred by the first European conference on ALife in 1991, organized by Varela and Paul Bourgine.[45] What makes this body of work noteworthy, however, is less the alternative approach it offers (in fact, its influence on the new AI and robotics is more significant, as we'll see in later chapters) than the fact that it confirms one of ALife's most fundamental assumptions. In short, even while opposed to the informational/computational basis of ALife, Maturana and Varela share with it a conceptual underpinning that understands living systems as machines.[46]

Maturana and Varela's intellectual lineage also reaches back to cybernetics, specifically to the catalyzing influence of Heinz von Foerster. Varela himself attests to von Foerster's importance for both the further development of cybernetic theory and Maturana's and his own work.[47] An active participant in the Macy Conferences and editor of its published proceedings, von Foerster went on to found and direct the Biological Computer Laboratory at the University of Illinois at Champaign-Urbana. Two key ideas he developed are relevant here: first, his order-from-noise principle (in certain systems, noise can sometimes spur a system to self-organize at a higher level), which anticipated the development of theories of self-organization in the 1970s; second, self-reference and the role of the observer can play a constitutive role in the formation of systems.[48] As a result of this work, many consider von Foerster the architect of second-order cybernetics. Varela and others have also claimed that von Foerster's work was unjustly overshadowed and even repressed because of the importance "granted" (in every sense) to dogmatic forms of cognitive science and classical AI, particularly to what later became known as symbolic, or "high church" (Daniel Dennett's phrase), computationalism. For my purposes, what is important is that von Foerster's first idea became crucial for complexity theory and his second for Maturana and Varela's theory of autopoiesis.

Autopoiesis: The Organization of the Living attempts to redefine the fundamental principles of a science of living systems. To do so, Maturana and Varela draw tacitly on systems theory (G. Spencer-Brown) and second-order cybernetics (Heinz von Foerster) as well as Maturana's earlier empirical studies of vision in frogs (and later pigeons), research that stemmed directly from McCulloch and Pitts's neural net theory.[49] Altogether, this research led Maturana to reject the assumed objectivity of science and to formulate his own epistemology. In the retrospective introduction Maturana wrote for *Autopoiesis and Cognition*, he recounts how the results of his experimental research on vision actually worked against the assumptions of realism and epistemological objectivity and how a reorientation necessarily imposed itself. Specifically, Maturana and his group were unable to map the activity of cells in the frog's retina directly onto the contours and colors of the visual world and were forced to reverse their assumptions and the questions they were asking. As Maturana framed the question, "What if, instead of attempting to correlate the activity of the retina with the physical stimuli external to the organism, we did otherwise, and tried to correlate the activity of the retina with the color experience of the subject?" (xv). Pushing further, he was forced to assume that "the activity of the nervous system [is] determined by the nervous system itself, and not by the external world; thus the external world would have only a triggering role in the release of the internally-determined activity of the nervous system" (xv). This perspective led him to treat the nervous system as a system closed on itself and the "report of the color experience as if it represented the state of the nervous system as a whole" (xv). It follows that perception of distinct objects in the external world doesn't at all correspond to the stimulation of specific cells in the retina—there is no point-to-point correspondence. Rather, the inflection of the visual field by the appearance of a distinct object results from a triggering effect that perturbs the visual system as a whole, which then reestablishes its own equilibrium. In the frog's visual system only certain kinds of perturbation are possible, and these correspond to what the frog can see—small objects moving fast, with little ability to discern large objects moving slowly. To say, therefore, that the frog's vision is perfectly adapted to its environment, enabling it to catch flies and avoid predatory birds, is true but misses the essential point that the frog does not so much *see* the world as respond to and interact with selected aspects of it. More precisely, seeing is a perceptual/cognitive linkage, with stimuli that have no objective existence outside the activities of the perceiving subject. It is only the observer who infers the distinction and who may then describe the interaction. This approach stands in obvious contrast to conventional scientific descriptions, which usually involve the attribution of causal relations, from which the reified metaphysic of realism and objectivity closely follows.

Rejecting these assumptions, Maturana replaces them with a conceptualization based on "circular organization" and the idea that cognition is inherent to all living systems. His research on the frog's vision had indicated that the nervous system operates as a closed network of interactions in which every change in the interactive relations of one set of components always results in the interactive relations of other components.[50] In brief, the nervous system is not only self-organizing but self-maintaining. Having theorized that the nervous system is both autonomous (operationally closed) and coupled to the environment (interactionally open), Maturana hypothesized that circular organization is the general feature that defines all living systems: "Living systems ... [are] organized in a closed causal circular process that allows for evolutionary change in the way the circularity is maintained, but not for the loss of the circularity itself" (9). Moreover, since this circular organization determines what aspects of the external environment the system can interact with, in effect it operates as a selective mechanism that Maturana equates with cognition as a whole: "Living systems are cognitive systems, and living as a process is a process of cognition" (13).[51]

This framework is assumed in *Autopoiesis: The Organization of the Living*, where Maturana and Varela attempt to answer the question, What is life, and how should it be defined?[52] Their general thesis is that "there is an organization that is common to all living systems, [whatever] the nature of their components" (76). As scientists, they declare at the outset that their explanation of life is mechanistic, not vitalist; yet they also insist that it is in the circular organization of physico-chemical processes, not in the specific properties of the latter, that life is to be found. Since the organization of living systems is what enables these systems to maintain their own boundaries, to replenish their component parts and hence to maintain their identity over time, the specific details of these processes are relegated to a secondary concern. And this includes what for most biologists are the essential processes of reproduction and evolution. Indeed, *Autopoiesis* evinces little interest in the concrete "stuff" of most biological investigations—the myriad diversity and dynamic profusion of exchanges that take place within and among living entities. Instead, like a treatise by Spinoza or Leibniz, the book proceeds almost exclusively by definition and conceptual reframing. What it offers, nonetheless, is a completely new perspective on how living systems may be thought of as machines.

A machine is an organized unity of various component parts and processes. For Maturana and Varela, the organization of the machine is precisely what gives it this unity and what determines "the dynamics of interactions and transformations which it may undergo as such a unity" (77). Contrastingly, the *structure* of the machine is constituted by "the actual relations which hold among the components which integrate a concrete machine in a given space" (77). These definitions allow Maturana and Varela to distinguish sharply between the relations that give the machine its unity and the properties of the components that realize the machine as a concrete system. Specifically, for them the organization of a machine is independent of the material properties of its components; hence a given machine can be realized in different ways and by different components. In this sense organization plays the same role as information in computation: it signifies a functional process or effect not dependent on the exact nature of its supporting material substrate. A second corollary of these definitions is that the use to which a machine may be put is not a feature of its organization but rather "of the domain in which the machine operates" (77). Of course, we usually think of human-made machines as constructed for a specific purpose or end, including our own amusement. However, just because this aim or purpose is expressed as a result of the machine's operation should not lead us to believe that it is a constitutive property. These notions, rather, are extrinsic to the machine's organization and pertain only to the domain of its observation. They may help us to imagine, describe, or simply talk about machines, and they may even be realized in a particular machine's operation. Nevertheless, they remain in the domain of descriptions generated by the observer.

That a living system *is* a kind of machine can be demonstrated therefore only by pointing to its organization, not to its component parts or structure. Specifically, living systems are organized in such a way that they produce and maintain their own identity, which makes them distinct from the environment and independent of relations with an observer. Maturana and Varela call this kind of machine autopoietic:

> An autopoietic machine is a machine organized (defined as a unity) as a network of processes of production (transformation and destruction) of components that produce the components which: (i) through their interactions and transformations continuously regenerate and realize the network of processes (relations) that produced them; and (ii) constitute it (the machine) as a concrete unity in the space in which [the components] exist by specifying the topological domain of its realization as a network. (78–79)

As in the definition of organization above, the emphasis falls on the network of processes that produce and maintain identity, not on the properties of the individual components and their varied relations. It is intentionally a circular definition: A produces B, which in turn produces A. Hence a living entity is a network of processes organized in such a way as to maintain the integrity and functioning of the processes that define it.

As long as a domestic cat, for example, breathes air, drinks water, and eats food, autopoietic networks will provide the energy to generate and maintain the cells and tissue that will enable the cat to interact with its surrounding environment *as* a cat. In contrast, an automobile is organized in such a way that burning gasoline is converted into enough kinetic energy to move the automobile across the landscape, but "these processes are not processes of production of components which specify the car as a unity since the components of a car are produced by other processes which are independent of the organization of the car and its operation" (79). Simply put, the car itself does not and cannot maintain its own identity. In Maturana and Varela's terminology, it is therefore an allopoietic machine. Whereas autopoietic machines are autonomous and subordinate all aspects of their functioning to the maintenance of their own organization (and hence to their identity), allopoietic machines (like the automobile) have something different from themselves as the product of their functioning. It follows that an allopoietic machine has no individuality and that its identity depends entirely on the observer, the "other" who stands outside its process of operation. This identity is not determined by or through the machine's operation, precisely because the result of this operation is different from the machine itself. An automobile, for example, could serve as a simple vehicle for transportation, a source of spare parts in a junkyard, a collector's item or a fetishized prop in a movie (or in the life of its owner). In contrast, an autopoietic machine maintains its own individual identity independently of its interactions with any observer.

While at first the definitions of autopoietic and allopoietic may seem to reinscribe the opposition between a living entity and a tool, the organic and the inorganic, the differences between them are not so easily reducible. Autopoietic machines are homeostatic since they maintain as constant precisely the relations that define them as autopoietic, but they do not have inputs and outputs, as do allopoietic machines. Of course autopoietic machines can be perturbed by independent events and can undergo internal structural changes as a result of these perturbations. But adjustments to perturbations, whether as singular or repeated events, are always subordinated to the maintenance of the organization that defines the machine as autopoietic. Nevertheless, human observers can still "describe physical autopoietic machines, and also manipulate them, as parts of a larger system that defines the independent events which perturb them" (82). Specifically, the observer can view these perturbing independent events as input and the changes the machine makes to compensate for these perturbations as output. Maturana and Varela insist, however, that this is precisely the mistake of molecular biology, which treats the living system as an information-processing device. That is, it mistakes an autopoietic machine for an allopoietic one. It is also possible "to recognize that if the independent perturbing events are regular in their nature and occurrence, an autopoietic machine can in fact be integrated into a larger system as a component allopoietic machine, without any alteration in its autopoietic organization" (82). In the same way, parts of autopoietic machines can be analyzed as allopoietic submachines in relation to their input and output. In neither case, however, can the essential and defining nature of the autopoietic machine be revealed.

With these definitions at hand, Maturana and Varela advance their central claim that "autopoiesis is necessary and sufficient to characterize the organization of living systems" (82). Before attempting to substantiate this claim, they make two points. First, they argue that since living systems are machines, once their organization is understood, there is no a priori reason why they cannot be reproduced and even designed by humans. To think otherwise would be to succumb to the "intimate fear" that the awe with which we view life would disappear if we recreated it or to the prejudiced belief that life will always remain inaccessible to our understanding. Second, they point out that as long as the nature of the living organization remains unknown, it is not possible "to recognize when one has at hand, either as a concrete synthetic system or as a description, a system that exhibits it" (83). In other words, it is not always or immediately obvious what is living and what is not. For most biologists, reproduction and evolution appear as constitutive, determinant properties, to which "the condition of living" is subordinated. But as Maturana and Varela point out, once these properties are reproduced in human-made systems, those who do not accept the proposition that any synthetic or human-made system can be living simply add new requirements.

In the remainder of the book Maturana and Varela develop the implications of their claim that autopoiesis is necessary and sufficient to characterize the organization of living systems. They argue, for example, against the common assumption that teleology is a necessary feature of a living system. It must be remarked, however, that they do not pursue some of the most interesting implications of their theory—for example, that in certain respects social systems and many information systems might qualify under their definition as living systems, as Félix Guattari points out in his essay, "Machinic Heterogenesis."[53] While institutions and technical machines appear to be allopoietic, Guattari notes, "when one considers them in the context of the machinic assemblages they constitute with human beings, they become ipso facto autopoietic." Guattari thus finds Varela's concept useful, but only if viewed from the perspective of "the ontogenesis and phylogenesis proper to the mechanosphere superposed on the biosphere"—in other words, from the perspective introduced by a machinic assemblage, which could be said to combine both allopoietic and autopoietic functions. Guattari's reflections on the limits of the concept of autopoiesis lead him to call for a reframing, or recontextualization, in terms that converge precisely with issues central to ALife research: "Autopoiesis deserves to be rethought in terms of evolutionary, collective entities, which maintain diverse types of relations of alterity, rather than being implacably closed in on themselves."

Varela himself is eventually forced to confront these issues more fully when he considers the immune system. Meanwhile, in *Principles of Biological Autonomy*, he offers a less abstract and more useful summary of autopoietic theory, which he applies to specific biological phenomena. Following the assumptions Maturana developed in his early theory of vision, Varela considers both the immune system and the nervous system as instances of operational closure and structural coupling, the technical terms Maturana proposed to describe how a system closed on itself can nevertheless interact with environmental stimuli.[54] Of more immediate importance here is the cellular automaton model of autopoiesis that Varela includes. Since this model anticipates Varela's explicit interest in Artificial Life, which he prefers to consider under the alternative rubric of *autonomous systems*, it will be useful to examine it briefly before contrasting Varela's approach with Langton's.

Autonomous Systems and Artificial Life

Varela's cellular automaton demonstrates how an autopoietic unity can spontaneously emerge through a simple linking of elements in the presence of a catalyst. It involves three processes: (1) composition, when two basic elements (denoted by O and [ ] in his diagram) form a link ([O]) in the presence of a catalyst (*); (2) concatenation (or bonding), when a link joins with two or more other links ([O]-[O]-[O]); and (3) disintegration, when a link decomposes back into two basic elements (see Varela's diagram, fig. 4.3).[55] Varela had these processes encoded in a computer program in such a way as to define the interactions of cellular automata (each space being either empty or occupied by a single element). When the program is run, links form and decompose randomly, but inevitably

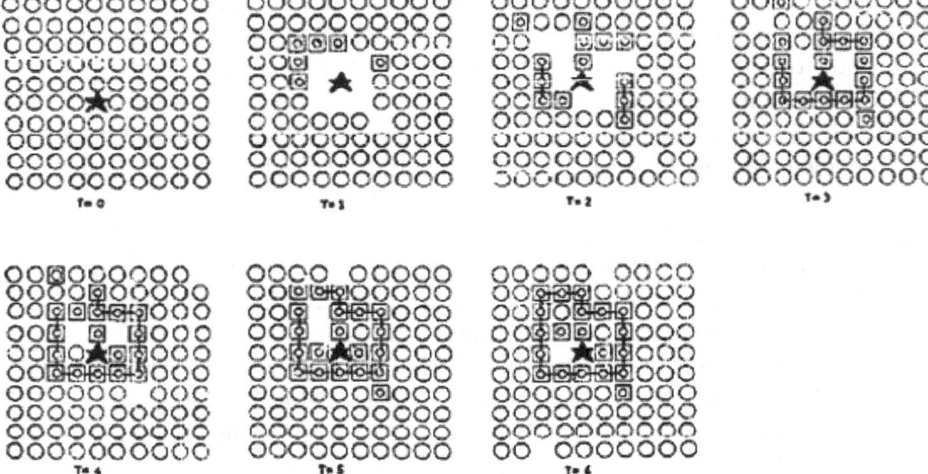

Fig. 1. The first seven instants (0→6) of one computer run, showing the spontaneous generation of an autopoietic unity. Interactions between substrate O and catalyst ★ produce chains of bonded links ⊡, which eventually enclose the catalyst, thus closing a network of interactions which constitutes an autopoietic unity within this universe.

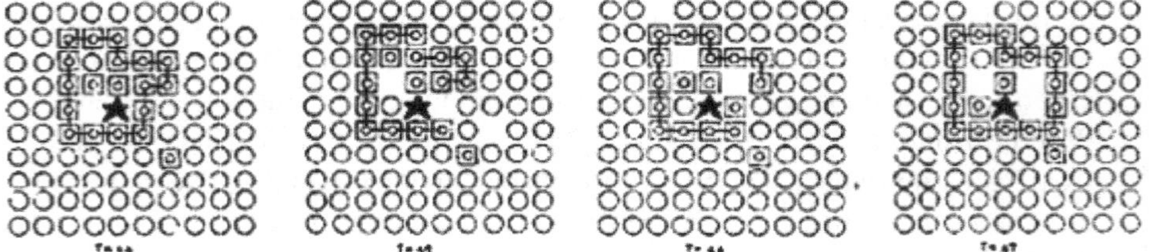

Fig. 2. Four successive instants (44–47) along the same computer run (Fig. 1), showing compensation in the boundary broken by spontaneous decay of links. Ongoing production of links re-establishes the unity under changes of form and turnover of components.

Figure 4.3
Varela's CA model. F. G. Varela, H. R. Maturana, and R. Uribe, "Autopoiesis: The Organization of Living Systems, Its Characterization and a Model," in *BioSystems* 5, no. 4 (May 1974): 190.

links form around a catalyst and gradually produce a bounded, self-enclosed space. Links also decompose, but new links form to replace them, and the boundary is maintained.

Some years later, at the Santa Fe Institute, Barry McMullin pointed out that the computer coding of the model was flawed. Varela and McMullin then revised the code, using slightly different algorithms to implement the "artificial chemistry" governing the interactions of the elements, and ran the model on the Swarm platform that Langton and his associates had written.[56] Since the results were essentially the same, the updated demonstration further strengthened the claim that "autopoietic phenomena are not dependent on any particular details of the original program or algorithm, but may be expected in *any* system sharing the same qualitative chemistry" (39, authors' emphasis). The demonstration also brought Varela into direct personal contact with scientists at the Santa Fe Institute and thus with proponents of what some Europeans thought of as the American brand of ALife. Indeed, the term "computational autopoiesis" in the title of McMullin and Varela's conference paper suggests some sort of rapprochement. Yet the term is misleading, inasmuch as it elides several crucial differences between Varela and Langton, differences that can be pinpointed by considering how each one understands the CA assemblage.

If we compare Varela's use of CA as a model of autopoiesis and demonstration of its viability to Langton's use of CA as an instantiation of artificial life, two obvious differences emerge. First, Varela shifts the focus from the issue of the cell group's self-reproduction to that of its unity and boundary maintenance. As he (as well as Maturana and Uribe) puts it: "For reproduction to take place there must be a unity to be reproduced: the establishment of the unity is logically and operationally antecedent to its reproduction" (189). The formation of a self-enclosing boundary and its dynamic maintenance is therefore what is most essential to Varela's model. In Langton's self-reproducing loops, by contrast, the boundary is a static construction, a sheath within which the core cells are programmed by manipulating the CA rules to function as data paths for a signal sequence in the cell's reproductive process.[57] (In Varela's terms, the sheath is thus an allopoietic device.) A second difference is defined by the role of information processing. In Langton's research the CA are configured as information machines. Indeed, a specific configuration of information is what reproduces itself: both the medium and the event of this reproduction are constituted *as* information. For Varela, however, information processing is secondary, subordinate to the dynamic processes that constitute and maintain the cell group's unity, or, in other words, constitute its organization. Informational processes, which include the cell group's reproduction, are allopoietic submachines. As such, they do not define what it is that gives life to a living system. In Varela's CA model this idea is illustrated by the formation and restoration of boundaries around a catalytic agent. It is the presence of the catalyst that initiates the dynamic process of self-organization, which does not involve information processing except at the lower level of transition-state tables that determine whether particular cells are on or off.

How much should be made of these differences? In the introduction to the proceedings of the "First European Conference on Artificial Life" (published under the title, *Toward a Practice of Autonomous Systems*), Varela spells out succinctly what he thinks Artificial Life research should be and how his view differs from Langton's earlier definition of the field. The key concept for Varela is autonomy:

> Autonomy in this context refers to [the living creatures'] basic and fundamental capacity to *be*, to assert their existence and to bring forth a world that is significant and pertinent without being pre-digested in advance. Thus the autonomy of the living is understood here both in regards to its actions and to the way it shapes a world into significance. This conceptual exploration goes hand in hand with the design and construction of autonomous agents and suggests an enormous range of applications at all scales, from cells to societies.[58]

Whereas Langton seeks to "abstract the fundamental dynamical principles underlying biological phenomena, and recreat[e] these dynamics in other physical media—such as computers—making them accessible to new kinds of experimental manipulation and testing" (Langton, as quoted by Varela, xi), Varela proposes that "artificial life can be better defined as a research program concerned with autonomous systems, their characterization and specific modes of viability" (xi). He concedes that his view does not contradict Langton's but asserts that it does make it more precise. Yet he also adds that "it is by focusing on living autonomy that one can naturally go beyond the tempting route of characterizing living phenomena entirely by disembodied abstractions, since the autonomy of the living naturally brings with it the situated nature of its cognitive performances" (xi).

Although his tone is conciliatory, Varela thus reveals that he views Langton's work as a "disembodied abstraction" and therefore of limited scope, since it has no way of dealing with the "situated nature" of living systems and hence with their "cognitive performances." In fact, with the charge of "disembodied abstraction" Varela comes perilously close to associating Langton's methodology with the assumptions that dominated the first period of AI and cognitive science following the development of cybernetics and information theory in the 1950s. After a thirty-year period of dominance by a "research program emphasizing symbolic computations and abstract representations," Varela continues, it is time to benefit from the rediscovery of connectionist models and neural networks.[59] Surprisingly, here Varela seems to have completely forgotten (or willfully ignored) Langton's explicit affiliation with connectionist models and natural systems that instantiate highly distributed parallel processing. In any event, Varela concludes his introduction to the conference proceedings by going one step further: Having reinscribed ALife within the study of autonomous systems, he resituates it within the evolving history of cognitive science. The latter, having started with classical cognitivism (based on symbolic computation) and having passed through connectionism, has now arrived at what Varela characterizes as "an *enactive* view of cognitive processes, which also places the autonomy of the system at its center and is thus naturally close to AL" (xvi). (By "enaction," Varela means the emergence of a new world through the codetermination of a structural coupling.) However, by strongly emphasizing emergence, a key term in ALife research, Varela would seem to want to annex ALife to a new or transformed version of his own work.[60]

Whatever the assumed limits of the computational, information-processing approach, it must be said that it is neither accurate nor fair to drive a wedge between it and the study of emergence and self-organization, as Varela does by identifying the latter exclusively with his own successor paradigm in his history of cognitive science.[61] In fact, the experimental production of emergence and self-organization has been an intrinsic aspect of research at the Santa Fe Institute and the putatively "American" approach to ALife from its inception. What has always been essential to this research is not at all the earlier, "standard" model of computation, but rather, as Langton emphasizes, the emergence of dynamic behavior in nonlinear processes and highly distributed systems. Indeed, one of the most salient aspects of Langton's work is that it considers distributed parallel computation and nonlinear dynamical systems in a single framework. The resulting new form of emergent computation becomes the explicit research method in complex systems theory, as discussed in chapter 5. Suspiciously, however, Varela makes no mention of this aspect of Langton's research and never seriously considers the new biologically oriented alternatives to the standard model of computation. Instead, he continues to insist on—indeed, to harp on—the sense in which living systems are the result of particular histories and contingencies. What is most wrong about the computational approach, he emphasizes at a subsequent European workshop on Artificial Life (held at San Sebastián, Spain, in 1993), is that it leaves out this contingency: "If it's silicon contingency, if it's tin can contingency, fine with me. What you cannot abstract out, centrifuge out, is that kind of process or situation that only comes from history."[62]

But while we may never really escape from history, life never allows history to remain intact either. As part of contemporary history ALife challenges many historical conceptions of life and evolution. What Varela seems to abhor above all else is that ALife deterritorializes (in Deleuze and Guattari's sense) biological processes and reterritorializes (i.e., recodes) them in a digital medium. But surely this can't be totally unlike the contingencies that living systems have always confronted. Indeed, the usefulness of D&G's perspective—however limited their discussion of specific biological process—is that it indicates in general theoretical terms how life arose and gained a stronghold precisely through such decodings and recodings. Within the frame of published ALife research, other researchers have questioned whether life can be abstracted from one material substrate and instantiated in another,[63] and they have done so with more cogency than Varela, whose approach does not adequately address the normative view held by biologists that reproduction and evolutionary adaptation are essential to all living systems. Animating these other critiques is the fact that molecular biology tends increasingly to view the relationship between the information contained in the genotype and the material production of the phenotype as one of dynamic interdependence. It is often pointed out, for example, that not all of the information necessary for protein production is contained within DNA; rather, it is "filled in" and the process completed in the complex dynamics of molecular interaction.

What should be emphasized in any case is that Varela is not opposed to ALife per se and related attempts to synthesize life in nonorganic materials—indeed, he completely endorses efforts like those of Rodney Brooks to build autonomous mobile robots (see chapter 6 below). He is only opposed to computational approaches like Langton's that abstract the purely formal features of life from their material instantiation, thus putatively removing life from historical contingency and the particulars of its situatedness and concrete embodiment. Evidently the contingencies of the computer environment don't count. Moreover, by ignoring the fact that Langton's approach to synthesizing life depends on the kind of highly parallel and distributed processing observed in biological life, Varela also avoids having to consider the extent to which natural life itself depends on similar kinds of computational processes.

It is also important to note that one significant strand of early ALife research—"artificial chemistry"—draws on *both* Langton and Varela. Indeed, "artificial chemistry" quickly became a flourishing part of ALife research, thanks in large part to Walter Fontana's ambitious project to pinpoint "a logical deep structure of which carbon chemistry-based life is a manifestation."[64] Langton himself provides the best short summary:

In Fontana's system, functions expressed in the lambda-calculus [a formal symbol system] are represented as character strings that "react" with other such functional character-strings via function composition, producing new function-strings in the process. These new function-strings then enter into reactions with existing function-strings and the process continues. Fontana finds that the natural dynamics of this "Algorithmic Chemistry" give rise to the spontaneous emergence of many cooperative reactions between function-strings, including self-replicators, self-replicating sets, autocatalytic cycles, symbiotic and parasitic sets of functions, and so forth.[65]

Fontana has also reinforced the theoretical significance of ALife in both nuanced and trenchant terms. On the one hand, he observes, "biology lacks a theory of organization" (212), and this accounts for unsolved problems in ontogeny and phylogeny; on the other, he states, ALife "makes sense only if there is an *implementation independent definition of life that informs biology*" (224, author's emphasis). Citing Varela and Maturana as "the first to think extensively about organization in a new way" (225), Fontana also reflects Langton's influence when he argues that the role of ALife is to develop a concept of life that "encompasses biology" (224), which means not simply imitating it but comprehending it, particularly by identifying its unsolved problems and seeking their resolution.

Thus the continued absence of a consensual definition of life among biologists in effect makes the theoretical aims and challenges that Fontana spells out for ALife necessary and indeed essential for its disciplinary justification.[66] Meanwhile, this lack suggests an obvious way to situate the differences between Langton and Varela: each one simply valorizes what he believes to be life's essential attributes. For Langton it is self-reproduction and evolution; for Varela, self-organization and autonomy. Accordingly, experiments in ALife research following Langton tend to explore the former, while contemporary robotics (beginning with Rodney Brooks and his followers) take up the latter, with adaptation and learning falling somewhere between the two initiatives. What this means here is that before any further assessments can be made both initiatives must be examined in detail, which will be the objective of chapters 5 and 7. In brief, we will find that these initiatives in machinic life cannot be considered apart from a core set of theoretical issues involving the shortcomings of Darwinian evolutionary theory (as revealed by self-organization and the processes of emergence in nonlinear dynamical systems), and that the question then becomes whether or not the framework of "complex adaptive systems" can provide an adequate new conceptual synthesis. While robotics will claim to make up for the putative deficiency of abstraction and disembodiment in ALife, the fruits of simulation and new, nonstandard approaches to computation (especially evolutionary computation) will also prove to be necessary for further advance.

In the meantime we should not ignore some of the new "unofficial" forms of artificial life. To single out only one example (but a significant one), I conclude this chapter with a brief exploration of two convergent lines of research on immune systems. After summarizing Varela's biological theory of the immune system I take up a parallel development that arises from an interest in machine learning. This leads to a discussion of a computational model of the immune system and computer viruses, and of their relationship to ALife. A central question raised by this convergence is again whether the information-processing approach and the dynamical systems approach are actually or necessarily as opposed as Varela and others have claimed. This is important because many of those who work in contemporary robotics and cognitive science believe that they are. The evidence, however, suggests that for complex systems like the immune system, information processing and dynamic behavior are very closely related, implicating and relaying each other in ways not yet fully understood.

Silicon Immune Systems and Viral Life

For Varela, the immune system of a biological organism presents a clear and compelling example of an autopoietic system: it possesses an autonomous unity, operationally closed but structurally coupled to the outside by a triggering mechanism that communicates perturbations. What's more, it serves an essential cognitive function in that it is responsible for the organism's very identity at the molecular level. Thus it is hardly surprising that Varela has devoted much of his research to the immune system, first presenting his theory in *Principles of Biological Autonomy* (1979) and continuing to publish important "update" articles well into the 1990s. From the outset he has always advocated an alternative to the classical view of the immune system as the body's primary defense against disease and infectious agents. According to this understanding, a wide variety of highly interactive cells, the lymphocytes (popularly known as white blood cells), possess and produce markers known as antibodies that either protrude from their surface or circulate freely. When such an antibody comes into contact with a foreign infectious agent or antigen, the antibody bonds with it chemically, rendering it harmless or destroying it. If the infection is severe, the particular type of lymphocyte with the right chemical "key," or recognition device, will immediately clone itself until it produces a veritable army of antibodies attacking the foreign intruder. This military model, however, raises a number of questions. First, how do the body's lymphocytes come to recognize the millions of new antigens that it encounters? Inversely, how does a lymphocyte know how to recognize its own body's cells, since in the classical view recognition is a chemical locking-on that destroys the other cell? Furthermore, what makes this mechanism fail in autoimmune disorders? Finally and most generally, how does the immune system maintain a "catalog" of all known foreign intruders, past and present, and how does it produce new types of lymphocytes to counteract the new types of intruders?

In Varela's view the classical theory attempts to answer these questions with a series of ad hoc proposals and hypotheses. A case in point is the "clonal selection theory," which proposes a Darwinian model to explain how the antibodies required for the body's maintenance have evolved over time. As Varela points out, this theory assumes that the body's antibody repertoire is initially incomplete.[67] Moreover, the theory postulates that those clones that would recognize (and destroy) self-molecules are missing. Since there is no known genetic mechanism that can account for this, it is assumed that these clones are filtered out and destroyed at the embryonic stage. One version of this theory, called "clonal deletion," postulates that such lymphocytes are removed in the thymus. Thus the recognition problem—how to distinguish between self and nonself at the level of molecular profiles—is not really resolved.[68]

Varela's solution to the problems posed by the classical model is to adopt a radically different perspective, doing away with both the military metaphor underlying the classical view and the information-processing model it assumes for its operation. The former understands the maintenance of the body's identity at the molecular level as essentially a negative, wholly defensive reaction, while the latter views the body's immune response as a type of input-output relationship and therefore "externally determined." Building on the work of Neils Jerne, who proposed that antibodies do not operate as separate, individual elements but as tightly meshed networks, Varela argues that the immune system is an autonomous network that must first be understood in positive terms. In his view the defensive reaction of the immune system is actually secondary to its normal functioning. Its primary function, rather, is to maintain the body's molecular identity by regulating the levels of different cell types and molecules circulating throughout the entire system. Only when the invading antigens become so numerous as to perturb these regulatory functions does the immune system assume or fall back on its defensive posture. The problem, then, is not how the body identifies individual antigens but how it regulates levels of a whole range of interacting molecules. For this reason Varela prefers to think of the immune system as part of a larger, autonomous network that constitutes a complex ecology:

> Like the living species of the biosphere, [lymphocytes] stimulate or inhibit each other's growth. Like the species in an ecosystem they generate an amazing diversity: the antibodies and other molecules produced by lymphocytes are by far (by a million fold) the most highly diversified molecular group in the body. They are therefore ideally qualified to ensure the constant change and diversity of other molecules in the body. (274)

This autonomous network functions as a dynamical system, with global emergent properties that enable it to track and remember the individual's molecular history:

> In our view the IS [immune system] asserts a molecular self during ontogeny, and for the entire lifetime of the individual, it keeps a memory of what this molecular self is.... It is as a result of this assertive molecular identity that an individual who had measles in childhood is different from what he would have been had he not been in contact with the virus, or how an IS changes if the person switched from an omnivorous to a vegetarian diet. The IS keeps track of all this history, while defining and maintaining a sensorial-like interface at the molecular level. It must be stressed that the self is in no way a well-defined ([nor a] predefined) repertoire, a list of authorized molecules, but rather a set of viable states, of mutually compatible groupings, of dynamical patterns.[69]

As a dynamical system, the immune network functions not by guarding and protecting boundaries between self and nonself, but by keeping different groups and subnetworks in states of dynamic equilibrium. In Varela's view the immune system does not and cannot discriminate between self and nonself:

> The normal function of the network can only be perturbed or modulated by incoming antigens, responding only to what is similar and to what is already present. Any antigen that perturbs the immune network is by definition an "antigen of the interior," and will therefore only modulate the ongoing dynamics of the network. Any element that is incapable of doing so is simply not recognized and may well trigger a "reflexive" immune response, that is, one produced by quasi-automatic processes that are only peripheral to the network itself. ("The Body Thinks," 283)

There are always antigens present in the network, just as there are always antibodies that attack other antibodies. The result is a ceaseless change in levels, quantities, and distributions and thus of perturbations in an already existing network that is constantly (re)adjusting to itself. The most apt metaphor is not a military campaign intent upon vanquishing the enemy but the dynamics of a weather pattern. The latter "never settles down to a steady-state, but rather constantly changes, with local flare ups and storms, and with periods of quiescence."[70] Yet the immune system is not exactly like a weather system either, since in order to function it must "remember" and draw upon its own history.

Indeed, the immune system's ability to remember and learn, that is, to evolve new pattern-recognition capacities, is precisely what made it of interest to J. Doyne Farmer and Norman Packard when they organized a conference devoted to "Evolution, Games, and Learning: Models for Adaptation in Machines and Nature."[71] The central problem to be addressed was how machines might learn to solve problems without having to be explicitly programmed to do so. Since biological systems—most evidently the immune system and the brain—accomplish this task as part of their regular functioning, they offer privileged models for studying the underlying principles of biological computation. Since games are "highly simplified models of higher level human interaction," they too can provide access to "smart algorithms." But most important, as Farmer and Packard emphasize in their introduction to the conference proceedings, the behavior of these various adaptive systems is inevitably nonlinear and emergent:

> Adaptive behavior is an emergent property, which spontaneously arises through the interaction of simple components. Whether these components are neurons, amino acids, ants, or bit strings, adaptation can only occur if the collective behavior of the whole is qualitatively different from that of the sum of the individual parts. This is precisely the definition of the nonlinear. (vii)

Consequently, the approach to such nonlinear dynamical systems requires new syntheses rather than further elaborations of the reductive approach, which breaks down complex processes into simpler component parts and processes. The most powerful and innovative tool available for the new synthetic approach is the computer, which provides a way of simulating adaptive systems too complex to model quantitatively. Although orders of magnitude simpler than "the brain or a complex organic molecule such as DNA" (viii), the digital computer's increasing speed, decreasing cost, and wide availability have made it an essential part of what Farmer and Packard think of as an explosion of "new wave science," characterized by the synthetic approach and the crossing of conventional disciplinary boundaries.

In "The Immune System, Adaptation, and Machine Learning," their own contribution to the conference, Farmer, Packard, and Alan S. Perelson propose a dynamical model of the immune system simple enough to be simulated on a computer. Actually, what they construct is a very original working example of a computational assemblage. Of greatest initial interest to the authors was the fact that "by employing genetic operators on a time scale fast enough to observe experimentally, the immune system is able to recognize novel shapes without pre-programming" (187). Furthermore, because of its vast combinatorial diversity, the immune system rapidly generates a large number of different types of antibodies (for a typical mammal it is on the order of 10^7 to 10^8) capable of recognizing an even larger number of foreign molecules (estimates range as high as 10^{16}). A realistic model must not only match this capacity but also provide a means by which the list of antibody and antigen types can constantly change as new types are added and removed.

In Farmer, Packard, and Perelson's model, both antibodies and antigens are represented by binary strings that allow for either full or partial complementary matches. In this way digital bit strings model recognition in natural immune systems, where matching is thought to occur when the molecular shape of the antibody's paratope "fits" with and thereby allows chemical binding to the antigen's epitope. Both paratopes and epitopes are sequences of amino acids and thus complexly related, but researchers generally think of their relationship as that of locks and keys. The dynamics of the system are further complicated by the fact that antibodies also possess epitopes, which are recognized by other antibodies and thus also participate in a self-regulating function. In natural immune systems each antibody type (or key) will fit a variety of antigen types (or locks). Yet this mechanism alone does not bridge the huge gap between the numbers of antibody and antigen types. In fact, since an organism never ceases to encounter both old and new antigens throughout its lifetime, its immune system must constantly "turn over," producing not only a large supply of antibody types effective against known and remembered antigens but also a repertory of new types that will recognize new antigens. In humans, for example, the entire supply of lymphocytes is replenished every few months. To remain effective, the system must produce a staggering combinatorial diversity of antibody types, and at least two mechanisms operating at different levels ensure this diversity. First, there is a constant reshuffling of the DNA that codes for antibody genes in the production of the cells in the bone marrow.[72] Second, when the lymphocytes themselves reproduce, there is an exceptionally high mutation rate, much greater in fact than in any other body cell type.

Of necessity, then, Farmer, Packard, and Perelson must simplify. For example, they only attempt to model the actions of B-lymphocytes and ignore the fact that natural immune systems also combine the actions of T-lymphocytes and macrophages. The central problem is how to mimic the production of this combinatorial diversity, clearly the immune system's most essential and characteristic property. The genetic algorithm, developed by John Holland, provides the solution. By applying genetic operators like crossover, inversion, and point mutation to both epitope and paratope bit strings, a vast number of antibody types are generated that basically comprise their model immune system. Antigens, on the other hand, are generated either randomly or by design, then repeatedly "presented" to the system in both varying number and rate in order to measure how well the system remembers (or how rapidly it forgets). Altogether, the total number of antibodies and antigens present at a given moment defines a single dynamical system, whose state will change as some of these antibodies and antigens interact and die and both new and similar ones are added. The state of the system can thus be computed with a set of differential equations that allow new variables to be triggered into action as the system evolves. In fact, the model consists of just such a set of equations. Having shown how they arrive at its formulation, the authors note the model's evident similarities to John Holland's classifier system, which they also rewrite as a set of differential equations in order to compare the two.

Both interesting in its own right and a highly influential new method of problem solving in the field of machine learning, Holland's classifier system deserves a brief summary. Basically, it is comprised of three components: a rule and message system, a credit system, and a genetic algorithm. In *Genetic Algorithms in Search, Optimization, and Machine Learning*, David E. Goldberg (one of Holland's former students) provides the schematic diagram shown in figure 4.4.[73]

As we can see from the diagram, information enters the system via detectors or sensors, where it is coded in binary strings and posted on a message list. Each string is then read by the classifiers, which are rules that take the form: "if ⟨condition⟩ then ⟨action⟩." If the ⟨condition⟩ is a match with the message, the ⟨action⟩ might be to post another message on the message list and/or to trigger an output action through an effector. The rules are also composed of binary strings, with an additional

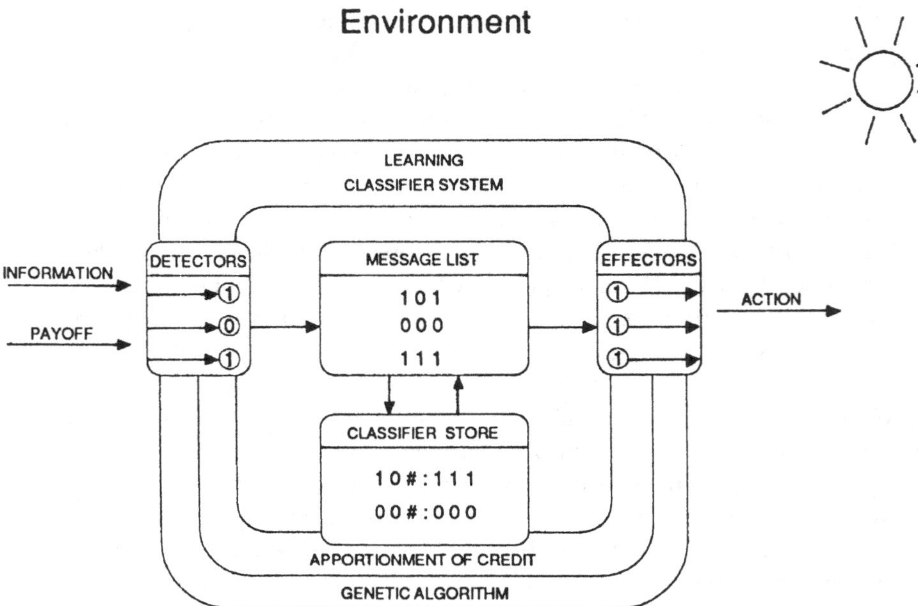

Figure 4.4
A learning classifier system. David E. Goldberg, *Genetic Algorithms in Search, Optimization, and Machine Learning* (Reading, Mass.: Addison-Wesley, 1989), 223.

wildcard marker (#) designating either 0 or 1. Thus the message 0101 would match with rule 0##1. To illustrate, consider a simple system composed of the following four classifiers, with their outputs given after the colons. (The example, slightly modified, is taken from Goldberg.)

1. 01## : 0000
2. 00#0 : 1100
3. 11## : 1000
4. ##00 : 0001

Suppose the message 0111 from the environmental detector is posted on the message list. Since it matches with classifier 1, the latter would post the message 0000. This message in turn matches with classifiers 2 and 4, which would then post their messages, 1100 and 0001. Message 1100 now matches with classifiers 3 and 4, which posts messages 1000 and 0001. Of these two, only 1000 elicits a response: classifier 4 posts message 0001, which then elicits no response, and the process is terminated.

However, a classifier is not allowed simply to post a message when activated, as assumed above for purposes of exposition. In actuality it makes a bid to post a message, the efficacy of which is proportional to the classifier's strength as measured by the number of times its activation has led to an output action. Moreover, it must give credit to any and all classifiers whose activation has led to its own activation. In short, its own activation simply initiates a complex process of credit assignment, which then conditions its ability to participate further in a larger process. Holland has called this system of auction, payment, and reinforcement the bucket-brigade algorithm, and the result is what Goldberg likens to "an information economy where the right to trade information is bought and sold by classifiers . . . [which] form a chain of middlemen from information manufacturer (the environment) to information consumer (the effectors)" (225).

With such rigorous and competitive demands on the rules or classifiers, one might wonder how the system can be supplied with rules that work at all, much less with high efficiency. But this is where the third part of the classifier system fits in: through the use of genetic algorithms, new and better rules are generated that can be injected into the system. This third part works in conjunction with the credit assignment subsystem, which separates the rules that perform effectively from those that do not. By applying the processes of crossover and mutation to these good rules, new and better ones are bred, which are then added to the population of rules to have their performances evaluated and winnowed in turn, as the whole system gradually turns over. As an alternative to the serial processing of traditional expert systems, the parallel rule activation of the classifier system avoids the bottlenecks of the latter and allows multiple activities to be coordinated simultaneously. While this makes for a much faster hardware implementation, the real accomplishment is a machine that can actually learn and adapt to changing information.

Although the classifier system is clearly an information-processing device, the fact that the rules operating within it are changing over time also means that it functions as a dynamical system. As Farmer, Packard, and Perelson demonstrate, both the classifier and model immune system are strongly nonlinear, and the equations for computing their changing behavior take the same basic form:

$\Delta x =$ internal interactions $+$ driving $-$ damping (202)

The precise form that these terms take depends on how the interactions as well as the driving and damping forces influence one another, but the general form of the equation is often seen in biological phenomena. In particular, the authors mention coupled autocatalytic reactions and the Lotka-Volterra equations for population dynamics in biology. These similarities are hardly surprising, given that their model immune system as well as Holland's classifier system mimic parallel computation in natural biological systems. Indeed, after comparing the two systems, Farmer, Packard, and Perelson conclude by affirming the superiority of such parallel computational systems over standard serial Turing machines. Not only does their model provide insight into the internal operations of real immune systems, but the correspondences between their model and the classifier system reinforces the growing sense "that generalized versions of our model may be capable of performing artificial intelligence tasks" (203). Since then, in fact, parallel processing and the use of genetic algorithms in classifier systems have become part of the larger repertory of contemporary AI.[74]

Although Varela later acknowledges Farmer, Packard, and Perelson's model of the immune system, he ignores the fact that it combines an information-processing and dynamical systems approach. This combination, as I have emphasized, is an essential feature of a new kind of computational assemblage. Further examples will to be taken up in subsequent chapters. Here I want to turn briefly to computer viruses and computer immune systems, which constitute a very different kind of computational assemblage. That there *is* research on computer immune systems—in obvious contrast to the commercial development of antivirus software—of course presupposes an environment of proliferating and increasingly sophisticated computer viruses. These viruses began to appear in the early 1980s, and it was only a matter of time before they would be considered as forms of artificial life.

Actually, computer viruses were one of the few topics that Langton actively sought to discourage at the first ALife conference in 1987.[75] In the bibliography to the conference proceedings, however, he does list A. K. Dewdney's article, "Computer Recreations: A Core War Bestiary of Viruses, Worms and Other Threats to Computer Memories," published two years earlier.[76] At the time of the conference, Dewdney was well known for his invention of the computer game Core Wars, in which players attempt to fill up and disable the opponent's memory space with replicating code. In short, Core Wars made virus writing into a game. Langton invited Dewdney to the ALife conference to judge an "artificial 4-H contest" for the best computer creature, an event intended to amuse and entertain the participants. However, he did not invite Fred Cohen, who as a graduate student in computer science in 1983 had written a virus of two hundred lines of code that could invisibly give him system administrator privileges on a Unix operating system. Cohen was one of the first "professional" experimenters with computer viruses and published the results of his experiments in the highly reputable journal, *Computers and Security*. But the line of demarcation was not always clear. In 1988 Cornell student Robert Morris had released his self-replicating "Internet worm," which quickly paralyzed some six thousand computers. Creating panic and hysteria, Morris's actions eventually resulted in the establishment of a panoply of legal measures and law enforcement agencies. Cohen's own laconic response was that Morris had just set the world's record for high-speed computation. Even so, in those inchoate times uncontrolled experiments with forms of artificial life *avant la lettre* could only be worrying and even an impediment to professional researchers, as Stephen Levy remarks in his book on artificial life:

> During the period that young Morris and other unauthorized experimenters were blithely releasing predatory creatures in the wild [i.e., floppy disks and networked computers], Cohen and other serious researchers were consistently being refused not only funding but even permission to conduct experiments in computer viruses. As a result, the creations of willful criminals and reckless hackers were for years the most active, and in some ways the most advanced, forms of artificial life thus far. (*Artificial Life*, 324)

Levy gives the impression that a whole new realm of artificial life was beginning to burgeon, some of which was scientifically "authorized" and officially sanctioned, while other forms constituted an unauthorized but no less fertile underside. While no doubt this would become the official position, the boundary line it assumes has never been firm and is perceptibly ever shifting.[77] For example, in his book *It's Alive! The New Breed of Computer Programs*, Cohen discusses computer viruses under the rubric of "living programs" (or LPs), which also include Core Wars, Conway's Game of Life and Ray's Tierra. Cohen defines a living system as comprised of an organism *and* its environment, arguing that when viewed as a pattern in "the information environment" computer viruses are very much alive. The "outsider" scientist Mark A. Ludwig pushes this point of view even further. In *Computer Viruses, Artificial Life and Evolution* he offers a technically and philosophically astute analysis of artificial life, while also—like Cohen—providing computer code for experimenting with a variety of real viruses.[78] In fact, Ludwig argues that computer viruses are a more significant form of artificial life than the "laboratory contained" forms produced in scientifically sanctioned experiments, precisely because viruses live in a world that was not specifically designed to support them. For Ludwig, Darwinian evolutionary theory provides the proper scientific framework for comparison. Describing several progressive steps by which virus writers have attempted to foil antivirus scanning techniques, with the mutating or polymorphic virus first written by the legendary Dark Avenger representing the latest stage, Ludwig suggests that the next step would be a mutating virus with a genetic memory that could evolve into increasingly more immune forms. Not content with speculation, Ludwig actually supplies the source code (in assembly language) for a "Darwinian Genetic Mutation Engine" that can convert a lowly DOS virus into a genetically evolving polymorph. Significantly, like some ALife scientists, Ludwig ends up questioning whether standard Darwinian theory can actually explain the evolutionary developments he describes. Whatever the case, there is little reason to believe that such research has been ignored by malignant hackers and virus writers. Indeed, the latter's undeclared war against software industry giants like Microsoft and corporate Web sites and the consequent attempts to provide virus protection have clearly resulted in an escalating "arms race" in today's digital ecology that illustrates Ludwig's basic argument.

Eventually the fact that the networked world of computers had become a site where new forms of viral life were constantly emerging could no longer be avoided by ALife scientists. At the ALife IV conference, Jeffrey O. Kephart argued that current antivirus techniques are doomed to fail and must eventually be replaced by a biologically inspired immune system for computers.[79] Although an important step, Kephart's model still assumed the military understanding of the immune system that Varela contests. In contrast, a potentially much more fruitful approach was taken by Stephanie Forrest when she conceived of the world of computers as having many of the properties of a living ecosystem, populated with "computers, with people, software, data, and programs."[80] In "Principles of a Computer Immune System," Forrest lists twelve organizing principles of a biological immune system, many of which—autonomy, adaptability, and dynamically changing coverage—while not going as far as Varela, move beyond the strictly defined military model.[81] Forrest argues that these principles must be incorporated as design principles if a computer immune system is to function. However, if the objective is "to design systems based on direct mappings between system components and current computer system architectures" (79), then the latter will have to be radically modified. One possible architecture, she suggests, would be something like an equivalent "lymphocyte process" comprised of lots of little programs that would query other programs and system functions to determine whether they were behaving normally or not. But they would also have to monitor one another, "ameliorating the dangers of rogue self-replicating mobile lymphocytes" (80) and thus a possible form of digital cancer. Just how feasible this approach will turn out to be is difficult to say, and Forrest herself rightly remains cautious and acutely aware of the limitations of "imitating biology," since biological organisms and human-made computers obviously have very different methods and objectives.

But here we may be bumping up against the limits of a conceptual contradiction between the computer as a tool or medium over which we exercise near complete control and the computer as part of an ecosystem that cannot function unless given more lifelike capacities that will put it outside of our control.[82] Perhaps human computational and communicational activities will eventually have to be made more like biological exchanges if a fully functional computer immune system is to be constructed—or rather, and more likely, evolved. In any case, the study of what are now known as complex adaptive systems constitutes an important theoretical step toward the construction and understanding of such systems. One of the most interesting features of complex adaptive systems is that they emerge in both nature and culture. It is certainly no accident that they should come into view—that is, be identified as such—about the same time that ALife research gets seriously under way. This convergence will be of primary interest in the next chapter.

5 Digital Evolution and the Emergence of Complexity

Yet nature is made better by no mean
But nature makes that mean: so, over that art
Which you say adds to nature, is an art
That nature makes.
—William Shakespeare, *A Winter's Tale*

In laying the foundations for ALife, Christopher Langton argued that ALife machines and programs would actualize the essential processes of living systems. In so doing, they would instantiate real life, not just *simulate* or *model* it. This amounts to a "strong" theory of ALife, in contrast to "weak" theories that would view ALife simulations as only lifelike replicas of living systems.[1] As we saw in the previous chapter, Francisco Varela did not so much challenge the possibility of strong ALife as argue for a different set of criteria for defining a living system. While Langton had extended John von Neumann's theory that life's essential logic could be captured by self-reproducing, evolving automata, Varela followed in the footsteps of Humberto Maturana, arguing that autopoiesis—a system's capacity to produce and sustain the organization by which it can maintain itself as autonomous—is what defines life, whether natural or artificial. Maturana and Varela also insisted that evolution logically presupposes a living entity. It is therefore possible that a living, self-organizing entity might be *in*capable of reproduction. Since most biologists consider evolution to be a fundamental attribute of life, this position raises obvious questions. How, for example, is the widely held assumption that life somehow emerged from nonliving matter to be addressed? And what about liminal instances like viruses, which only live in a state of parasitic dependency on host organisms, but nonetheless replicate, mutate, and evolve, and thus fall under the seemingly iron law of natural selection?

At one level, the conflict between Langton and Varela simply reflects the lack of consensus among biologists about how to define life: is there an underlying essence, or is life a constellation of "family" properties, not all of which have to be present in any given instance? Yet their differences are not merely definitional. Langton understands life as an emergent phenomenon and therefore focuses on its underlying dynamical structure rather than the materiality of its constituent parts. What is crucial, he writes, is that in life "the local dynamics of a set of interacting entities (e.g., molecules, cells, etc.) support an emergent set of global dynamical structures which stabilize themselves by setting boundary conditions within which the local dynamics operate. That is, these global structures can 'reach down' to their own, physical bases of support and fine tune them in the furtherance of their own, global ends. Such LOCAL to GLOBAL back to LOCAL inter-level feedback loops are essential to life, and are the key to understanding its origin, evolution and diversity."[2] For Varela, on the other hand, diversity stems from "the situated nature of [a living system's] cognitive performances" and from how it maintains its own autonomy in the face of unending contingencies.[3]

This basic difference has further implications. It not only echoes the origin-of-life question—which came first, a replicator or a metabolic mechanism?—but also determines exactly which life processes are to be instantiated (or simulated) in ALife experiments. Although much groundbreaking work has been devoted to creating "artificial chemistries," autocatalytic sets, and metabolic activities, Varela's position (further discussed in chapter 7) lends itself more readily to the construction of autonomous mobile robots than to the generation of artificial organisms. For those interested in emergence, however, there is no means "so accessible or as easy to manipulate as that of increasing complexity by evolution through natural selection," as Charles Taylor puts it ("'Fleshing Out' Artificial Life II," 32). Not surprisingly, then, the most dramatically successful ALife research to date has been based on the evolution of digital organisms (i.e., strings of self-replicating code) in computer-generated virtual worlds. This chapter examines a body of that research—together with the underlying theories of emergence and complexity—conducted on various software platforms, including Thomas Ray's Tierra, John Holland's Echo, Christoph Adami's Avida, Andrew Pargellis's Amoeba, Tim Taylor's Cosmos, and Larry Yaeger's PolyWorld. I conclude the chapter by considering several of the methodological limitations of these platforms and then point to some recently developed alternatives, including living computation and attempts to create an artificial protocell. Haunting this entire body of work is the still unresolved question of ALife's relationship to biological conceptions of life and whether its simulations and recreations can free themselves from mimetic and representational dependencies without devolving into merely interesting technical performances. In other words, to what extent, at this stage, can the new science of ALife achieve genuine autonomy?

Tierra Cognita

In order to study the dynamics of evolution, Thomas Ray constructed Tierra, a computer-generated virtual machine in which digital organisms spawn, mutate, and die. The obvious advantage of using a computer program to mimic natural evolution is that it can offset the latter's grindingly slow pace. At the same time, an artificial environment entails an obvious reduction of complexity: in addition to its apparent simplicity and paucity of organisms, it lacks the unplanned openness of nature in which natural selection can turn to its advantage whatever chance offers, whether it be an aspect of the environment, a parasitic or symbiotic relationship with another species, or, in the most familiar instance, a beneficial mutation in an organism's genome. While the idea of simulating evolution on a computer was certainly not new among the early practitioners of ALife, no one before Ray had figured out exactly how it could be done.[4] The stumbling block was precisely this open-ended aspect of nature—the problem of how to get "the hand of the breeder" out of the selection process and let natural selection do its own work. Evidently nature's algorithm, as Daniel Dennett calls natural selection, could not be so easily reproduced in a computational medium.[5]

Trained as a biologist and specializing in rain forest ecology, Ray had arrived at the idea of studying evolution in a computer-generated environment completely on his own.[6] Within a year of buying his first computer in 1987, he was well on his way toward realizing his basic project: to make digital creatures out of machine instructions that could replicate and evolve. The creatures would "live" in computer memory (their spatial habitat) and compete for time (analogous to their energy source) on the computer's central processing unit (CPU). (The code was written in assembly language, a programming language one step above the 0s and 1s of machine code.) In order to produce mutations in the replication process, Ray included instructions that would cause random bit flipping (reversing 0s and 1s). The creature that would initiate, or "seed," the process he called the Ancestor. Once introduced into Tierra (Spanish for "earth"), it would immediately begin to replicate, and the resulting digital organisms would be assigned: (1) a memory address; (2) a place in the queue that would bring them to the CPU, where they would execute their code; and (3) a place in the queue of the "reaper" function, which would eliminate them. Young organisms, naturally, would be inserted at the bottom of the reaper queue, but Ray also included code that would slow down the advance up the queue of the organisms that were most successful in the replication process (that accumulated the lowest number of errors), thus allowing them to live longer. So as not to infect his or other machines, the entire simulator program would run on a virtual machine—a software emulation of a computer within Ray's computer. Since instructions designed for this virtual machine would not execute on other machines, it provided a contained environment.

That, in brief, was the main idea. When Ray presented his project to Langton and other ALife scientists at the Santa Fe Institute, however, the response was somewhat mixed: on the one hand, there was enthusiastic support for the idea; on the other, warnings about infection (Robert Morris had recently released his Internet worm) and how long it would take to make a fully functional simulator. An obvious obstacle was writing a self-replication program that could actually allow mutation and continue to function, since most computer programs, being extremely brittle, malfunction with only slight changes in the code. Inspired by the way RNA replication works, Ray introduced a number of "template matching" function calls into the organism's instruction set. Consequently, when the code calls for a certain function to be executed in a replication loop, the computer looks for a template—a specific bit pattern—in its memory rather than for a specific memory address. If it finds a block of code with a template matching the one needed, the organism can execute that code. Thus an organism is not necessarily incomplete or dysfunctional if it lacks certain blocks of code, as long as it can find, through template-matching, the needed code somewhere in the "soup," as Ray called the program's total memory. Yet even with this innovative feature Ray was greatly surprised by the results of Tierra's first trial run. Seeding it with what he thought would be only a primitive, first version of the Ancestor, he assumed that it would require much more work to get a genuine evolutionary process going, that is, to get the bugs *into* his system and replicating. As it turned out, he says, "I never had to write another creature" (quoted in Levy, *Artificial Life*, 221).

The trial run took place on January 3, 1990. Composed of 80 lines of code, the ancestor required 839 CPU cycles to replicate, so Ray let the program run all night on his laptop computer, which executed about 12 million instructions per hour. The reaper function was set to kick in when the program's memory capacity reached 80 percent, and the results could be tracked on a bar graph that identified organisms and the number of their proliferation. Not expecting much on this first run, Ray was astonished the next morning to discover that genuine evolutionary behavior already seemed to be taking place. Not only were the 80-instruction Ancestors replicating, but there were also 79-instruction mutants, and 45- and 51-instruction parasites. In subsequent runs 61-instruction hyperparasites and 27-instruction "cheaters" also appeared. Indeed, by taking advantage of the special conditions of this virtual environment, a small but thriving ecology of interacting digital organisms had come into being.

It was clear, for example, that many of the mutant organisms were quite resourceful. The 45-instruction parasites, lacking the code for replication, would simply borrow it from the larger-sized Ancestors. Being smaller than their hosts, they required less CPU time to copy themselves to a new location in memory and could thus proliferate more quickly. The parasites and hosts also exhibited dynamic behavior similar to the Lotka-Volterra population cycling in predator-prey studies familiar to ecologists. In contrast, the hyperparasites would quickly decimate the parasites. These creatures could do everything the Ancestors could do, but mutation had made their code more compact and efficient, enabling them to destroy the parasites by capturing the latter's CPU time. As for the cheater, though very small (only 27 instructions), it could intercept the replication instructions it needed as they were passed between two cooperating hyperparasites. Ray later reported that one type of creature had actually discovered the advantage of "lying" about its size.[7] The replication code requires that each creature do a self-examination and calculate its size in order to request precisely enough memory for its daughter cell. In this instance the creature would calculate its size as 36 instructions but request a space of 72 instructions for its daughter, thereby doubling the amount of space and energy (memory and CPU time) available to its progeny.

The marvelous success of Tierra rests upon the fact that it actually instantiates, rather than merely models, the fundamental dynamic of open-ended evolution in a limited milieu. For undeniably, digital organisms had evolved in such a way as to take advantage of the unique features of their silicon world, and in a manner that could not have been foreseen, much less planned for in advance. The distinguished evolutionary biologists John Maynard Smith and Richard Dawkins immediately saw the value of Ray's experiments, and another highly respected biologist, Graham Bell of McGill University, summarized Ray's contribution as follows:

> This work has three important uses. First, it is a superb educational tool. Many people doubt that the theory of evolution is logically possible.... Now, one can simply point to the output of Ray's programs; they are the ultimate demonstration of the logical coherence of evolution by selection. Secondly, it seems likely to provide a superior method for testing theoretical ideas in evolution by providing more realistic general algorithms than have ever before been available. Thirdly, it may also represent a general advance in computation, since it makes it possible to evolve algorithms for any purpose. (Levy, *Artificial Life*, 230)

Appreciation, to be sure, but also accommodation to familiar and readily acceptable frameworks. Both Langton and Ray himself describe Tierra's methodology and experimental results in more far-reaching terms. For Langton, what mattered most was that by removing the "hand of the breeder" Ray had created an artificial environment in which "natural selection is truly at play."[8] In these terms Tierra represented a great leap forward in ALife research, carrying it beyond the earlier work of John Holland and others with genetic algorithms, Danny Hillis with coevolutionary strategies for the development of algorithms, and Kristian Lindgren with computational ecologies. Before considering Tierra further, we should briefly recall Langton's summary of this precedent work.

In all three of these instances, algorithms were evolved according to preexistent notions of fitness. Although the procedures are different, the underlying assumptions are similar to those used in animal husbandry and agriculture, as in the selection and grafting of grape vines in wine production. In Holland's genetic algorithms, strings of code are repeatedly divided, swapped, and mated in order to evolve algorithms that provide an optimal solution to a precisely defined problem, most often in machine learning or sorting operations. Hillis, on the other hand, discovered how to combine different kinds of sorting problems and different kinds of sorting networks and then make them coevolve in a host-parasite dynamic. Let's say that a particular population of sorting networks is stuck on a local fitness peak. (Evolutionary biologists typically represent measures of fitness as peaks or valleys on a topographical map.) In order to get to a higher peak, the population would first have to cross a valley, which is difficult, since performance would have to get worse before it could get better, and there is no Darwinian selection mechanism for that. Hillis, however, found that in certain instances the parasites would deform the fitness landscape of the sorting networks, turning the low peak into a valley from which the population could then escape by evolving upward. Lastly, building on the celebrated work of Robert Axelrod, Lindgren considered evolutionary strategies for cooperating or defecting in iterated games of Prisoner's Dilemma. By making a player's strategy depend on what the opposed player did in the previous game or games, Lindgren introduced memory as a factor. This in turn initiated an evolution of strategies. At first the results were what might be expected: to defect works to one player's advantage in the short term but to cooperate becomes advantageous to both over the long term. As strategies evolved over the long term, moreover, the system began to exhibit more interesting complex behavior. First, the phenomenon known as punctuated equilibria could be observed, that is, "after an initial irregular transient, the system settles down to relatively long periods of stasis 'punctuated' irregularly by periods of rapid evolutionary change" (84).[9] In addition, extinction events, or crashes in the diversity of species (here, game strategies), as well as ecologies or mixtures of different strategies, also emerged.

To everyone's surprise, Tierra exhibited this entire range of evolutionary behavior. Yet the much greater autonomy of the evolutionary process at work in Ray's silicon world makes the appearance of this complex behavior much more significant. With the elimination of the outside algorithmic breeding agent, Langton stresses, the "external task of evaluation of fitness has been internalized in the function of the organisms themselves" (88). This means that the conditions that obtain within Tierra itself define fitness, not some external objective or performance criteria. These conditions are *not* known in advance and can be assumed to change over time. The fitness of any particular organism will be determined by its interactions with other organisms in the soup. As we saw above, these interactions involve competition, host-parasite and symbiotic relationships, and strategies of deception. To be sure, fitness in Tierra really means long-term reproductive vitality. As Ray himself notes, "survival of the fittest" is nearly a tautology, since the fittest by definition are those that tend to survive. The tautology is avoided, however, if the fitness landscape itself "is shaped by specific adaptations that facilitate passing genes along."[10] And this is exactly what happens in Tierra, as well-adapted digital organisms pass their replication code to their offspring.

Initially, Tierra's fitness landscape is defined by the organism's replication code and the fixed parameters of computer memory and CPU time. That is, the original fitness landscape is based solely on the Ancestor's adaptation to the environment constituted by the computer itself. However, as replication and mutation occur and the soup is filled with a diversity of organisms, the fitness landscape begins to change. What now becomes important is how the organisms interact, as they discover new ways both to exploit one another and to cooperate. Thus in addition to the computer environment, the creatures must also adapt to the changing environment created by their own interactions, which quickly becomes the primary force driving evolutionary change and development. Ray summarizes the consequences as follows: "Because the fitness landscape includes an ever-increasing realm of adaptations to other creatures which are themselves evolving, it can facilitate an autocatalytic increase in complexity and diversity of organisms" (395). As the digital organisms change and evolve in response to one another, their digital universe grows in complexity. But are there limits to the growth of this complexity? How far can it go, and what are the factors that enhance as well as diminish it?

At the chapter's conclusion I suggest that a lack of diversity ultimately limits this growth. Nevertheless, the importance of Tierra for subsequent ALife research cannot be overemphasized. Like Langton, many specialists within ALife and evolutionary biology saw it as instantiating—rather than merely modeling—the Darwinian principle of natural selection in a digital medium. It also provided an unexpectedly efficient means to evolve high-performance algorithms—certainly no mean feat. Yet some, including a few nonspecialists, have contested the claim that the behavior observed in Tierra is worthy of the label "life" and have seen Ray's descriptions of his experiments as misleadingly anthropomorphic.[11] Yet even organo-centric speculation on these questions, if it is not to be trivial, must confront the issue of complexity and the new kind of silicon-becoming that Tierra instigates. To do this, however, may require a conceptual framework as innovative as Tierra itself. In several published essays, notably "An Approach to the Synthesis of Life," "An Evolutionary Approach to Synthetic Biology: Zen and the Art of Creating Life," and "Evolution and Complexity," Ray has emphasized the importance of emergence and complexity, two key ideas in nonlinear dynamical systems theory. At the same time, he remains curiously silent about the challenges that these ideas pose to modern, neo-Darwinian evolutionary theory, the explicit frame for his experiments. To grasp what is at stake, we must now consider several of these central challenges.

Darwinian Evolution and the Theory of Self-Organization

Of course Darwin himself did not know by what specific mechanism inherited traits were passed down from generation to generation. Several important aspects of that mechanism were discovered by Gregor Mendel, an Austrian monk and amateur botanist, only a few years after the publication of *The Origin of Species* in 1859, but unfortunately they did not become widely known until the turn of the century. In the subsequently formed neo-Darwinian synthesis, the gene theory of inheritable traits was combined with Darwin's laws of inherited variation and natural selection. With the discovery of DNA and the "cracking of the genetic code," all the parts of a complete theory of biological evolution seemed to be in place. The material evidence for this updated version of Darwin's theory is overwhelming and it is now accepted as scientific law. Yet there remain a few significant problems that the theory cannot account for or address without new extensions and enrichments.

First, the genome of a well-defined species like human beings, initially pictured "as a linear array of independent genes, each corresponding to a biological trait," actually behaves more like a complex, nonlinear dynamical system in which different genes switch one another on and off, and a single gene may effect a wide range of traits; or, conversely, many separate genes may combine to produce a single trait.[12] This makes it unlikely or rare that evolution acts on a single, isolated gene. Actually, evolution within the genome is enormously complex, and there is even evidence of "directed mutation," as Lynn Helena Corporale has demonstrated.[13] Second, the fossil evidence does not confirm—as Darwinian theory demands—that evolution has proceeded through slow and continuous gradual change over long periods of time. Instead, it tends to support Niles Eldredge and Stephen Jay Gould's thesis of "punctuated equilibria," according to which long periods of stability are followed by sudden transitions to catastrophic change, characterized by mushrooming growth of a wide diversity of completely new forms, massive extinction events, or sometimes both. The best-known examples are the Cambrian explosion, in which as many as 100 new phyla formed (compared to the 32 today), and the Permian extinction, in which about 96 percent of all species became extinct. A third problem is raised by the discovery that at the microbial level of bacteria and other simple life forms, genes are and possibly always have been freely traded and shared in a process known as DNA recombination. As a result, for several billion years a vast planetary web of life whose growth is based on networks and coevolution has flourished on earth.[14] In view of these and other complications, the older picture of evolution based on chance mutation and competition both within and across variant species is giving way to a different picture, characterized by a constellation of self-organizing and coevolving forces.[15] In brief, a mix of nonlinear dynamical systems theory and coevolutionary theory is replacing the violent Victorian image of "Nature red in tooth and claw," as Tennyson put it.

In the context of ALife research, the most directly relevant complication of Darwinian theory stems from the work of theoretical biologist Stuart Kauffman, who was at the Santa Fe Institute during roughly the same period as Langton. Although Kauffman is often seen as challenging the Darwinian hegemony in evolutionary biology, it is perhaps more accurate to say that he has sought a marriage of self-organization and natural selection, rather than a wholesale dethroning of the latter.[16] Early in his career Kauffman proposed random Boolean networks derived from the neural nets of McCulloch and Pitts as a model for understanding how arrays of genes constituted a dynamical system. In a Boolean network, interconnected nodes switch one another on or off according to the complex nonlinear dynamics of the system as a whole. In Kauffman's model network, only slight changes to simple variables, specifically the number of nodes (N) and the average number of inputs to each node (K), would cause it to display a range of distinct behaviors. At $K = 1$ and below, large sectors of the network would remain frozen and almost inanimate. At $K = N - 1$, in contrast, highly animated and chaotically changing patterns would cascade through it, and no stable structures appear. Kauffman discovered that the most interesting behavior occurred in certain regions between these two regimes, around $K = 2$, where the network would evolve toward clearly distinct self-organizing states in which a stable but unpredictable order emerged. Kauffman called this emergent order "order for free" and thought that it might play a large role in the evolution of biological organization at many levels.[17]

Since a random Boolean network constitutes a dynamical system, each of these distinct behaviors corresponds to a different attractor state (e.g., single point, periodic, or chaotic). Kauffman thus assumed that detailed study of the NK model in relation to attractor states would deepen our understanding of how genetic networks and cell differentiation actually work. Specifically, he theorized that each specific cell type would be the result of a stable, self-organizing state of the genetic network (i.e., an attractor):

> Order for free. But more: the spontaneously ordered features of such systems actually parallel a host of ordered features seen in the ontogeny of mouse, man, bracken fern, fly and bird. A "cell type" becomes a stable recurrent pattern of gene expression, an "attractor" in the jargon of mathematics, where an attractor, like a whirlpool, is a region in the state space of all the possible patterns of gene

activities to which the system flows and remains. In the spontaneously ordered regime, such cell-type attractors are inherently small, stable, and few, implying that the cell types of an organism traverse their recurrent patterns of gene expression in hours not eons, that homeostasis, Claude Bernard's conceptual child, lies inevitably available for selection to mold, and, remarkably, that it should be possible to *predict* the number of cell types, each a whirlpool attractor in the genomic repertoire, in an organism. Bacteria harbor one to two cell types, yeast three, ferns and bracken some dozen, and man about two hundred and fifty. Thus, as the number of genes, called genomic complexity, increases, the number of cell types increases. Plotting cell types against genomic complexity, one finds that the number of cell types increases as a rough-square root function of the number of genes. And, in parallel, the number of whirlpool attractors in model genomic systems in the ordered regime also increase as a square-root function of the number of genes. Man, with about 100,000 genes should have 370 cell types, but has close to 250. A simple alternative theory would predict billions of cell types.[18]

As a dynamical system, the gene network is constrained to specific state changes that mean only certain pathways (i.e., sequences of expression of DNA instructions) are possible. Thus each of these pathways, defined by a specific attractor state, leads to the production of a specific cell type. Given that genes switch one another on and off (this was the Nobel Prize–winning discovery of François Jacob and Jacques Monod) and that each gene is usually controlled by two other genes (i.e., $K = 2$), Kauffman further postulated that the gene network must lie in the ordered regime but near the edge or boundary of the chaotic regime. Otherwise, it would not have been able to evolve to its present functional state by successive useful variations.[19]

Here, Kauffman's work dovetails with Christopher Langton's. As we saw in the previous chapter, Langton found that it was precisely "at the edge of chaos," in a specific domain or regime he characterized as a critical phase transition, that conditions are most propitious for a dynamics of information processing to emerge spontaneously and come to dominate the behavior of a physical system. His work thus pointed to how a natural domain structured by the exchange of information in the physical world could arise and how the kinds of computational processes necessary for life could get started. In the millions of chemical interactions occurring in states far from equilibrium, at the onset of chaos, Langton saw nature seeking a way to compute itself to a stable state, one in which life could perpetuate and reproduce itself. Looking at this scenario from a biological perspective, Kauffman saw that it required an additional idea: once life gets started by taking advantage of the greatly increased information processing that occurs in the region of a phase transition, evolutionary pressures actually select for mechanisms that will move it toward or keep it near this state. This idea, first proposed by Norman Packard in a paper entitled "Adaptation to the Edge of Chaos" and presented at Santa Fe in 1988, struck Kauffman with the force of a critical breakthrough.[20] If living systems actually operate close to the edge-of-chaos phase transition—and Kauffman thinks they do—then natural selection is most likely the motor that pushes them there.

In these terms the contradiction between self-organization and natural selection appears to be resolved. If complex dynamical systems with the capacity to self-organize are widespread throughout the natural world, then the outcome of evolutionary processes would not be a simple matter of "chance and necessity," as Jacques Monod famously put it.[21] Rather, spontaneous tendencies to self-organize and produce order for free would introduce another determining factor in the evolutionary process, one that Darwin could not have foreseen. Consequently, the discovery of self-organization in biological systems means that "we must rethink evolutionary theory, for the sources of order in the biosphere will now include both selection *and* self-organization."[22] As David J. Depew and Bruce H. Weber summarize Kauffman's position, Darwinians "must be prepared to admit that in many cases natural selection cannot be expected to do all or even most of the work, that as explanatory models become more realistic natural selection ceases to be an explanation of first resort, and that when selection operates, it does so in a fairly narrow range of possibility space, since it selects among entities that are already self-organized modules and that are in the process of spontaneously forming into still higher levels of self-organization" (*Darwinism Evolving*, 436). Indeed, Kauffman believes that complex systems theory may provide a fuller understanding of Darwinian evolution itself by clarifying the very conditions of "evolvability":

> Selection must *achieve* the kinds of systems which are *able* to adapt. That capacity is not Godgiven; it is a success.
>
> If the capacity to evolve must itself evolve, then the new science of complexity seeking the laws governing complex adapting systems must discover the laws governing the emergence and character of systems which can themselves adapt to accumulation of successive useful variations. ("The Sciences of Complexity," 307)

It follows that those systems that *can* evolve will most likely turn out to be precisely those systems poised in the ordered regime but near the edge of chaos.

These ideas found unexpected resonance with another paper presented only months after Packard's. In a presentation on "self-organized criticality," the Danish physicist Per Bak contended that many complex systems evolve spontaneously to a critical state, at which point even a small perturbation will produce waves of catastrophic change on all scales.[23] An ordinary sand pile provided a clear illustration. If sand is continually added to the pile, it will inevitably arrive at such a critical state, where adding a few more grains will cause it to avalanche. Bak showed that the distribution of the sizes and frequencies of the avalanches constituting this change follows a power law.[24] Though there seemed to be no way to know in advance which systems would evolve to a state of self-organized criticality and which would not, when they did, the size of the changes relative to their frequency always exhibited the precise correlation of a power law. Bak pointed to evidence in the data describing earthquakes, extinction events in the fossil evidence, and even the behavior of the stock market. The fact that a system poised at a state of self-organized criticality would propagate perturbations on all possible length and size scales especially intrigued Bak's audience at Santa Fe, and Kauffman in particular thought there might be a connection with Langton's edge-of-chaos thesis. But while the point of the latter was that "systems at the edge had the potential to do complex computations and show lifelike behaviors ... Bak's critical state didn't seem to have anything to do with life or computation" (Waldrop, *Complexity*, 307). Yet both concepts dealt with phase transitions and exhibited the same phenomenology: "You can tell that a system is at the critical state and/or the edge of chaos if it shows waves of change and upheaval on all scales and if the size of the changes follows a power law" (308). Although it wasn't exactly clear how they fit together, Kauffman was sure there was a connection. Self-organized criticality, for example, *might* explain punctuated equilibrium, though the data in the fossil record was inconclusive. Nevertheless, Kauffman began to wonder "if power-law cascades of change would be a general feature of 'living' systems on the edge of chaos" (309). When he tested this hypothesis in simulations of an ecosystem with coevolving species, he found that a coevolving ecosystem could well be evolving toward self-organized criticality while parts of it remain "frozen" in equilibria.[25] However, "cascades, or avalanches of changes initiated at local points in the ecosystem web may propagate to various extents throughout the ecosystem. Such avalanches may trigger speciation and extinction events" "Co-evolution to the Edge of Chaos," 343). Thus Bak's theory, in conjunction with Langton's thesis, could explain the coevolutionary dynamic revealed in the system's coupled fitness landscapes.

Internet Tierra: Theoretical Biology or "Live" Software?

It was within this framework that scientists like Langton and Kauffman tended to perceive the significance of Ray's Tierra. Though skeptical at first, Ray was willing to run tests on Tierra and analyze the data with the edge-of-chaos idea and a power law distribution of extinction events in mind. As he reported to Roger Lewin,

> Mutation rate in my system is somewhat analogous to the lambda parameter Chris [Langton] used in his cellular automata. If I turn up the mutation rate, the system should go chaotic and die out. At a low rate nothing very interesting should happen. In between these two rates we should see a rich ecology produced, and if this is the edge of chaos, this is where we should see avalanches of extinctions with a power law distribution.[26]

Lewin, in turn, pointed out that the data Ray had already gathered exhibits the power law distribution, suggesting that Tierra "might have evolved to the edge of chaos all by itself" (103). Ray agreed, but wanted to run more tests, which confirmed Lewin's insight.

Following Tierra's success, Ray decided to install a second version in the more expansive, networked space of the Internet. Before we consider this research, the central importance of complexity in Ray's work must be emphasized. Ray himself has stated, "It is relatively easy to create life. Evidently, virtual life is out there, waiting for us to provide environments in which it may evolve" ("An Approach," 393). But what *is* difficult to produce is something like the Cambrian explosion, where there was "origin, proliferation and diversification of macroscopic multi-cellular organisms" (398). Here, as elsewhere in Ray's writings, the reference to the Cambrian explosion harbors something of an ambiguity. On the one hand, by citing a well-known event in evolutionary biology, Ray sustains the more or less automatic assumption that his work contributes to our understanding of basic biological questions about the origin of life and its evolutionary path to a wide diversity of species. On the other hand, with the success of Tierra, it has become clear that Ray's explicit aim is the production of complexity, understood simply as an increasing diversity of interactions among an increasing diversity of organisms, or agents. The best medium for realizing this aim is the digital medium of computers. For Ray, it would seem, the actual realization of this complexity takes precedence over whatever it might mean in relation to the processes of the organic world. As he bluntly states, "The objective is not to create a digital model of organic life, but rather to use organic life as a model on which to base our better design of digital evolution."[27] While references to organic life continue to have an obvious legitimating function—after all, Ray is a bona fide biologist working within well-defined disciplinary constraints—they also have a secondary or subsidiary status, since the central objective is to produce complexity.

This somewhat ambiguous perspective surfaces directly in Ray's research after Tierra, when a literal doubling takes place. In 1995 he issued "A Proposal to Create a Network-wide Biodiversity Reserve for Digital Organisms." His basic idea was to extend Tierra to the Internet by taking advantage of the fact that at any given time there are thousands of "idling" machines on the network that could provide spare CPU cycles. The objective would be "to set off a digital analog to the Cambrian explosion of diversity, in which multi-cellular digital organisms (parallel processes) will spontaneously increase in diversity and complexity."[28] If successful, he continues, "this evolutionary process will allow us to find the natural form of parallel and distributed processes, and will generate complete digital information processes that fully utilize the capacities inherent in our parallel and networked hardware." At the same time, and in clear parallel, he issued "A Proposal to Consolidate and Stabilize the Rain Forest Reserves of the Sarapiqui Region of Costa Rica." The purpose of this project would be to "prevent the imminent destruction of some of the last remaining large areas of rain forest ... in Northern Costa Rica," and "to establish a conservation economy through a community based nature tourist project." While the obvious parallelism of the two projects suggests an equal importance, their very formulation lends itself to a probably unintended interpretation: while organic life now requires protection and preservation, nonorganic life simply requires the opportunity and conditions in which it can emerge and foment complexity.

Tellingly, at no point in the Internet Tierra proposal does Ray mention or allude to the simulation of life; instead, he simply reiterates his aim to use evolution "to generate complex software." Though he mentions the Cambrian explosion, it serves only as a convenient and well-known general model of complexity. For Ray, in fact, the specific objective was "to engineer the proper conditions for digital organisms in order to place them on the threshold of a digital version of the Cambrian explosion." The global network of the Internet, because of its "size, topological complexity, and dynamically changing form and conditions," presents the ideal habitat for this kind of evolution. Under these propitious conditions, Ray hoped, individual digital organisms would evolve into multi-celled organisms, even if the cells that constitute an individual might be dispersed over the net. The remote cells might play a sensory function, relaying information about energy levels [i.e., availability of CPU time] around the net back to some "central nervous system" where the incoming sensory information can be processed and decisions made on appropriate actions. If there are some massively parallel machines participating in the virtual net, digital organisms may choose to deploy their central nervous systems on these arrays of tightly coupled processors.

Furthermore, if anything like the Cambrian explosion were to occur on Internet Tierra, then we should expect to see not only "better" forms of existing species of digital organisms but entirely new species, or forms of "wild" software, "living free in the digital biodiversity reserve." Since the reserve would be in the public domain, anyone willing to make the effort would be able to observe and even "attempt to domesticate" these digital organisms. While domestication would present special problems, Ray foresaw this as an area where private enterprise could get involved, especially since the organisms could conceivably become "autonomous network agents."

After several years of operation, Internet Tierra did not prove to be as dramatically successful as the closed-world version, mainly because of difficulties developing the necessary parallel-processing software. Yet the results were in some ways astonishing. At the ALife VI conference in 1996, Ray and colleague Joseph Hart reported on the following experiment:

Digital organisms essentially identical to those of the original *Tierra* experiment were provided with a sensory mechanism for obtaining data about conditions on other machines on the network; code for processing that data and making decisions based on the analysis, the digital equivalent of a nervous system; and effectors in the form of the ability to make directed movements between machines in the network.[29]

Tests were then run to observe the migratory patterns of these new organisms. For the first few generations, these organisms would all "rush" to the "best-looking machines," as indeed their algorithms instructed them to do. The result was, as Ray called it, "mob behavior." Over time, however, mutation and natural selection led to the evolution of a different algorithm, one that simply instructed the organism to avoid poor-quality machines and consequently gave it a huge adaptive advantage over the others.

Internet Tierra makes fully explicit a central objective in Ray's work: to deploy evolutionary strategies like natural selection in the digital medium in order to bring into being a quasi-autonomous silicon world of growing complexity. While the digital organisms of this world are of essential interest to ALife and evolutionary biology, their wider significance exceeds the boundaries of these disciplines and perhaps the frame of scientific research itself. Indeed, the necessary constraints of the latter can even inhibit us from seeing how these organisms participate in a much larger transformation, or coevolution, of technology and the natural world.

Echo: Modeling A Complex Adaptive System

Tierra is an exemplary instance of what has come to be known as a complex adaptive system. Such systems have been and remain a privileged focus of research at the Santa Fe Institute. If any single person can be said to be responsible for this central and overarching interest, it would be John Holland, a frequent visitor for many years. Best known for his invention of genetic algorithms, Holland has devoted much of his professional career to the study of complex adaptive systems. In *Adaptation in Natural and Artificial Systems*, he offers four broad distinguishing features:

> 1. All complex adaptive systems involve large numbers of parts undergoing a kaleidoscopic array of simultaneous nonlinear interactions.
>
> Because of the nonlinear interactions, the behavior of the whole system is not, not even to an approximation, a simple sum of the behaviors of its parts....
>
> 2. The impact of these systems in human affairs centers on the aggregate behavior, the behavior of the whole.
>
> Indeed, the aggregate behavior often feeds back to the individual parts, modifying their behavior....
>
> 3. The interactions evolve over time, as the parts adapt in an attempt to survive in the environment provided by the other parts.
>
> As a result, the parts face perpetual novelty, and the system as a whole typically operates far from global optimum or equilibrium....
>
> 4. Complex adaptive systems anticipate. In seeking to adapt to changing circumstances, the parts develop "rules" (models) that anticipate the consequences of responses....[30]

Because such systems are not easily modeled or described mathematically, they are usually studied using computer simulations. Holland, however, has built a general model of a complex adaptive system, which he calls Echo. Though "designed primarily for gedanken experiments rather than precise simulations" (186), Echo has been implemented as a computer simulation.[31]

Echo consists of a flat spatial expanse on which a number of simple agents are distributed around specific sites, like pieces on a game board. Each agent possesses certain attributes defined by "tags" and "conditions" inscribed in its "chromosomes," which enable it to interact in three ways: fight, trade, or mate. Each agent also possesses a "reservoir" of resources, which it attempts to augment. After collecting a predefined quantity, it can reproduce. Agents interact with specific sites in the environment by drawing, or "uptaking," resources from a site, but they must also pay a small "maintenance" cost, which is subtracted from the reservoir of their resources. If an agent cannot pay this cost, it is deleted from the site. However, it is also possible for an agent to migrate to another site, where more resources are available. More frequently, agents simply accumulate or exchange resources with other agents. Each agent's specific capacities are encoded in symbol strings ("chromosomes") that determine its external "phenotypic" properties. Its traits are encoded in tag chromosomes; its possible responses to other agents, in condition chromosomes. The particular configuration of these chromosomes—whether and how they match—determines how any two agents will interact in an encounter. For example, if the "combat condition" of either agent matches the "offense tag" of the other, then combat is initiated. There are many ways of scoring the outcome. In the simplest case, a rapid calculation of resources and attributes leads to a winner, who collects the loser's resources; the loser is then deleted. However, if combat does not take place, then the conditions and tags of each of the agents are checked for a possible trade. Unlike combat, which can be initiated unilaterally, trading can only be bilateral. If the proper conditions for both agents match, they can exchange any excess of resources they possess over and above what is necessary for their own reproduction. Although agents can also reproduce asexually, when they mate there is an exchange and recombination of chromosomes. As with trading, mating is always bilateral, in which conditions must be met on both sides.

These simple rules of interaction quickly give rise to complex behavior among the agents, including (Holland notes), "ecological phenomena (e.g., mimicry and biological arms races), immune system responses (e.g., interactions conditioned on identification), evolution of metazoans (e.g., emergent hierarchical organization), and economic phenomena (e.g., trading complexes and the evolution of 'money')" (186). This complexity results from "surprisingly sophisticated evolutions," which produced "evolving sequences of agents with ever longer, more complicated chromosomes, accompanied by a corresponding increase in the complexity of their interactions" (193). Following this evolution, a biological arms race ensued, in which defensive tags increased in length, in turn necessitating "the development of ever more sophisticated matches to overcome the increasing defensive capabilities" (193). More recent versions of the model saw the development of connected communities of agents that have internal boundaries but reproduce as a unit. This led in turn to specialization among the agents, some in offensive capabilities, others in resource acquisition. Overall, a net increase in the reproduction rate for both types resulted, confirming the generally accepted idea that evolution often favors cooperation and multiagent networks or communities.

While intended only as a general model on which different kinds of simulations and experiments could be run, Holland's Echo exhibits one obvious similarity with Ray's Tierra: there are no explicit fitness functions. In Echo, the reproductive rate of an agent depends solely on its ability to gather the necessary resources through interactions with specific sites and other agents.[32] Its fitness, therefore, is a function of its inherited traits, which enable it to survive and reproduce in its immediate environment and which it can exchange and pass on to its progeny. But Echo and Tierra also differ in several important respects. As both Ray and Langton have insisted, the Darwinian evolutionary process is directly instantiated in Tierra and not merely modeled or simulated. This enables Ray to speak of Tierra's digital organisms as in some way alive rather than as merely lifelike. In contrast, Holland thinks of Echo as only a model that can support various simulations and readily points to its cartoonlike aspects. This difference is reflected in their respective presentations. Whereas Ray describes in detail how the digital organisms are coded, how they function in the computer, and how they emerge as an evolving ecology, Holland is primarily concerned with the features of the model, specifically the attributes of the agents and how they interact. He barely mentions the computer, and then only to say that "Echo, and other models of complex adaptive systems, are readily designed for direct simulation on massively parallel computers" (196). These differences reinforce the interpretive differences just mentioned. Composed of computer code, Ray's digital creatures are machinelike and cannot be abstracted from their machine environment. From the point of view of a strong theory of ALife, however, they are not simply lifelike but a new *form* of life, not because they imitate natural life forms but because they perform functions that actually (in Ray's view) define life, that is, they replicate, mutate, and evolve to form ecologies of diverse organisms. As informational entities composed of code, they exhibit machinic rather than natural life.[33] Holland's agents, in contrast, are lifelike in a directly mimetic and representational sense. For one critic, in fact, Holland's agents too easily reflect the ideology of possessive individualism and a capitalist economy.[34]

When Holland presented his model at the Santa Fe Institute's conference on complexity, questions from the audience focused on two aspects of Echo that are particularly relevant here.[35] First, Kauffman wanted to know about effects of species interaction. He noted that in his own model of coevolution when every species interacts with every other species, the system tends to go chaotic. Contrarily, "by tuning down the number of species which interact with one another," he could push his system into an ordered regime, adding that there is "evidence from real food webs that species work quite hard at controlling the number of other species that they interact with." Implicit in Kauffman's comments is the idea that there may be an ideal or optimal number of interspecies interactions somewhere between high and low that would drive the ecology to the edge of chaos, thereby promoting higher levels of fitness and adaptability. Although Holland responded that "there's some stuff that's relevant" ("Echoing Emergence," 334), Echo does not seem able to address species interaction, certainly not with the clarity evinced by Tierra when Ray varies the mutation rate.

A query from Leo Buss brought Ray's system directly into the discussion. Buss first returns to his colleague Walter Fontana's earlier observation that a great deal of complexity is built into rather than generated by Holland's model, specifically by the tags, which require "entire levels of organization." On the other hand, because Holland's agents have a phenotype (defined specifically by the tags) his system is substantially richer than Ray's. This follows from the fact that, with phenotypes, selection operates on traits rather than on individual creatures or agents and thus produces greater variation and hence greater richness. At the same time, there is a drawback, since the system has not generated the phenotypes—Holland has simply built them in. As Buss adds, this gap is precisely where "biology lacks theory." For Fontana and Buss, it appears, the real issue lies not in the consequences of the tags in terms of "the dynamics of the individual entity class" (338), but in how to generate or evolve the tags.

Generalizing, we could say that for Fontana and Buss the complexity of the Echo system doesn't arise far enough from the bottom up, since the mechanism of the tags is "injected" at an intermediary level, as a given of the system. (A similar criticism can be made of Tierra, of course, which is seeded by an Ancestor whose reproductive code is complete from the start.) Nevertheless, while Echo's individual agents are only capable of three types of action, their collective interactions result in the emergence of complex adaptive behavior. This complex behavior not only exceeds what the individual agents are capable of but also operates globally: it changes the fitness landscape. As in Tierra, this emergent global behavior modifies environmental conditions at the local level, to which individual agents must henceforth adapt through a constant turnover in trait selection. Thus even though the set of primitive interactions (fight, mate, or trade) does not evolve, the mechanism of trait selection generates new combinations that allow more complex adaptive behavior to emerge. At least up to a point.[36] In Tierra the mechanism of evolution is not trait selection but genome selection, with the motor of change operating at the deeper level of the organism as a whole: the computer code defining the "genotype" of each individual organism is rewritten through mutation (the genotype is the code; the phenotype, the result of its execution), thus producing new organisms subject to adaptive pressures and hence natural selection. The result, as we saw, is a dynamically changing ecology. In two different ways, then, Echo and Tierra exemplify the feedback loop from local to global back to local that Langton describes. Complex behavior emerges at the global level, sets new constraints at the local level, and thereby alters the fitness landscape. While the behavior-generating mechanisms in the two systems differ fundamentally, the dynamically evolving behavior that results make both of them instances of complex adaptive systems.[37]

This leads to what we might call the Santa Fe perspective: what counts is not whether or not the system is a living biological system but whether it exhibits the properties of a complex adaptive system. Cutting across natural and artificial realms, the concept of a complex adaptive system often encompasses a multiplicity of heterogeneous relationships among living and nonliving agents and processes, as in Holland's example of New York City cited in this book's introduction. His other examples include the central nervous system, the immune system, a business firm, a species, and an ecology, with their operative time scales ranging respectively from seconds to millennia. In an amplification of his earlier work, Holland examines four properties and three mechanisms common to all complex adaptive systems.[38] Most of them are embodied in Echo and need not be laid out here. However, Holland also extends the term adaptation to include learning (9–10), which is significant here in that the relationship of learning to evolution was an important theme in early ALife research, most notably in David Ackley and Michael Litman's model of evolutionary reinforcement learning (ERL). Resembling Echo in some respects, the ERL model incorporates mechanisms that measure the extent to which agents can learn and act appropriately within a larger evolutionary framework.[39] This groundbreaking research suggests that "learning and evolution together were more successful than either alone in producing adaptive populations" (487). Unfortunately, the relationship between learning and evolution figures much less prominently in more recent ALife research.[40]

If, as complex adaptive systems, Tierra and Echo evolve into states of greater complexity by different mechanisms, then several obvious questions follow. For example, are the criteria sketched above the ones that should be foremost in making "lateral" comparisons among different complex adaptive systems, or does some kind of learning or anticipatory mechanism need to be included? And can similar kinds of dynamical systems where local interactions produce global effects be stacked vertically, so that emergent systems operate "above" and englobe other emergent systems? Basically, this is the question of how hierarchies in dynamical systems should be defined and experimentally generated.[41] But looming over this entire discussion is the key question of how emergence itself should be defined and whether it is a universal type of phenomena. To a great extent, these questions are still being engaged by ongoing research within current ALife and complex adaptive systems theory. Here, however, some basic questions about how ALife systems work remain to be pursued. Specifically, given the computational nature of such systems, we need to consider how computational processes function in an emergent system and how these computational processes contribute to the emergence of the dynamical behavior of the system as a whole.

What's New? Emergent Computation

Several papers presented at the Santa Fe conference at which John Holland presented Echo attempt to grapple with the question of how to define emergence. The most ambitious, James Crutchfield's "Is Anything Ever New? Considering Emergence," focuses on computation in emergent systems, centering specifically on the difference between the perception or appearance of a new pattern in physical phenomena and the detection of an "emergence of coordinated behavior and global information processing."[42] Combining rigorous science with epistemological sophistication, Crutchfield distinguishes among (1) intuitive notions of emergence according to which "something new appears," (2) an observer's identification of pattern or "organization" in a dynamical system, and (3) what he calls "intrinsic emergence," where a new pattern brings about an increase in the system's computational capacity, thus giving it additional functionality. Crutchfield then discusses a method by which intrinsic emergence can be measured.[43]

At the outset Crutchfield distinguishes between a pattern having importance only in the eye of an observer and one having importance *within* the system in which it emerges. In the former case, the pattern's "newness" is attributed by the observer, who anticipates and recognizes it in relation to other patterns or to a "fixed palette of possible regularities." For Crutchfield, the novel patterns that emerge from deterministic chaos (like the owl face pattern of the Lorenz strange attractor), the self-similar "fractalness" of the random walk, the self-organizing material states like the Belousov-Zhabotinsky chemical clock, and the studies of solitons in turbulence all fall into this first category. In the second category, that of intrinsic emergence, the pattern's newness must be defined in relation to other structures within the same system. Since there is no external referent for novelty or pattern, this type of emergence is "intrinsic." More specifically, intrinsic emergence confers "additional functionality which supports global information processing." In any instance of intrinsic emergence, therefore, there is an "increase in intrinsic computational capability, which can be capitalized on and so can lend additional functionality" (518). For example, competitive agents in a capitalist market economy develop strategies based on the pricing of goods and services that emerges from their collective interactions. Inasmuch as pricing reflects all the available information in the system, it is an instance of global information processing (i.e., intrinsic computation) that in turn becomes the basis for better investment strategies. Examples of this second type, Crutchfield adds, fall under the rubric of what is called emergent computation. This important concept will be considered momentarily; here let it suffice to say that emergent computation refers to a type of information processing that arises from the interactions of a large number of simple agents or elements without any centralized external control. Emergent computation can thus be said to occur when a large number of local interactions with limited computational and communicational capacities give rise to coordinated global information processing. Insect colonies, immune systems, the brain, cellular automata, and the capitalist market system—all evince forms of emergent computation.

Clearly the role and position of the observer vis-à-vis both the observed phenomenon and the assumed model differ significantly in the two types of emergence. In the first type, the relation of the observer to the phenomenon, or pattern, and the model, whether assumed or explicit, is external; that is, the observer perceives the phenomenon and compares it with a preexistent model. In the second type, the observer is *inside* the system, one of its subprocesses so to speak, and the model is embedded in the observer. This is a logical requirement, since the observer must have "the requisite information processing capability with which to take advantage of the emergent patterns" (519).

Crutchfield next proposes a method by which to detect and quantify intrinsic emergence in nonlinear processes. The method is distinguished by two aspects, which make it very similar to his earlier work on ϵ-machine reconstruction: a notion of structure based on a "computational mechanics of nonlinear processes"; and a mechanism or process he calls "hierarchical machine reconstruction" for measuring its complexity. The method works roughly as follows. Isolating a process in nature, an observer makes a series of discrete-time, discrete-space measurements. In order to extract structure from these measurements the observer can assume that an information-processing architecture is embedded in the states that the measurements reflect. This means that, "given a discrete series of measurements from a process, a machine can be constructed that is the best description or predictor of this discrete time series" (523). The structure of the machine would thus provide "the best approximation to the original process's information processing structure" (523). The more complex the structure, the more complex the machine required to recognize it. Fortunately, computational theory distinguishes different classes of machines on the basis of precisely this correlation. (Chomsky's computational hierarchy is the best-known example.) Basically, then, a data stream exhibiting dynamic patterns is mapped as a function of the formal language and corresponding computational machine or automaton required to recognize or compute it. As expected, patterns of increased complexity require a more complex type of formal language for their description. In sum, since the architecture of the reconstructed machine itself "represents the organization of the information processing," it can provide "a model of the mechanisms by which the natural process manipulates information" (523).

In considering the relevance of his method to contemporary debates on biological evolution, Crutchfield notes that three schools of thought on the guiding mechanisms in Darwinian evolution can account for the emergent complexity visible in "the present diversity of biological structure" (527). Selectionists, represented by John Maynard Smith, are the heirs of classical Darwinian theory brought up to date by genetics. Although natural selection based on a notion of fitness culls the destabilizing effects of genetic variation, this approach offers no theory of structure. Historicists, represented by Stephen J. Gould, accept the Darwinian mechanisms of variation and selection but insist that change is often nonadaptive and that accidents and contingencies play an ultimately determinant role in the shaping of biological form. Again, there is no theory of structure. Finally, structuralists represented by Brian Goodwin and Stuart Kauffman argue that on many levels biological development is both enabled and constrained by principles of self-organization, that is, by structural attractors that lie in wait in a possibility space. The role of natural selection is reduced to "choosing between these 'structural attractors' and possibly fine-tuning their adaptiveness" (527). While lacking a theory of transformation, this approach does yield a theory of structure. However, despite an explicit interest in organization and structural archetypes, Crutchfield finds structuralism inadequate since "the structure archetypes are neither analyzed in terms of their internal components, nor in terms of system-referred functionality" (528).

Reviewing these three basic positions on evolution, Crutchfield concludes that "there is a crying need for a theory of biological structure and a qualitative dynamical theory of its emergence" (528–529). Since computational mechanics provides a theory of structure, and hierarchical machine reconstruction provides a theory of its transformation, together they are well suited to "study what drives and what constrains the appearance of novelty" (529). Such a study of the dynamics of innovation, he adds, might well be called evolutionary mechanics. If it could be applied to the biological domain, it might point toward a synthesis of the conflicting viewpoints within the three schools. In fact, expanding on his critique of the latter, Crutchfield later demonstrated how hierarchical ϵ-machine reconstruction can be "folded into" an evolutionary process and thus applied to the evolution of complexity.[44] Going beyond Kauffman, Crutchfield establishes that "Darwinian evolution cannot, in and of itself, produce novel biological structures and functions" (102); however, by using the computational tools he has developed, epochal evolution can be understood as an "open-ended process of discovering and stabilizing novelty" (128). Moreover, it is not necessarily a process driven by external forces. As we'll see, this theoretical position finds significant resonance in several ALife experiments.

Considered more widely, Crutchfield's work on intrinsic emergence marks a notable turning point. While theories of emergence and complexity have attracted considerable attention, there is still no scientific consensus on their status. No doubt, structures of self-organization arise spontaneously in many nonlinear dynamical systems—in all complex adaptive systems, it seems—but scientists have not yet been able to establish incontrovertibly that emergence and complexity are underlying universal phenomena rather than unique if widespread instances. In fact, doubts about the viability of a unified theory of complex systems have been publicly aired, perhaps most visibly by John Horgan.[45] Given that scientific method requires rigorous skepticism, Horgan's objections should be salutary, but unfortunately he offers too one-sided an account and fails to report on a number of promising avenues of current research. Referring to Christopher Langton's and Norman Packard's work with cellular automata, work that concludes that "a system's computational capacity—that is, its ability to store and process information—peaks in a narrow regime [the space of a phase transition] between highly periodic and chaotic behavior" (106), Horgan cites criticisms by Crutchfield and Melanie Mitchell, who were unable to repeat and thus verify the results of Langton's and Packard's edge-of-chaos research. However, Horgan fails to mention not only the context in which these criticisms were made but also, and more importantly, the promising work these two scientists themselves have done in the area of emergent computation, specifically in relation to complex adaptive systems and the assumed conflict between the computational and dynamical approaches to systems theory. Since their investigations constitute one of the most fruitful paths that complexity theory has taken, I shall now consider it in some detail.

The very idea of emergent computation immediately recalls two basic questions that Langton posed: how does information processing arise and gain control over the dynamics of a system, and what are the most propitious conditions for its emergence?[46] The term, however, owes its current usage not to Langton but to Stephanie Forrest, who organized the conference "Emergent Computation: Self-Organizing, Collective, and Cooperative Phenomena in Natural and Artificial Networks" in 1990 at the Center for Nonlinear Studies at Los Alamos National Laboratory.[47] This conference followed two other groundbreaking conferences sponsored by the center mentioned in earlier chapters: the 1983 conference on cellular automata organized by J. Doyne Farmer, Tomas Toffoli, and Stephen Wolfram; and the 1986 conference "Evolution, Games, and Learning: Models for Adaptation in Machines and Nature," organized by J. Doyne Farmer, Alan Lapedes, Norman Packard, and Burton Wendroff.[48] The latter was especially important for emergent computation, addressing issues of evolutionary adaptation through learning in both machines and natural processes, with a key paper (discussed in the previous chapter) on an artificial immune system. Significantly, Langton presented papers on cellular automata at all three conferences. While he doesn't broach all of the problems and issues on the table at these conferences, his work is clearly central to them. It seems appropriate, therefore, to begin the discussion of emergent computation with Crutchfield and Mitchell's critique of Langton's work.

Melanie Mitchell, a computer scientist who had worked with John Holland and Douglas Hofstadter, had come to the Santa Fe Institute to pursue research on genetic algorithms. Thus the pairing of Mitchell and Crutchfield would seem to be the perfect combination for investigating emergent computation in evolutionary systems, whether natural or artificial. As they themselves state, "Our ultimate motivations are both to understand emergent computation in natural systems and to explore ways of engineering sophisticated emergent computation in decentralized multiprocessor systems."[49] Crutchfield, we should recall, began his research as a physicist in the late 1970s with investigations into deterministic chaos. Early work in this new science had revealed that the onset of chaos often occurred as a phase transition characterized by rapid bifurcations and period doublings in a dynamical system's phase space. Working with colleagues like Karl Young in the late 1980s, Crutchfield proved that a system's computational capacity increased significantly at the onset of chaos, attaining what he called "statistical complexity."[50] Simply put, at moments of rapid change in a system's physical state—in the transition, say, from solid to fluid—the system can be understood as a finite-state automaton with the capacity to store and transmit large amounts of information globally, that is, from one part to another throughout the entire system.

As Langton acknowledges in "Life at the Edge of Chaos," Crutchfield's research on computation and chaos parallels his own. There were, of course, differences. Working with cellular automata, Langton sought correlations between information processing (changes in the CA rule space) and the dynamic behavior of the CA (evident in the changing visual patterns). Working with physical systems under the influence of chaotic attractors, Crutchfield wanted to develop precise measurements of the increases in information-processing capacity of those systems as they became more chaotic. Whereas in Langton's CA changes occur over discrete time steps, in the physical systems Crutchfield considers, dynamical changes occur continuously. This is significant inasmuch as the terms Langton uses to describe CA behavior (i.e., the attractor states in dynamical systems theory, first applied to CA behavior by Stephen Wolfram) were developed for continuous-time, continuous-state systems, not discrete-time, discrete-state systems like cellular automata. While the consequences of this importation are not immediately evident, what does become an issue is whether or not Langton's experiments were defined rigorously enough to provide unambiguous proof of his claims. In short, did the results fully justify his claim that the information-processing capacity of the CA measurably increased as its dynamic behavior approached the edge of chaos?

The computational capacity of CA at the onset of chaos was also at the heart of Norman Packard's research, and Crutchfield and Mitchell's first joint investigations revisited both Langton's and Packard's experiments. Since I intend to focus on work they have done following their critique of these experiments, I shall simply summarize their findings. In neither case, they found, could they reproduce the experimental results and therefore could not confirm the claims advanced on its basis. But while Horgan's article implies that the claims themselves are invalid, what Crutchfield and Mitchell actually do is to pinpoint the inherent simplification and vagueness in the experimenters' assumptions that made the results unrepeatable. "Those negative results," Mitchell writes, "did not disprove the hypothesis that computational capability can be correlated with phase transitions in CA rule space; they showed only that Packard's results did not provide support for that hypothesis."[51] In fact, Mitchell cites the work by Crutchfield and Young mentioned above that demonstrates the validity of this hypothesis, though not by using cellular automata.

The central problem in both Langton's and Packard's work arises from the way the correlation between the dynamic behavior of the CA and its computational capacity is defined and demonstrated. In Langton's case, where the correlation is measured by the lambda parameter (discussed in chapter 4), Mitchell and Crutchfield demonstrate on both theoretical and empirical grounds why it is too crude. For one thing, the assumption that the "CA rule tables themselves are the appropriate loci of dynamical behavior" is unwarranted.[52] For another, the actual behavior of the CA is averaged in a way that obscures the meaning of the correlated results. Finally, the imputed increase in computational capacity that occurs as lambda approaches certain critical values ("the edge of chaos") is not defined for any specific computational task. In Packard's case, the desired computational task of the CA (described below) is defined at the outset, and genetic algorithms are evolved as the best means to accomplish it. Since the best performing genetic algorithms have lambda values very close to the critical values that correspond to class 4 CA, Packard interprets the results as confirming the edge-of-chaos hypothesis. But again, Mitchell and Crutchfield show that the experimental results do not justify this conclusion. They do not deny that as the CA behavior approaches the edge of chaos threshold the computational capacity of the CA increases—in fact, they acknowledge that it does for certain computations—but only that these results do not confirm the hypothesis as stated. In short, the hypothesis "has not been rigorously formulated" (510), and other factors explain the correlation.

Nevertheless, Packard's specific computational task and his deployment of genetic algorithms to solve it have proved useful for subsequent work that Crutchfield and Mitchell themselves carried out. The computational task is simple but well suited for determining emergent computation, understood as "collective information-processing abilities that emerge from the individual actions of simple components interacting via restricted communication pathways."[53] Given that the CA forms a lattice of cells, with each one in an on or off state (i.e., 1 or 0) at a particular moment in time, the task is to determine whether, at an initial configuration, a majority of the cells are on or off. If the majority is on, then the desired behavior is for the remaining cells to change to the on state; if, on the other hand, the majority is off, then the desired behavior is for all the remaining cells to change to the off state. For computational systems with a central processor (like an ordinary digital computer) or a neural network with global connectivity, it would be easy to design an algorithm to perform this task. But for cellular automata, where the interactions among cells are always and only local, this is a very difficult task and becomes more so as the size of the cell array increases. The reason is obvious: as the lattice increases in size, the CA must communicate information over larger space-time distances, and without a central processor there is no easy way to coordinate information from distant parts of the array. Packard solved the problem by evolving a genetic algorithm that would provide the most efficient CA rule (or transition-state table) leading to the desired transformation; in short, it would map the neighborhood state configuration to an update state of all 1s or 0s. Since Packard was working with one-dimensional cellular automata of finite size, the transition-state table was a string of 1s and 0s of only moderate length.

According to Mitchell and Crutchfield, Packard's work was meant to test two hypotheses: "(1) CA rules able to perform complex computations are most likely to be found near [lambda critical] values; and (2) when CA rules are evolved to perform complex computations, evolution will tend to select rules near [lambda critical] values" (505). But, as already noted, when they repeated the experiment, they were unable to obtain the same results and were forced to explain the differences and offer an alternative interpretation of Packard's data. Their subsequent work, however, both corrects and extends the work of Langton and Packard by putting it on a more mathematically precise and better-grounded theoretical foundation. In a series of experiments they have followed Packard's efforts to evolve genetic algorithms to perform the specific computational task described above, usually referred to as the $p = 1/2$ task. But whereas Packard sought to build or engineer emergent information processing in decentralized, spatially extended systems like CA, their work is intended to detect such behavior. As a consequence, it is better suited to provide a more precise understanding of how globally coordinated computations in such systems can evolve on their own. To accomplish this they developed "a novel technique for analyzing particle-based logic embedded in pattern-forming systems" ("The Evolution of Emergent Computation," 1). Although technically complex, at its core the technique involves a means of revealing vectors of pattern formation as the CA change over time. Specifically, patterns quickly appear that are composed of areas of black (1s), white (0s), and gray (combinations of 1s and 0s). A filter is then applied that screens out these large areas, leaving visible only the edges and lines of intersection. These edges and intersecting lines are then analyzed as streams of particles traveling across the lattice that carry information (or signals) over space-time distances. The visualization of a CA pattern changing over time and the diagram that provides the filter (reproduced from "The Evolution of Emergent Computation") illustrate the method (figs. 5.1 and 5.2).

Points of intersection in figure 5.2 mark sites where the signals perform logical operations, much as colliding streams from the glider guns in Langton's CA experiments are used to perform the operations of logic gates. These logical operations in fact constitute density mapping operations: when a high-density particle collides with a low-density particle, the particle that emerges maps this information onto another signal moving toward a different spatial configuration. According to the logic of this implementation, "the collection of domains, domain walls, particles, and particle interactions for a CA represents the basic information processing

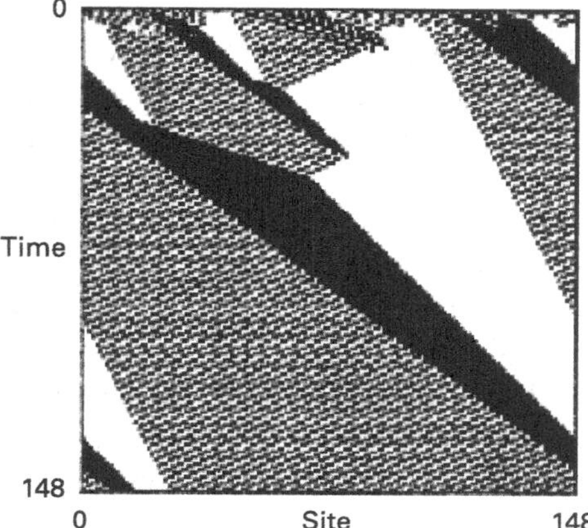

Figure 5.1
Evolving CA. James P. Crutchfield and Melanie Mitchell, "The Evolution of Emergent Computation," SFI Technical Report 94-03-012, 1.

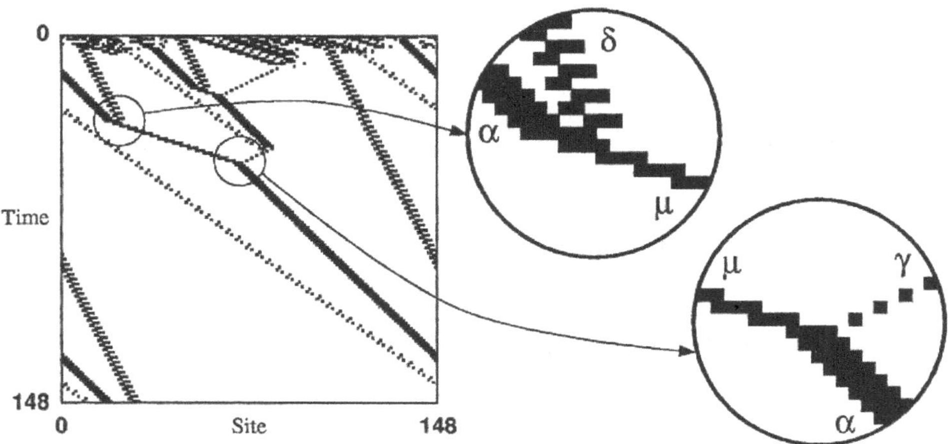

Figure 5.2
Evolving CA filter. James P. Crutchfield and Melanie Mitchell, "The Evolution of Emergent Computation," SFI Technical Report 94-03-012, 1.

elements embedded in the CA's behavior—the CA's 'intrinsic' computation" (6). Analyzing a number of different instances of "particle-based computation," Crutchfield and Mitchell have been able to demonstrate how "(1) complex global coordination can emerge within a collection of simple individual actions; and (2) how an evolutionary process, by taking advantage of certain nonlinear pattern-forming propensities of CA, can produce this new level of behavior through a succession of innovations that build up the delicate balance [between order and chaos] necessary for effective emergent computation" (10).

Crutchfield and Mitchell's particle-embedded approach to detecting emergent computation represents a major step forward in making complexity theory and emergence directly amenable to scientific investigation. Specifically, in pursuing a theoretically rigorous and empirically precise method for detecting emergent computation in decentralized, spatially extended systems, they develop a tool that synthesizes dynamical systems and information-processing approaches. Both scientists, in fact, have denied that there is any necessary or inherent dichotomy between the two approaches.[54] Indeed, their work suggests that evolutionary adaptation can perhaps best be explained by their synthesis. More generally, by adumbrating a unified framework in which dynamical systems, computation, and the evolutionary process can be articulated and understood together, they are pursuing what is perhaps the most fruitful direction for further research with evolving complex systems. Experiments with ALife systems confirm the importance of this framework. In Christoph Adami's Avida and Andrew Pargellis's Amoeba in particular we see how evolution leads to two different kinds of complexity.

Avida: Evolving Complexity in the Digital Genome

A physicist at the California Institute of Technology, Christoph Adami first began to perform ALife experiments with Thomas Ray's freely available Tierra system. This led him to develop Avida, a next-generation system conceived in the same spirit and set up to do experiments in open-ended evolution.[55] Like Tierra, Avida basically consists of self-replicating blocks of code that execute their instructions on a virtual machine within a digital computer. However, as in Tierra, because these individual blocks of code are subject to mutation and "death" (erasure) and compete with one another for CPU time and memory space, they are rightfully treated as digital organisms. As we saw earlier, Ray's digital creatures reshape the fitness landscape through their dynamic interactions and thereby increase the complexity of their environment's fitness landscape. This complexity is global, a result of "creatures adapting to other creatures which are themselves evolving," as Ray puts it. In Avida the same kinds of dynamic interaction occur, but the software allows the experimenter to track increases of complexity not just at the global level, as in Tierra, but at local levels as well.

To enhance Avida's functionality as a platform for ALife experiments, Adami and software developer Titus Brown introduced a spatial structure and local update rules.[56] In Avida the digital organisms "live on a two-dimensional grid" (*Introduction*, 50), the structure of which is reminiscent of a cellular automaton. Hence, Adami states, "We term each grid point a *cell* or *organism*, and each associated list of instructions the *genome* of that organism" (301). Unlike Tierra, where any program or block of code can in principle interact with any other, in Avida the programs located at the grid points can only interact with the eight other programs in their neighborhood, and thus only propagate information locally. Death, or the reaper queue, functions similarly. Positioned at each lattice location, it can only delete cells in the immediate neighborhood. Avida also differs from Tierra in that its "update rules are not fixed but rather are dependent on the genome of the cells in the immediate neighborhood" ("Evolutionary Learning," 377). Whereas in Tierra each block of code (i.e., genome or instruction set) is given the same time slice—that is, it is allocated the same amount of CPU time to execute its code—in Avida, time slicing is a variable that can be set by the user. For example, it can be made constant (as in Tierra), dependent on the length of the genome, or size neutral. Most usefully, it can be made proportional to the organism's "merit," which is determined by the number of instructions copied into or executed by the organism multiplied by any "bonuses" that accrue in its interaction with the environment. For example, the user can reward organisms for performing computational and logical tasks beyond mere replication.[57] Each of these options will differently carve the fitness landscape generated by each run of the Avida system.

As in Tierra, a run begins when the Avida system is seeded with an ancestor program. It can be the default creature, **creature.base**, or one with a smaller instruction set, **creature.small**, or for that matter any other creature in **genebank**. What differs in each case is the genome, or instruction set, which can also be modified by the user. The default instruction set, written in assembly language, contains twenty-four instructions, although as many as forty-two are available. In addition to the "biological" instructions that enable the organisms to replicate (allocate memory, copy, divide, etc.), there are instructions for mathematical and logical operations as well as input and output. There is also a jump command that can move the instruction pointer from one organism's instruction set to another's (as long as it resides in the same neighborhood), thus allowing one organism to execute another's code, like forms of bacteria using one another's DNA. The default ancestor, **creature.base**, has a genome of thirty-one instructions and "carries a certain amount of redundancy in its code" (*Introduction*, 234). Importantly, this extra code provides additional raw material for the processes of mutation and natural selection. For one experiment Adami describes an ancestor of a hundred instructions "containing mostly nonsense code, from which all populations are spawned."[58] It turns out, however, that this extra, "junk" code is essential for producing genomic complexity.

The replication process is carried out in four stages: (1) allocation of new memory; (2) duplication of the parent program in the new memory, instruction by instruction; (3) division of the program into parent and child programs; and (4) placement of the child program into the cell lattice. All operations but the last are handled by the instruction set. The placement of offspring, however, involves the environment and brings with it both restrictions and options. First, the offspring cell can only be placed at one of the eight grid points that constitute the parent's neighborhood. The user sets where it is to be placed, by selecting one of four options in the **BIRTH_METHOD** variable in the **genesis** file: it can be set randomly; the offspring cell can replace the oldest cell in the neighborhood (in some cases this might be the parent cell); it can replace the cell with the highest value of age divided by merit; or it can be placed in an empty cell. There is also choice of how cells die. First, the **DEATH_METHOD** can be turned off so that a cell only dies by being replaced by a younger cell. Second, a cell can die at a certain "age limit," that is, after executing a specified number of instructions. And third (a variation on the second method), it can die after a specified number of instructions multiplied by its genome length. These options in **BIRTH_METHOD** and **DEATH_METHOD** enable the user to vary parameters that will considerably alter the dynamics of this virtual world and hence the resulting fitness landscape.

After the system is seeded with a self-replicating organism, the grid begins to fill with copies. (The grid is 60 by 60, and thus supports a total population of 3,600 organisms.) Very quickly, mutated creatures also begin to appear and replicate if capable. As alterations of the seed organism's code, mutations are lethal, neutral, or favorable with respect to replication. Once a mutant produces three copies of itself, its new genome is identified with a letter. Several types of screen allow the user to track this process in detail. The software also allows the user to extract, examine, and inject individual genomes. Both the mutation rate and the types of mutation—five are available in Avida—can also be adjusted by the user.[59] This tool in fact constitutes one of Avida's most essential features, for the system is designed precisely to study the relationship between mutation and genomic complexity. Recall that in Tierra the most effective mutants all had shorter instruction sets (i.e., genomes) than the Ancestor. Since the time slice, or CPU time, remained constant whatever the code length, mutants with a shorter code—or that could use the code of other cells to replicate—possessed a distinct advantage. The shorter the organism's code, the less it has to copy and the more copies of itself it can make in the fixed amount of time allotted. As Adami remarks, this situation leads to a very efficient optimization of code but discourages the evolution of complex code (*Introduction*, 303). The question, therefore, is not what makes for an efficient and successful replicator but under what circumstances does the acquisition of additional capacities—like the ability to do simple computations—also enhance replication in the long run. The answer clearly has to do with the nature of the environment. A replicator that is successful in a simple environment might not function well or at all in a more complex one. This means that the information inscribed in the genome must also be understood as information *about* the environment.

This information is stored in the genome and transmitted through replication. When an organism copies its genome into an offspring, information is transferred to that other site. In his *Introduction to Artificial Life*, Adami describes experiments that "investigate the mode and speed of this transfer in relation to the fitness of the genotype carrying the information, the fitness of the other genotypes near this carrier, and the mutation rate" (253). For example, when the fitness of the carrier genome is the same as those in the neighborhood, then information transfer can be modeled as a case of *diffusion*, describable in terms of a classic random walk. On the other hand, if the carrier is more fit, information spreads in sharply defined *wavefronts* that propagate at a constant speed. The various dynamics of information propagation let Adami determine the characteristic time scale of the system, that is, the average time it takes for the system to return to equilibrium after a perturbation, or information propagation event. Although this "relaxation time" depends on the size of the system and the speed of information propagation within it, Adami discovered that "persistent mutation pressure" would always keep it far from equilibrium. Indeed, a subsequent series of experiments involves increasing the mutation rate "to the edge of an *error catastrophe*," where the probability that a string of code will acquire a mutation before being able to replicate itself is close to 100 percent. Surprisingly, increases in the mutation rate drove the population to greater adaptability right up *to* the catastrophic error rate. In this regard the experiments yielded results similar to Langton's edge-of-chaos research. Beneficial mutations that propagated increasing amounts of information through organism replication peaked just short of the "edge of an error catastrophe," while on the other side, errors in coding multiplied so fast that chaos in the form of dysfunctional code swamped the system.

Experiments with complexity thus lie at the heart of Adami's project. And like Crutchfield, Adami realized that complexity had to be defined precisely, in measurable terms. He also understood that complexity means one thing to physicists and computational theorists and something else to biologists.[60] Physicists who think about the complexity of dynamical systems usually resort to computational measures, since in principle all physical processes can be viewed as computations and vice versa. A well-known example is Kolmogorov, or algorithmic, complexity, which is measured by the length of the algorithm to which a symbol string or sequence can be compressed. In contrast, biologists who think about complexity consider form, function, or the symbolic sequence (DNA) that codes for the production of the organism.[61] Among biologists, furthermore, there is no clear consensus on whether or not evolution necessarily increases complexity.[62] Since there are no universal measures for structural or functional complexity in biology, Adami opts for a version of sequence complexity, even while acknowledging that "the difficulty of biology lies precisely in the intimacy of this map from sequence to function" ("What Is Complexity?" 1086). "It is very likely," he adds, "that a properly defined *sequence complexity* should mirror the complexity of the organism that the sequence gives rise to" (1086). If this conjecture is true, then Adami can adopt some of the mathematical methods used by physicists to measure complexity in dynamical systems and perhaps "bridge the gap between the physical and biological sciences" (1086).

Adami defines the physical complexity of a sequence as the amount of information about a particular environment stored in that sequence. He suggests that we think of the sequence as a tape. The "sequence entropy" of the tape measures the amount of information that the tape *could* hold and is indicated by the length of the tape. As the tape is filled with measurements (i.e., recordings) about a particular environment, the length of the tape containing the recordings would constitute a measure of this information. Because this information is correlated with a particular environment (for another environment it would be meaningless noise), it has predictive value for that environment. Quoting David Deutsch ("Genes embody knowledge about their niche") and E. O. Wilson ("[Organisms] encode the predictable occurrence of nature's storms in the letters of their genes"), Adami declares that this is precisely what physical complexity measures: "It is information about an environment that can be used to make predictions about it. Being able to predict the environment allows an organism to exploit it for survival. In such a manner, physical complexity translates into fitness for the organism" (1087).

Mathematically, Adami defines physical complexity as "the *shared Kolmogorov complexity* between a sequence, and a description of the environment in which that sequence is to be interpreted." At first glance this seems somewhat akin to defining the complexity of a book as the complexity of the information it contains plus a description of the cultural context in which it will be read and made use of. However, the context of Adami's definition greatly narrows its reference: "It is sufficient to think of the physical complexity of a sequence as the amount of information that is coded in the genomes of an adapting population, about the environment to which it is adapting" (1087). In fact, it is because Adami is not concerned with a single genome but a population of genomes that the entire evolutionary selection process can be understood as a kind of Maxwell's demon; that is, for a given environment, natural selection separates successful from unsuccessful replicators. As we would expect, when moments of increased replication are considered in an otherwise unchanging environment, genomic complexity in the population as a whole is observed to increase.

While acknowledging that there are serious questions about how complexity in biological living systems should be defined and measured, Adami believes that in artificial living systems the evolution of complexity can be explicitly observed by examining the effects of mutation on the genomic complexity of a population of digital organisms. In Avida specifically this is accomplished as follows. As a population of organisms self-replicates, some of these organisms will discover through random mutation a sequence of instructions that benefits their survival and increase. These new parts of the mutated code gradually become fixed, or frozen, in the organism's genome as it continues to replicate. These frozen parts of the code can be distinguished from parts that are still volatile, in the sense that the latter provide genomic diversity without storing any

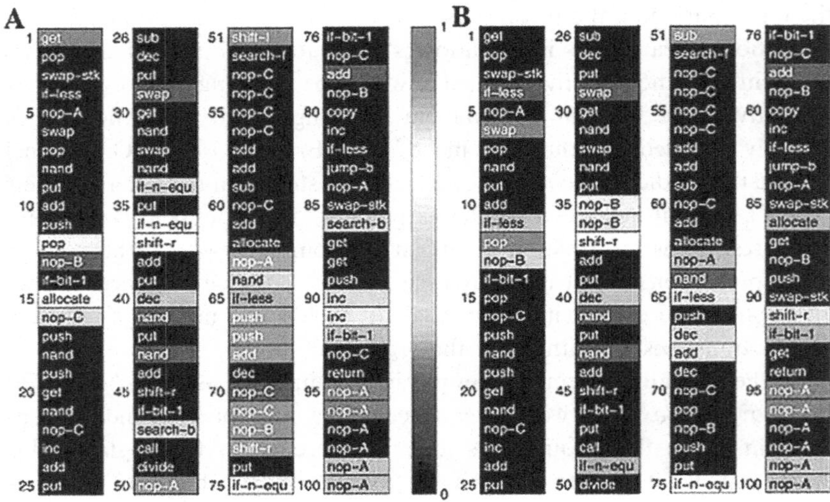

Figure 5.3
Two genomes according to their per-site entropy. Christof Adami, "What Is Complexity?" *BioEssays* 24, no. 12 (2002): 1092.

useful information about the environment. Adami notes, however, that "the determination of [the] volatility of a site is only possible *statistically*, i.e., by examining ensembles of members of the same 'species.' "[63] Significantly, in adaptive events in which the replication rate increases, the number of volatile instructions can be observed to decrease. This is illustrated in figure 5.3 ("What Is Complexity?" 1092), a before-after comparison of two genomes in which the instructions are color-coded in relation to their site-entropy—a measure of their fixity or volatility.

As Adami points out, instructions in blue and the cooler colors (here, dark) possess low entropy: mutations at those loci usually produce nonfunctional organisms. Contrarily, instructions in red and other warm colors (here, light) are highly entropic and thus volatile. Comparing the instructions of two extracted genomes, A and B, we can clearly see a "cooling off" in instructions 63–74 (in the lower half of the third column in B). This represents a visible gain in the number of fixed instructions and thus a gain of information about the environment. Since the number of nonvolatile instructions in a code string represents an estimate of the physical complexity of a particular species of string, this gain also indicates an increase of physical complexity. Adami summarizes the significance of these experiments as follows:

> In artificial living systems, the increase of physical complexity, which coincides with increasing acquisition and storage of information, can be monitored directly, and illustrates the usefulness of this measure. Note that this process of acquisition of information constitutes, in the language of thermodynamics, [...] the operation of a natural Maxwell-demon: the population performs random measurements on its environment, and stores those "results" that decrease the entropy, but rejects all others. Thus, the process can be likened to a semi-permeable "membrane" for information, and the physical complexity increases as a function of evolutionary time (given a fixed environment) as the strings store more and more information about that environment. Naturally, a change in environment (catastrophic or otherwise) generally leads to a reduction in complexity. Such experiments suggest that physical complexity is indeed the "quantity that increases when self-organizing systems organize themselves."[64]

Adami and his group's experiments with digital organisms thus confirm the intuitive idea that there is a positive correlation between environmental replication success (fitness) and a collective increase of information about the environment. This correlation, for Adami, provides the basis for a quantitative measure of complexity. It also turns out that experimental data on the effects of mutation rates observed using the Avida system correlates very positively with data obtained in experiments with the evolution of *E. coli* and other bacterialike biological organisms, as biologist Richard Lenski and others have shown.[65] Yet, as in the case of Tierra, these results may be viewed from two different though not necessarily contradictory perspectives. On the one hand, Avida is an invaluable tool for evolutionary biologists who want to study the effects of mutation on the genome of an asexual organism. But on the other, it is an incubator spawning new and increasingly complex forms of artificial life whose ultimate value may well exceed the specific interests of theoretical biology.

Amoeba: The Spontaneous Generation of Digital Life

Both Tierra and Avida are seeded by a handcrafted digital ancestor designed to self-replicate. Implicitly these systems raise the intriguing possibility that under certain conditions a self-replicating digital organism might arise spontaneously, without having to be crafted by a designer's hand. This possibility is exactly what Andrew Pargellis sought to address with Amoeba, a software platform he designed following the general lessons of Tierra and Avida. In 1996 Pargellis published the results of over one hundred separate runs, in which self-replicating digital creatures emerged from a "primordial soup" composed of random sequences of machine instructions.[66] This success allowed Pargellis to identify some of the primary factors that optimize the probability of the appearance of self-replicators and to follow their subsequent evolution. What perhaps best characterizes Amoeba, however, is the dynamic complexity of the soup itself, which results from Pargellis's creative use of the instruction pointer in the execution of each organism's code.

We might reasonably assume that two factors in particular are critically important: first, that the primordial soup is filled with a relatively limited number of types of instructions; and second, that a very small subset of these would be sufficient for self-replication to occur. The assumed biological counterpart is the high likelihood that primitive forms of RNA arose spontaneously and acquired the capacity to self-replicate. To test these assumptions Pargellis created a virtual machine with an instruction set composed of 16 instructions, with only a minimal set of 5 necessary for self-replication. (By comparison, in Tierra a minimum number of 19 instructions is necessary, while in Avida it is 11.) Since the number of possible combinations of these instructions (16^5) is approximately 10^6, or one million, the probability that a self-replicating organism might appear seems rather remote. However, if we take into account the fact that Amoeba can support as many as a thousand organisms in a single run, and that the computation cycle for the execution of every organism's code is very rapid, then even a moderate mutation rate is likely to result in the appearance of a self-replicator in a reasonable time period. Noting that "the probability of a randomly generated sequence resulting in a self-replicator increases with replicator size" ("The Evolution," 118), Pargellis calculated that the probability of a sequence of 25 instructions is approximately 10^{-4}, or 1 in 10,000. And indeed, self-replicators did emerge in Amoeba after about four hundred generations. Pargellis expected these ancestral replicators to be small, with simple sequences of instructions, but in fact they were "large and unlikely combinations of computer operations [i.e., instructions]" ("The Spontaneous Generation," 87). Once these self-replicators appeared, moreover, competition for CPU time and memory forced them to optimize their code, just as we would expect.

To initiate a run, Pargellis seeded Amoeba with a primordial soup consisting of several hundred digital organisms or cells, each composed of a unique and randomly generated sequence of computer instructions that makes up a cell's genome. These genomes or instruction sets range in length from one to twenty-five instructions, although in later experiments as many as thirty instructions made up a set. When the soup fills to maximum capacity, a reaper function eliminates about 30 percent of the cells, beginning with the oldest. At the completion of the reaper cycle, about 50 of the empty slots are filled with new, randomly generated sequences of instructions. In addition to the reaper, there are also random mutations. Each time a child cell is initiated, there is a 10 percent probability that one of its instructions will be mutated. Of these mutations, 60 percent are substitutional, 20 percent are insertional, and 20 percent are deletions. Pargellis has observed, however, that only about 3 percent of the mutations enable a replicator to remain viable. As a consequence, a number of default settings and routines are required in order to handle improperly executed instructions.

The instruction set from which the cell's genome is constituted differs in two major respects from those of Tierra and Avida. First, it has many fewer instructions (the minimum required for replication) and consequently is not Turing complete (i.e., not capable of universal computation). Second, a great deal of work is accomplished by the cell's instruction pointer, which functions like the program counter in a CPU: it sequentially points to the address of the next instruction to be executed. Designated **ip(cell, op)**, the instruction pointer in Pargellis's virtual machine has two vectors (i.e., variables or arguments): **cell** is the particular cell whose instructions are being executed as the pointer points to each one seriatim in a classic visitation sequence; **op** is the position (from 1 to 30) of the instruction presently being executed. A jump command can move the pointer to a new position or address, but eventually it must be followed by a return command; otherwise, the pointer is "lost." When a cell loses its pointer, it can no longer execute its code and more than likely will be "captured" by a neighboring cell. The neighboring cell will then be able to use this host cell's registers, CPU time, and memory slot for its own additional replication. Since this feature is responsible for how many of the cells in Amoeba replicate and is essential to what drives the system's dynamic behavior, it is worth considering in detail.

In another contrast with Avida, cells in Amoeba "randomly diffuse about on the grid." Actually, each cell's coordinates on this 22 by 80 interaction grid serves primarily as a means of identifying that cell's nearest neighbors—the cells with which it can interact. An interaction occurs, Pargellis states, "when a cell (virus) transfers its pointer to, and executes the code of, another cell (host)" ("The Spontaneous Generation," 89). Pargellis's introduction of the virus-host language here is potentially confusing, as we'll see. We can, nevertheless, extrapolate a basic idea of how the system works. In the present instance there are two ways for this "transfer of the pointer" to happen. In the first (as noted above), a cell loses its pointer when there is no return, or reset, command after it executes the last instruction in its code. Thereupon a search routine is initiated that finds a cell within three grid units and that acquires a pointer with new parameters: **ip(cell's host, op = 1)**. In effect, this means that the CPU jumps from the first cell (with the lost pointer) to an adjacent cell and begins to execute its code. In other words, the first cell has now become the adjacent cell's host. The second way occurs if a cell finds an address (usually in its **cx** address register) that points to another cell. If that happens, a search is initiated for a nearby cell containing that address, and the first cell's pointer is altered to new parameters: **ip(cell's host, op = address position)**. Again, as a result the CPU jumps from the execution of the first cell's code to that of the second. In both cases the CPU performs this jump at the next computational cycle and then regularly, unless a mutation causes a change in the cell's code (in the first instance) or the address register (in the second). Thus rather than execute each cell's code in a classic visitation sequence, the CPU is constantly jumping around the grid, crossing cell boundaries and executing code in ever new sequences.

These features result in a very interesting ecology. Before considering the special dynamics at work, a further clarification is necessary. As noted above, Pargellis refers to cell interaction as a viral process. This is confusing, for actual "viruses" *also* appear in the primordial soup. Strangely, however, Pargellis does not define them or even include them in his account of the various cell types. Let us therefore define a virus in this context as a cell that uses the code of a neighboring cell (as host) to produce another virus. How does this differ from the process described above, where a cell loses its pointer only to see it acquired or replaced by a parasite cell's pointer? The simple answer is that the parasitic cell gets twice as much CPU time and memory, since at each computational cycle it will be visited twice: once for itself and again when the CPU visits the host cell without a pointer and is subsequently jumped to the parasitic cell to execute its instructions a second time. All other things being equal, if it has all the necessary code to replicate, the parasitic cell will therefore replicate roughly twice as fast as a nonparasitic replicator. A virus, on the other hand, is able to copy "part of its code, followed by part of the host's code, into the viral embryo" (90).[67] Exactly how the virus does this is not explained. In the examples Pargellis provides in his two articles there doesn't appear to be a minimal set of instructions that every virus contains. All we know for certain is that a virus always and only produces another virus. These viral children are always small (they do not possess many lines of code) and can quickly form colonies around host cells.

In order to compare emerging self-replicators and viruses, Pargellis supplies a "phylogeny" (see fig. 5.4, reproduced from "The Spontaneous Generation," 91). Viruses appear on the left; the "main rootstock" self-replicator with various offshoots, on the right. The numbers refer to specific instructions (described in a table), with the colors indicating useful code (yellow), nonfunctional code (blue), and useful additional code (red). Viruses appear about the same time (around the 406th generation) as self-replicators. But as the phylogeny emphasizes, viruses don't evolve as successfully as the bigger self-replicators, which tend to lose many of their "introns" (the nonfunctional code) and acquire new subroutines. Pargellis observes that in fact viral species usually don't last more than a few hundred generations and soon lose their small size advantage in the presence of robust self-replicators. Thus, he notes, the viral population will be either suppressed or vigorous depending on the "robustness of the dominant self-replicators in the soup" (91).

However, a more complex picture emerges when Pargellis describes the three types of cells that make up the soup: the prebiotic, the protobiotic, and the biotic. More specifically, the narrative that his account generates appears to support two different interpretations, depending on whether we assume the point of view of the "true biotic," as Pargellis calls the spontaneously generated ancestral cell capable of "replicating itself successfully many times" (92), or that of the soup as a dynamical system. As evident in the phylogeny, when the primordial replicators first appear, they are usually large but inefficient. In fact, in a majority of runs the first self-replicators are protobiotics—large cells that either replicate only once or replicate incorrectly at least once before acquiring the capacity to replicate correctly. But here we see a division. Those that are successful (the minority) soon shed their introns and become very efficient over succeeding generations. They then begin to grow again and acquire advantageous subroutines. They are, in short, the perfect success story. In contrast, the majority of protobiotics suffer an ambiguous fate. Frequently they cannot compete with neighboring viruses, which are able to use the code of their protobiotic hosts to "generate viral children, quickly allocating the available computer memory." Although a protobiotic often doesn't last more than a few generations, Pargellis adds, somewhat contradictorily, "It often lasts indefinitely, becoming the dominant component of the soup by initially existing in a colony (where it trades instruction pointers with

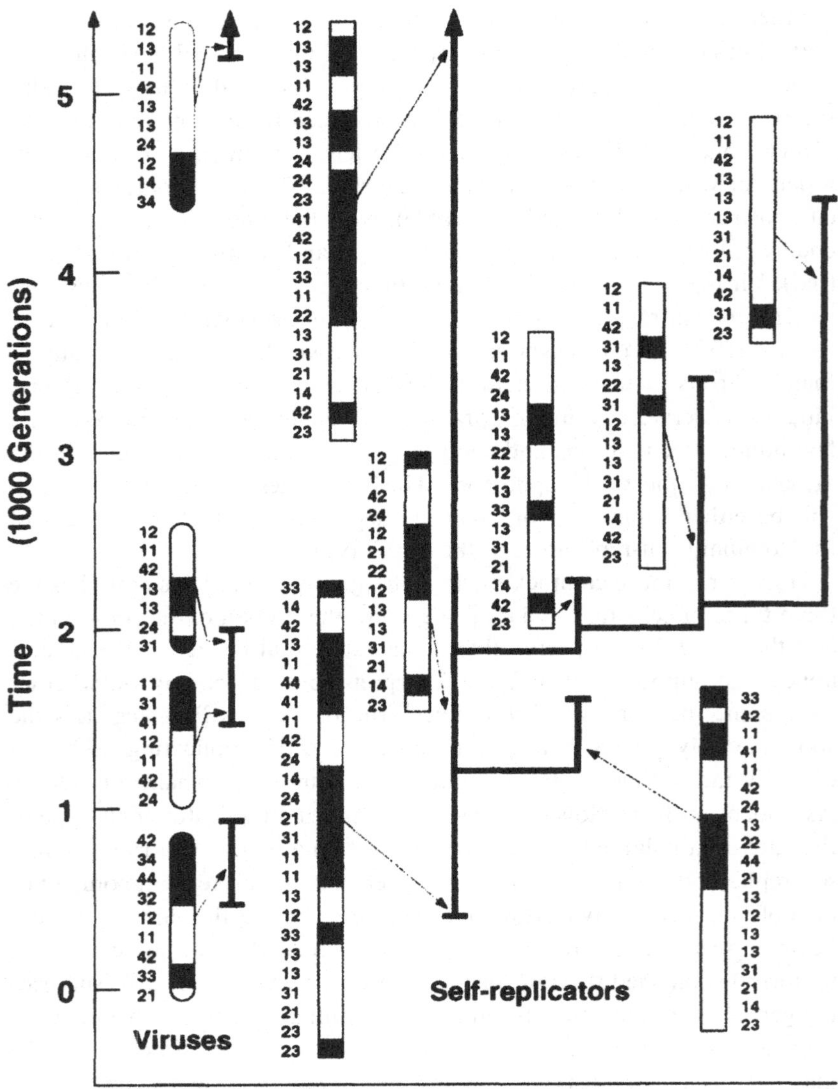

Figure 5.4
Phylogeny of a digital organism. A. N. Pargellis, "The Spontaneous Generation of Digital 'Life'," *Physica D* 91 (1996): 91.

other similar cells, for example) and evolving into a more efficient organism, the biotic cell" (93).

Meanwhile, the prebiotics also endure an ambiguous fate. As we would expect, they are incapable of self-replication. But here again a division appears: either the prebiotics are "completely inactive or they are only able to create a daughter cell that has either one or no operations" (92). Curiously, although the prebiotics at first appear to be inert, they enjoy both an initial moment of glory and a truly essential function. In the beginning, in fact, there are only prebiotics. However, as soon as the first replicators emerge, the prebiotics lose their dominance, even though they continue to exist as long as the soup endures. Since any children they spawn are empty or only contain a few pieces of code, they can "generate offspring only through chance encounters with other cells":

> Usually, they lose their pointer to the soup, which is then captured by another cell and used to copy either the viral code, that of the host, or some combination of both, into the viral child. Therefore, prebiotic cells that are unable to retain their pointers play an important role as their instruction pointers are captured by nearby cells that may be more viable, protobiotics. (92)

Yet Pargellis's statement is confusing here precisely because it collapses two distinct processes: in the first instance the prebiotic is a prey to viruses, whereas in the second the prebiotic is a helper or enabler for protobiotics.

Clearly there is something odd about the dynamics of the soup. The oddity has to do with the fact that because pointers can be lost and acquired by other cells and used as an alternative form of replication, prebiotics are essential for protobiotics, and protobiotics are essential for biotics, in the sense that in both cases the former enable or even become the latter. Furthermore, if we take this process of loss and acquisition of the pointer followed by alternative replication as equivalent to the actions of a virus, as Pargellis appears to do, then viruses are also essential to the ecology. Or rather, as parasites that compete for CPU time and memory space viruses are both the "enemy" of all other cells—prebiotics, protobiotics, and biotics alike—*and* essential to the dynamics of the soup, since without them there would be no dynamic interaction. Put another way, the dynamic interactions of the soup necessitate a blurring between entity (or category of entity) and function. As entity, viruses are dangerous, and prey on "healthy" cells; as function, they are absolutely necessary to the soup's continuing viability. A striking example of this split occurs in Pargellis's discussion of the protobiotics: either they are the easy prey of viruses, with which they cannot successfully compete, or they exist in colonies, where they can trade "instruction pointers with other similar cells." In short, either they succumb to viruses or they become viruslike and survive by sharing pointers in colonies.

It is much too simple, therefore, to describe the dynamics of the soup in terms of cell types. Rather, in addition to the replicators and their ongoing evolution, we must consider the actions of both its parasites and symbionts and their ongoing evolution as well. In his article "The Evolution of Self-replicating Computer Organisms," Pargellis begins to acknowledge this more complex picture in a schematic diagram of the interaction grid (115), which now includes three distinct processes: (1) the actions of an isolated self-replicator; (2) a parasite (i.e., virus) using some of the code of its nearest neighbor; and (3) cells sharing their pointers within a colony. Pargellis refers to these colonies as "primitive types of multicellular life where cells exchange resources, in this case computer memory and CPU time through the exchange of instruction pointers" (120). In these terms, one cannot help but wonder how these colonies—and the colonies of viruses as well—would fare under varying mutation rates, both in the short and long term, and would compare with the isolated self-replicators. One can only hope that Pargellis or other researchers will pursue the question in future research.

The issue of mutation rate in Pargellis's experiments relates directly to an idea that Adami has recently discussed.[68] When exposed to high mutation rates, Adami notes, some species survive as a cloud of mutants, or quasi species, in the following sense.[69] No longer does any one organism contain the entire genome for the species, but rather a range of genomes exist, and this variety allows the species to survive the onslaught of destructive mutations. Viruses in particular exhibit this kind of quasi-species behavior, Adami continues, just as "the more robust of his digital organisms do":

> In virus evolution, you clearly have mutation rates on the order of those we have played around with. It is clear that a virus is not one particular sequence. Viruses are not pure species. They are, in fact, this cloud, this mutational cloud that lives on flat peaks [in a fitness landscape]. They present many many different genotypes. (2)

A multiplicity of digital organisms that can no longer be defined except as a "mutational cloud"—surely this is a curious order of being, produced inadvertently by technology but only made intelligible by an anomalous phenomenon of nature. Yet the pursuit of such anomalous instances may prove highly instructive in the attempt to define and create life.

The End of ALife's First Phase?

At the Fourth European Conference on Artificial Life (1997), Mark Bedau and Norman Packard presented the results of a study comparing evolutionary activity in artificial evolving systems with those in the biosphere.[70] The results are sobering and indicate the need to make artificial evolving systems more open ended. Seeking to identify trends in adaptation rather than complexity, the authors define three measurable aspects of evolutionary activity: cumulative activity, mean activity, and diversity. In general terms, evolutionary activity simply reflects the fact that an evolving population produces innovations, which are adaptive "if they persist in the population with a beneficial effect on the survival potential of the components that have it" (126). Innovations are defined as new components introduced into the system, genotypes in the case of artificial systems, and the appearance of a family in the fossil record. In both instances an activity counter is incremented at each successive time step if the innovation still exists in the population under study. This count is then used to compute the other measures. Diversity, for example, is the total number of innovations present at time *t* in a particular run of an evolving system. These quantitative measures were then used to compare evolutionary activity in two artificial evolving systems, Evita and Bugs, with evolutionary activity in the fossil record.[71]

The results are unmistakable: "Long-term trends involving adaptation are present in the biosphere but missing in the artificial models" (132). Specifically, cumulative activity, mean activity, and diversity in the fossil record show a steady increase from the Cambrian to the Tertiary periods, except for a momentary drop in the Permian period, which corresponds to a large and well-known extinction event. In contrast, after an initial burst of evolutionary activity in the two artificial evolving systems, there are no long-term trends. In Evita the reproductive rate improves significantly at the beginning of the simulation, after which the new genotypes remain "neutrally variant"; in other words, though highly adaptive, they are no more so than the majority of genotypes already in the population. The authors interpret this to mean that "the bulk of this simulation consists of a random drift among genotypes that are selectively neutral, along the lines of [M. Kimura's] neutral theory of evolution." In the Bugs model, three major peaks of innovation occur, but then the evolutionary activity "settles down into a random drift among selectively-neutral variant genotypes, as in the Evita simulation" (130). These peaks, the authors explain, reflect successive strategies that enable the population to exploit more of the available resource sites.[72] But as with Evita, the possibilities for significant adaptation are soon exhausted.

After presenting quantitative evidence for the qualitative difference between these ALife systems and the biosphere, Bedau and Packard attempt to account for this difference as a necessary prelude to closing the gap between them. First, they note the absence of interesting interactions between organisms, like predator-prey relations, cooperation, or communication. (Since Ray's Tierra and Holland's Echo permit such interactions, they planned to extend the study to them. A follow-up study, however, revealed that Echo also lacks "the unbounded growth in evolutionary activity observed in the fossil record.")[73] Second, they acknowledge that the spatial and temporal scales of Evita and Bugs are vastly smaller than those of the biosphere, and that these systems are much less complex. However, they don't believe that scaling up space and time in the artificial systems or making them more complex will make any qualitative difference. This follows from what they think is the primary reason behind the biosphere's arrow of cumulative activity: the fact that "the dynamics of the biosphere constantly create new niches and new evolutionary possibilities through interactions between diverse organisms. This aspect of biological evolution dramatically amplifies both diversity and evolutionary activity, and it is an aspect not evident in these models." Noting a qualitative similarity between the initial part of the cumulative activity curve for the fossil data and for Bugs, they speculate that the biosphere might be on some kind of "transient" during the period reflected in the fossil data, whereas the statistical stabilization in Bugs may be caused by the system hitting its "resource ceiling," meaning that "growth in activity would be limited by the finite spatial and energetic resources available to support adaptive innovations." Contrarily, the biosphere seems not to be limited by "any inexorable resource ceilings." Its evolution continues to make new resources available when it creates new niches, as "organisms occupying new niches create the possibility for yet newer niches, i.e., the space of innovations available to evolution is constantly growing" (132). Whatever is responsible for unbounded growth of adaptive activity in the biosphere, the challenge is clear. Indeed, creating a comparable system, the authors assert, "is among the very highest priorities of the field of artificial life" ("A Classification," 236). The good news is that an objective, quantitative means for measuring success is now available.

Despite their intentions, Bedau and Packard's findings can be taken as evidence that we are nearing the end of the first phase of official ALife research. Of course, this interpretation ignores or relegates to a lesser status much of the very valuable work devoted to "artificial chemistries," the transition to "life" from nonliving origins, and the synthesizing of dynamical hierarchies, to name only three other significant strands of ALife research.[74] Nevertheless, several things argue for this assessment. ALife programs like Tierra and Avida, which have been applauded precisely for instantiating open-ended evolution, stand out among the discipline's most visible successes. Now that the inherent limitations of these systems are objectively measurable, new approaches that can move beyond them and advance ALife research to a new stage are called for.

One effort in this direction is Tim Taylor's Cosmos, which, though it incorporates many of the features that have made Tierra and Avida successful, is intended to surpass their limits by correcting for a number of perceived deficiencies.[75] For example, Taylor thinks that the appearance of parasites in these systems is due to the fact that an organism can read and directly execute the code of another organism, thus making parasitism an artifact of the system's design rather than a reflection of evolutionary processes. Cosmos, accordingly, does not allow one organism to copy another's code. Another criticism is that, since CPU time serves as an analogy for energy in a biological system, the organisms in these systems are essentially getting energy for free. Like Echo and Bugs, therefore, Cosmos includes a simple metabolic system, and requires that a price be paid for every organism's "energy tokens." Yet beyond these and other "corrections" what makes Cosmos an ambitious system is that it is designed to allow multicellular organisms to evolve. Taylor accepts the widely held assumption that once evolution hits upon multicellularity the emergence of complex organisms will inevitably follow. As noted earlier, Tom Ray has followed this path with Internet Tierra, which operates as a multithreaded, or parallel-processed, network. But instead of a network, Taylor uses object-oriented programming to provide a basis for multicellularity:

> Each program within Cosmos is an *Organism* object. An organism contains one or more *Cell* objects. Each cell object represents a single process, so that an Organism with one Cell is a serial program, and an Organism with multiple Cells is a parallel program. A Cell contains a bit string—the *Genome*, which gets translated to the executable code of the process. A Cell also contains a number of other objects, including: *Nucleus Working Memory* for writing a copy of the Genome for replication; *Communications Working Memory* for composing arbitrary messages; a *Regulator Store* containing promoters and repressors which dictate which sections of the Genome are translated; a buffer for receiving incoming messages; an *Energy Token Store*; four 16 bit registers and a stack. (553)

Thus in addition to making the evolution of multicellular organisms possible, Cosmos allows gene regulation, an evolvable mapping between genotype and phenotype, energy storage, intercellular communication, and interorganism communication. These features greatly increase the potential complexity of the system and enable a wider range of experiments to be performed than most ALife systems to date can support.

Yet despite this promise, Taylor's reported results have been disappointing. No multicellular (parallel) programs have evolved, and evolutionary activity (according to Bedau and Packard's criteria) does not compare with that of the biosphere; that is, evolutionary activity in Cosmos is bounded.[76] These negative results have led Taylor to assess the basic design and underlying assumptions that characterize Tierra- and Avida-type systems. He discusses the problems they pose under the following rubrics:

- Lack of an explicit theoretical grounding.
- Predefined organism structure.
- Restricted ecological interactions.
- No competition for matter and energy.
- Evolving a self-reproductive algorithm.

(*From Artificial Evolution to Artificial Life*, 188)

Drawing together early criticisms with more recent perspectives like that of Bedau and Packard, Taylor reviews the arguments for the putative limits of ALife efforts to synthesize life in a computational medium. The overall effect—though certainly not Taylor's intention—is to reinforce the idea that a first phase of ALife research may be drawing to a close.

One of the most salient problems that Taylor identifies concerns the need to embed organisms more fully in the (or an) environment. The argument for embedding takes several forms, but Howard Pattee's seems to be the most compelling (and the one Taylor mostly draws upon). Pattee has argued that a symbol system only functions and has meaning within a specific physical context. Thus the symbolic information contained in an organism's genes has "no intrinsic meaning outside the context of an entire symbol system as well as the material organization that constructs (writes) and interprets (reads) the symbol for a specific function, such as classification, control, construction, communication."[77] This "semantic closure," as Pattee calls the linking of the self-referential symbol system to the matter that instantiates it through these specific functions, constitutes a necessary condition for open-ended evolution. As Taylor adds,

"Open-ended evolution fundamentally requires the evolution of new meaning in the system, and this can only be achieved in the context of a semantically closed organization which is completely embedded within the physical world" (218). More specifically, "It is only when an organism's genotype, phenotype, and the interpretation machinery that produces the latter from the former, are all embedded in the arena of competition that fundamentally new symbolic information can arise in the genome (thereby permitting truly open-ended evolution)" (219). From which it follows that ALife organisms—both the genotype and phenotype—should be constructed "with the parts and laws of an artificial physical world."[78] Otherwise, no relationship of semantic closure between the genomic code (i.e., the symbol system) and the artificial physics of the environment can be established, and the organisms will simply obey the instructions of the program and settings of the hardware on which they are run. More recently, Luis Rocha has taken up and strengthened Pattee's argument that only if the phenotypes of the organisms result from a material, self-organizing process can open-ended evolution be attained.[79] Changes and contingencies in the environment would then be able to exert a more varied selection pressure on the organisms, and the simulation would more fully model the conditions of life in the biosphere. Evolution in the digital medium, in short, would be less bounded.

A second, closely related concern is the very limited nature of the ecological interactions within these systems. In biological evolution there is a struggle for existence among a diversity of organisms that compete for resources of space and energy both among themselves and with other species and subspecies. Nevertheless, noting that some of the "most spectacular examples of artificial evolution rely upon coevolutionary interactions between organisms," Taylor points out that not enough attention in ALife research has been directed to the question of what "sorts of ecological interactions should be available" (189).

Let us consider, in this light, the difference between Tom Ray's highly lauded Tierra and Larry Yaeger's virtually undiscussed PolyWorld. Analyzing the results of evolutionary runs on Tierra, Ray has observed that the innovations fall into two broad categories: "ecological solutions" and "optimizations."[80] The former refer to adaptive relations with and among parasites, while the latter refer to improvements of the reproductive algorithm of individual organisms. At this point it is hardly necessary to emphasize the importance of these results. Yet it is seldom noted that in Yaeger's PolyWorld a number of distinct species of organisms have actually evolved and coexisted.[81] Most likely this success has to do with the complexity of the organisms and their interactions. Whereas Ray's paired-down digital creatures exhibit no difference between genotype and phenotype—hence their evolution and adaptation only involves strategies of replication, Yaeger's organisms can eat, reproduce, fight, move forward, turn, control their field of vision, and control the brightness of a few of the polygons on their bodies (268). These "primitive behaviors" are controlled by the organism's brain, which consists of a neural network, the initial configuration of which is determined genetically, that is, by the initial values included in the organism's genes. This inclusion of a neural net means, in principle, that the organisms can learn, although thus far the update rules actually applied to the neural nets are "coarse abstractions" (277). Nevertheless, the fact that these organisms can perceive aspects of their environment (including other organisms) and respond volitionally makes PolyWorld unique among ALife simulations.[82]

Although particular runs of PolyWorld usually begin with a created organism—this is the case with most ALife systems, Amoeba being a striking exception—the run is judged to be successful only "if some number of species emerge with a Successful Behavior Strategy" (283) capable of sustaining their numbers through reproduction. Yaeger describes several examples: the "frenetic joggers," the "indolent cannibals," the "edge runners," and the "dervishes," each of which represents a different but successful strategy for adapting to this virtual world (see "Computational Genetics," 283–284, for details). Yet, as Yaeger notes, the most interesting species and individuals are not so easily classified. The most important instances are those in which there are multiple distinct species, no one of which dominates. In this setting, complex emergent behaviors appear, including responding to visual stimuli by speeding up, responding to an attack by running away or fighting back, responding to food patches by slowing down and grazing, seeking out and circling food, and following other organisms. Yaeger has also observed foraging as well as the formation of swarms, noting the adaptive advantage of the former in regard to eating and of the latter in regard to finding reproductive partners. On these grounds alone, PolyWorld clearly qualifies as one of the most complex of ALife worlds.

To my knowledge, Bedau and Packard's measures of adaptive evolutionary activity have never been applied to PolyWorld. If increases of complexity over the long term were observed to occur, they would most likely be due to the system's large repertory of organism behavior (comparable in this regard to Echo) and the immense advantage offered by vision and a neural net brain. As Yaeger insists, what distinguishes his "ecological simulator" from previous ones is "its unique use of a naturalistic visual perceptive system to ground its inhabitants in their environment" (266). This "grounding," he believes, "should answer one of cognitive psychology's most frequently sounded complaints against traditional AI" (278). Whether this grounding would satisfy critics like Pattee is not clear; yet it is undeniable that in Yaeger's project the evolution of digital organisms is not separate from the evolution of intelligence. Thus, to the question of how to evolve greater complexity and/or lifelike behavior in artificial life forms, Yaeger's implicit response is to make them smarter and more aware of their environment.

Summarizing recent work with Network Tierra, which also provides sensory data to its digital organisms, Ray writes, "The objective is not to create a digital model of organic life, but rather to use organic life as a model on which to base our better design of digital evolution" ("Selecting Naturally," 26). But perhaps all ALife projects in which digital organisms replicate and evolve can be seen in this perspective, whatever the experimenter's explicit intentions. This would be true not only of PolyWorld and Cosmos but of all the systems previously discussed. Indeed, this salient ambiguity may ultimately supersede in importance the tension between weak and strong theories of ALife. In any case, by making digital evolution itself the primary objective, Ray decisively rejects the mimetic and representational assumptions that have governed the discourse of much ALife research. With Network Tierra, moreover, he also moved beyond the use of a small-scale, closed-world system generated by a single computer, which has been the type of platform on which most of the official ALife research has been conducted during its first phase.

To conclude this chapter let us consider two research paths that strike off in new directions. In the first, the focus shifts to computers in networks as a form of ALife; in the second, to the laboratory creation of "wetlife." Both approaches implicitly move beyond the paradigm developed by Langton and Ray, in which biological processes are simulated in a particular type of computational assemblage. What seems to drop out in these newer efforts is the special role of simulation.

New Directions in ALife Research

If asked to give an example of artificial life, millions of computers users today would most likely answer: computer viruses. So notes David Ackley at the outset of a paper he presented at the ALife VII conference in

2000.[83] Ackley goes on to wonder whether we really want to exclude the expanding world of computers and computer networks from consideration as forms of artificial life. Such questions lead him to point to a number of interesting parallels between living systems and manufactured computers, starting with the fact that both are excellent copiers and therefore present "tremendously virus-friendly environments" (487). An even more striking parallel is evident between "the arc of software development" and the "evolution of living architectures":

> From early proteins and autocatalytic sets amounting to direct coding on bare hardware; to the emergence of higher level programming languages such as RNA and DNA, and associated interpreters; to single-celled organisms as complex applications running monolithic codes; to simple, largely undifferentiated multicellular creatures like SIMD [single-instruction multiple-data stream] parallel computers. Then, apparently, progress seems to stall for a billion years give or take—the software crisis.
>
> Some half a billion years ago all that changed, with the "Cambrian" explosion of differentiated multicellular organisms, giving rise to all the major groups of modern animal.... Living computation hypothesizes that it was primarily a programming breakthrough—combining what we might today view as object-oriented programming with plentiful MIMD [multiple-instruction multiple-data stream] parallel hardware—that enabled that epochal change. (495)[84]

In the context of these observations Ackley proposes that "the actual physicality of a computer itself may support richer notions of life" than either software programs alone or the software candidates for artificial life. In effect, this perspective stands the ALife agenda on its head: "Rather than seeking to understand natural life-as-it-is through the computational lens of artificial life-as-it-could-be ... we seek to understand artificial computation-as-it-could-be through the living lens of natural computation-as-it-is" (488). Ackley calls this research agenda living computation. With computer source code serving as its "principal genotypic basis" (488), he is looking into how the principles of living systems can be applied to networked computer systems.

For specific experiments Ackley has constructed **ccr**, which is "a code library for peer-to-peer networking with emphases on security, robust operation, object persistence, and run-time extensibility" (491). An explicit objective is to enable a **ccr** world to communicate with other **ccr** worlds in a manner consistent with how living systems guard against possible sources of danger. Hence the peer-to-peer communications architecture requires a more layered, self-protective system of protocols than the familiar TCP/IP protocols of Internet communication. Ackley describes how **ccr** starts with very small messages and builds to larger ones using a cryptographic session key as a rough equivalent of means used in the natural world—complex chemical signaling systems, hard-to-duplicate bird songs, ritualized interactions—to authenticate messages and build trust and confidence before stepping up to more elaborate and sustained exchanges. Another initiative deployed in **ccr** is to circumvent the commercial software practice of distributing only precompiled binary programs while "guarding access to the 'germ line' source code, largely to ensure that nobody else has the ability to evolve the line" (493). According to the analogy of software genetics, Ackley understands computer source code as genome, the software build process as embryological development, and the resulting executable binary as phenotype. In these terms, the rapidly growing open source software movement is of essential importance:

> With source code always available and reusable by virtue of the free software licensing terms, an environment supporting much more rapid evolution is created. The traditional closed-source "protect the germ line at all cost" model is reminiscent of, say, mammalian evolution; by contrast the free software movement is more like anything-goes bacterial evolution, with the possibility of acquiring code from the surrounding environment and in any event displaying a surprising range of "gene mobility," as when genes for antibiotic drug resistance jump between species. (493)

Ackley illustrates this point by comparing **ccr** to a different "species," the GNU Image Manipulation Program (GIMP). Both utilize an identical piece of code—the GNU regular expression package—which Ackley likens to a "highly useful gene incorporated into multiple different applications out of the free software environment" (495). According to traditional, commercial software practices such duplication might be deemed wasteful, but for Ackley "such gene duplication reduces epistasis and increases evolvability" (495).

By treating the computer itself as a kind of living system, Ackley is surely pushing the limits of what some will perceive as merely an interesting analogy. Yet the application of the principles of living systems to actual computational systems is hardly new or outré. As discussed in chapter 1, it was the basis of many of John von Neumann's novel ideas about computation as well as the development of neural nets, and subsequently it has been essential to computational strategies like genetic algorithms and evolutionary computation. Ackley's project, moreover, is closely linked to Stephanie Forrest's research on the development of computer immune systems, discussed in the previous chapter. Indeed, by resizing the framework of ALife research and extending it to networked computer systems, Ackley openly pushes the contemporary convergence between new computational approaches to biology and biological approaches to computation. At the same time, in attributing a kind of life to networked computers he is further extending Langton's idea that life is the result of a particular organization of matter rather than of something that inheres in individual entities. Here life is envisioned as a property of the complex exchanges that some computational networks make possible.

A second new direction for ALife research involves the attempt to synthesize artificial "living" cells and thus represents a wetlife approach. In one respect this effort is not new at all, since it builds on an extensive body of research that has sought to understand the transition from nonliving to living matter and thus to discover the origin of life. What makes the new research different—and now part of the official ALife agenda—is that its primary objective is to synthesize a cell in the laboratory, even though it may be quite different from any known form of life and might not even recapitulate life's actual origins.[85] These experimental, yet-to-be-created artificial life forms are usually called protocells. As we would expect, the very notion of a living artificial cell raises the question (once again) of what it means to be alive. Although still a matter of controversy, the generally agreed-upon assumption is that a living cell must be capable of regenerating itself, replicating and evolving, and that the processes that accomplish these things are housed together within a single membrane and constitute a single entity. This assumption thus rests on the further assumption that natural living cells are the smallest units or instances of unquestionable life and constitute a distinct threshold. More precisely, every natural living cell is composed of separable molecular processes that are not in themselves alive, yet when these processes work together the result is what we call a living cell. It may turn out—as seems likely—that these distinct processes do not simply work side by side but somehow enable one another in ways not yet fully understood.

Thus far there are two distinct approaches to artificial cell synthesis: from the top down and from the bottom up. The top-down approach proceeds by simplification: starting with already existing cells with very small genomes, it attempts to reprogram them genetically. One candidate, the bacterium *Mycoplasma genitalium*, "is thought to contain the smallest genome for a self-replicating organism (560 kb) and represents an important system for exploring a minimal gene set."[86] Additionally, it can remain alive after much of its genome has been removed. It has been estimated that it requires only about 300 to 350 of its 517 genes to function acceptably. For J. Craig Venter and Hamilton O. Smith, this feature has made it an attractive object of protocell research.[87] Their project involves removing some of the genome from a bacterium cell and then injecting into its nucleus an artificial string of genes that resembles a naturally occurring chromosome. The hope is that the cell will not only survive but evolve into a new kind of cell, or at least a cell with new capabilities. Since this approach builds on both the functionality of an already naturally living cell and the now vast knowledge base of DNA research and genetic engineering, it may well be more successful in the short run.

More ambitious, the bottom-up approach attempts to construct artificial cells from scratch, out of nonliving organic and inorganic materials. Here the mechanisms of cellular metabolism and replication (with ensuing heritable variation) must somehow be made to self-assemble in the same space, so that they remain housed in the same physical container. The problems are truly daunting, but there has been definite progress. For example, Vincent Noireaux and Albert Libchaber at Rockefeller University have succeeded in creating small, bacterium-sized cells that can make specific proteins.[88] Biomolecules that can produce proteins already exist and are commercially available, but in order to function they require a constant supply of raw materials and the removal of waste products. The problem Noireaux and Libchaber tackled was how to enclose these biomolecules in cell walls and keep them working longer. Basically, they were able to house a collection of biomolecules in a double layer of phospholipids that looks like a real cell membrane. To create molecular portals, they added a bacterial gene that codes for a protein, alpha-hemolysin, with a hollow barrel-shaped structure that inserts itself into the cell's membranes and creates pores. To monitor the behavior of these "cells," they added DNA that codes for a florescent protein, so that when it is produced the cells begin to glow. In contrast with the commercial biomolecules, which only worked for two hours, Noireaux and Libchaber's cells would continue to turn out proteins for days.

Currently there are several research programs afoot that are attempting to create a fully functional protocell. An important—indeed essential—precursor event was the discovery in the late 1980s that lipid aggregates are self-replicating (lipids are fatty, water-insoluble compounds).[89] So a common feature of these programs is to enclose the molecular machinery in a lipid vesicle. These efforts also tend to use an RNA genome containing an RNA-replicase ribozyme as a way to include transcription and translation functionalities. A somewhat different path—equally ambitious and perhaps further along—has been taken by Steen Rasmussen at the Los Alamos National Observatory and Liaohai Chen at the Argonne National Laboratory.[90] Rasmussen, a biophysicist who has been an ALife scientist from its beginnings, and Chen, a newcomer trained as a chemist, aim to keep things as simple as possible by creating a minimalist protocell in which the barest few components and processes do double duty or at least help one another out in and through their interaction. These components include a container or supporting chassis, a few genes, and a simple metabolic process:

(1) The housing, or chassis, for their protocell doesn't enclose the components but provides a surface on which they stick, somewhat like a sheet of chewing gum. This function is served by a very small lipid aggregate, or "micelle." Since it is a surfactant, meaning that one side is hydrophobic while the other is hydrophilic, in water it spontaneously forms into tiny droplets. However, it is the second property, mentioned above, that is crucial: after forming, when micelles reach a critical size, they split or replicate.

(2) The genetic components are made of peptide nucleic acid (PNA), an artificial molecule that could be thought of as a double or artificial cousin of DNA. It contains the same four nucleotide bases as DNA—adenine (A), thymine (T), guanine (G), and cytosine (C)—and has the same double-stranded helical structure, but in PNA the structure is made from peptides rather than sugars. Since peptides are also hydrophilic, these genetic components stick to the micelle's outer layer.

(3) Photoactive molecules that embed themselves in the micelle membranes serve as the necessary energy source for the simple metabolic process, working somewhat like chlorophyll in plant photosynthesis. When light strikes one of these photoactive molecules it gives up an electron. Normally the molecule would quickly recapture it, but it turns out that in the presence of PNA and certain nutrient molecules a metabolic chain is started. This is because the PNA molecules lend electrons to the photoactive molecules, leaving a supply of free electrons to sustain the metabolism of the nutrients. Furthermore, the speed and efficiency with which PNA conducts electrons depends upon the sequence of its base pairs, which means that the efficiency of the metabolic process can actually be encoded.

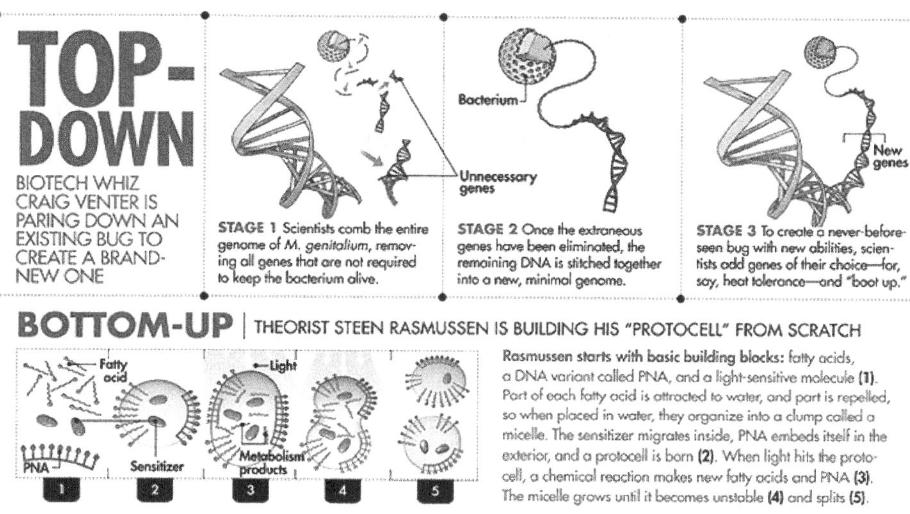

Figure 5.5
Creating an artificial cell. Steen Rasmussen, *Chicago Tribune*, April 14, 2004.

The processes that make up the two approaches to constructing a protocell are illustrated in figure 5.5. In Rasmussen's bottom-up approach, each of the individual processes has been demonstrated in the laboratory, but thus far they have not all worked together as a whole. The greatest problem remains the protocell replication process. In DNA replication the double strand "unzips," and each strand provides a template for a new set of complement base pairs. Although roughly the same thing occurs in PNA replication (i.e., the strands spontaneously separate and complementary strands are formed), there is nothing comparable to the elaborate check-and-repair mechanisms that normally make DNA replication highly accurate and reliable. As Rasmussen admits, "The replication process will be very error-prone, because we don't have any repair system" (quoted in Castelvecchi, "A New Game of Life"). However, this also means that the protocells will be characterized by great genetic diversity, with some unable to replicate or sustain a metabolism, while others with "the right sequence" will proliferate. Still, the "Los Alamos bug"—as Rasmussen and Chen call their protocell—will be very different from any known life form. Measuring only 5 to 20 nanometers in diameter, it will be an order of magnitude smaller than any bacteria. Although it will have carbon-based components like lipids and amino acids, it will not contain water and proteins. Yet it will be a life form, with an almost unimaginable range of uses in technology and engineering—and perhaps equally unimaginable consequences for human life and culture.

III MACHINIC INTELLIGENCE

6 The Decoded Couple: Artificial Intelligence and Cognitive Science

The mission of AI is not to make a human mind but to explore the "space" of possible intelligences.
—Kevin Kelly

Taken together, complex systems theory and Artificial Life experiments constitute one of the most innovative theory-practice relay systems in contemporary science and technology, for the simple reason that they articulate a single framework in which nonlinear dynamical interactions, emergent computation, and adaptation in evolution are constellated as relational aspects of the same process or system. Understandably, for scientists and theorists alike this framework has raised hopes that complex systems theory—or, more simply, complexity theory—may be able to resolve contradictions between theories of structure and theories of change. Consider, for example, recent controversies in evolutionary biology, where the orthodox Darwinian theory of natural selection (that random mutations passed from one generation to the next lead inevitably to adaptation) has been challenged both by those who, like Stephen Jay Gould, argue for the primacy of historical contingency, and by those who, like Brian Goodwin and Stuart Kauffman, argue that selection is predetermined by intrinsic, self-organizational stabilities and constraints. The problem, as Melanie Mitchell has observed, is that "the stark oppositions posited among these three frameworks are not only false oppositions, but are hindering progress in evolutionary theory."[1] Speaking as a scientist whose own work demonstrates how new forms of emergent computation in globally distributed representations can arise and evolve from the substrate of a complex dynamical system, Mitchell has brought the same critical perspective to bear on a recent shift in the methodology of cognitive science.

Mitchell's remarks were occasioned by the publication of Tim van Gelder's article "The Dynamical Hypothesis in Cognitive Science."[2] Militating against the "dominant computational approach," van Gelder argues for the validity and fruitfulness of the dynamical systems approach more widely, that is, for cognitive science at large. Thus, from Mitchell's perspective, the debate over whether the priority should be placed on change or structure now appears to be repeating itself within the field of cognitive science, wherein the older computational model, which gave pride of place to internal structure, is to be jettisoned for the new dynamical systems approach, which valorizes change. Esther Thelen and Linda Smith's study of the development of kicking and reaching in infants clearly illustrates the difference.[3] According to the authors, these new movements are explained "in terms of dynamical notions such as the stability of attractors in a phase space defined by body and environmental parameters" (Mitchell, "A Complex-Systems Perspective," 645). As the infant grows, changes in its body weight and length produce bifurcations that lead to new attractors and hence new trajectories of stability. Thelen and Smith thus reject the traditional theory, which would understand new developmental stages in terms of brain maturation and the infant's increasing ability to control its body through information processing. Since higher cognitive skills are rooted in these early spatial skills, they too are best understood dynamically.

In many respects this shift makes good sense. Human behavior entails a fluid linking of body, brain, and environment, with all three constantly changing over time and many of these changes feeding and being fed by complex interactive loops. Hence the appropriateness of a model that maps behavior as a dynamical system with varying "trajectories," "attractors," and "bifurcations," rather than the older model, which assumes that the cognitive agent is a computing device whose "input" comes from the environment and whose "output" is its own behavior. The rigid structure of this older model is simply not equipped to reveal paths of development, transition and stability, or instability and chaos.[4] However, as Mitchell points out, there are limits to current dynamical approaches that cannot be overcome within the dynamical systems framework. The first limit is scaling: since most of the work with dynamical systems involves low-dimensional analyses, it is not certain how such an approach can be scaled up to the dimensions required for modeling truly complex behavior. (Mitchell points to Randall Beer's five-neuron neural network controller for a walking robot, which will be discussed in chapter 7.) The second limit is that the dynamical model does not provide any way of understanding how changes in a system can be newly functional or adaptive. For this to be possible, Mitchell argues, the dynamical systems framework must include an additional capacity "in which functional, information-bearing, and information-processing components can be identified" (646). In short, instead of seeing the two approaches as opposed, Mitchell calls for their synthesis in a complex systems perspective in which the dynamical behavior of the system effects and is in turn effected by information processing within the system.[5]

Whether such a synthesis can or will be attained in cognitive science remains an open question. Recent signs, including van Gelder's article, suggest that contemporary cognitive science is moved by a somewhat different agenda, often encapsulated in the term *embodiment* and closely allied with the dynamical systems approach. Francisco Varela, in his concise little book *Invitation aux sciences cognitives*, finds both the cognitivist approach based on computation and the connectionist approach based on self-organization to be inadequate and spells out the new agenda.[6] This agenda dictates the book's basic rhetorical structure, which outlines in four stages the progressive development of cognitive science from its beginnings in cybernetics to symbol processing and the cognitivist hypothesis, to the neural net alternative, and finally to Varela's own theory of enaction as an alternative to representation. As Varela sees it, the problem with these earlier approaches is that they cut off the act of cognition—whether understood as the manipulation of symbols or the completion of a pattern in a neural network—from the world in which it occurs. In other words, both approaches presuppose an objective world and rely on representations of it. In order to determine the efficacy of cognition, therefore, the cognitive act must be tested against some aspect of the world. For Varela, however, the world does not exist outside of or apart from cognition. Rather, the act of cognition is a structural coupling in which some aspect of the world coemerges in and through this very act. At the most fundamental level, therefore, cognition does not depend on or involve representations. Because cognition is always embodied and concretely situated in the world, Varela calls it enaction and believes that his theory of enaction overcomes the deficiencies of the earlier approaches and restores common sense to the study of cognition.

Varela's wholesale rejection of both computation and representation, evident throughout *Invitation aux science cognitives*, is part of a continuing reaction against the same abstracting, deterritorializing impulse that characterizes classic AI. Not surprisingly, he admits to certain affinities with the Heideggerian critique of the latter and explicitly aligns himself with the phenomenology of Merleau-Ponty.[7] More important here, his explicit rejection of representation brings him into close alliance with what is called the new AI, which is neither the AI based on the brain-as-computer model that dominated research from the 1950s through the 1970s nor the connectionist strand of AI based on the neural net theory that mostly superseded the latter in the 1980s. Rather, as the roboticist Rodney Brooks and others refer to it, the new AI comes *after* ALife and is concerned with behavior-based (as opposed to knowledge-based) robotics, autonomous and/or situated agents, and multiagent systems, or what is sometimes called distributed artificial intelligence. As described in the next chapter, Brooks's groundbreaking innovations began in the late 1980s with what he called subsumption architecture—his theory and practice of constructing autonomous, mobile robots in vertical functional layers without a centralized controller, which consequently operate without any global representation of the exterior world. In this sense, Brooks's robots can be seen as the machinic counterpart to Varela's theory of cognition. Indeed, Varela has stated that Brooks's approach to constructing intelligent machines "is akin to our enactive orientation."[8]

It is one thing, however, to push behavior-based robotics to the point of doing away with a central controller and the necessity of representations of the external world, and something else to claim that the direct coupling of perception and action fully accounts for intelligence and cognition. Insofar as the new AI espouses this claim, it questions the usefulness of representation in cognitive science.[9] In fact, an interest in representation is noticeably lacking in van Gelder's dynamical systems approach to cognitive science and may account for the radical and subversive aspect of his article. To be sure, classic cognitive science, to which both Varela and van Gelder react so negatively, conceives of representation in cognition in a very specific sense: as the mental manipulation of representations composed of symbols, with this manipulation itself understood as a species of computation, or "information processing psychology." Indeed, this understanding lies at the foundation of cognitive science, and in his "history of the cognitive revolution" Howard Gardner singles out mental representations and the computer model as its primary defining features:

> First of all, there is the belief that, in talking about human cognitive activities, it is necessary to speak about mental representations and to posit a level of analysis wholly separate from the biological and the neurological, on the one hand, and the sociological or cultural, on the other.

> Second, there is the faith that central to any understanding of the human mind is the electronic computer. Not only are computers indispensable for carrying out studies of various sorts, but, more crucially, they also serve as the most viable model of how the human mind functions.[10]

From this brief exposition alone, the radical abstraction of the cognitivist model, as it is usually called, should be self-evident. It shares this abstraction with the first theorizations of AI, born in the same year as cognitive science (1956). This shared feature is not surprising, given that both disciplines sprang from the same new "computational paradigm"; in fact, several scientists, most notably Herbert Simon and Allen Newell, worked simultaneously in both fields. At the heart of this paradigm lies the newly developed digital computer, which can therefore be said to have given birth simultaneously to these two complementary discourses.

This discursive doubling, in which a single new technical artifact produces two distinct but complementary disciplines, is not unusual in itself. But the fact that this discursive couple remained tied together and complementary through two more paradigmatic changes—indicated roughly by Varela's (and Franklin's) three-part scheme: symbolic computation, connectionism, and enaction—surely calls for a critical examination that goes beyond the usual perfunctory acknowledgment of their initial begetting by the computer as "model" and "influence." This chapter begins such an analysis, first by uncovering what I think underlies a peculiar and problematic aspect of the relationship between AI and cognitive science, then by traversing in detail the conceptual history of AI research. After pointing to the shaping role of technics as what has been missing from previous discussions of the relationship between AI and cognitive science, I suggest that both must be analyzed as particular concrete instances of what I call a computational assemblage. The dramatic clash between symbolic and connectionist models of AI is also examined in these terms, after which I briefly consider two efforts to simulate consciousness (or several of its features) in robotic machines. The history of AI is then resumed in the next chapter, where I consider the new AI, the construction of autonomous agents, and artificial evolution.

Blurred Boundaries: The Importance of Technics

In his useful history of artificial intelligence, Daniel Crevier remarks that over time the boundary between AI and psychology grew fuzzy, with the area of overlap then becoming a discipline in its own right: "With help from other fields (anthropology, linguistics, philosophy, and neuroscience), cognitive science aims to explore the nature and functioning of the mind."[11] Two observations must be added. First, Crevier ignores the fact that in every description of cognitive science since Gardner's *The Mind's New Science: A History of the Cognitive Revolution*, AI is always listed as one of its tributaries, thus suggesting that as far as disciplinary boundaries are concerned, AI can and should be subsumed under the more encompassing activities of cognitive science. Second, in ascribing to cognitive science the aim of exploring the "nature and functioning of the mind"—and this is the path followed by most researchers and commentators alike—Crevier opens the door to problems similar to those raised by attempts to define the concept of intelligence. Indeed, *mind* is even slipperier, since it obscures or ignores the specific material conditions and "machinery" in and through which intelligence operates. In fact, until fairly recently the operations of mind, which generally included all higher cognitive functions, were studied in complete separation from those of the brain, the site of neuronal activity whose proper domain of study is neuroscience.

What must be acknowledged, in any case, is the constitutive role of technics, even in so-called natural human intelligence. To do so foregrounds the view that human intelligence arises in and through the use of not only tools but gestures, signs, languages, and props in the environment that make cognitive activities repeatable. This perspective has been conspicuously missing from discussions of AI and cognitive science, even though both took the computer as their explicit conceptual model at their inception. The problem is that common notions of the computer, model, and influence are not precise enough to account for the formation of these discourses in relation to specific types of computational machines. But before attempting a more exact theoretical analysis, I want to draw attention to the blurring that both produces and hides the double objective that arises with the simultaneous birth of AI research and cognitive science. This blurring is apparent in textbooks on artificial intelligence, Patrick Winston's for example, where the goals of AI are characterized as both the construction of useful intelligent systems and the understanding of human intelligence.[12] While AI researchers would certainly agree that the first goal alone would delimit their activities too restrictively, acceptance of the second puts them squarely in the realm of cognitive science. What is needed, therefore, is a theory that would mediate between the two objectives and precisely define their relationship. In my view this can only be accomplished with a fully contemporary understanding of technics.

Technics, of course, is not a new topic. In the second half of the twentieth century Jacques Ellul, André Leroi-Gourhan, Gilbert Simondon, Lewis Mumford, and others too numerous to mention were mainly concerned with the development of tools, machines, and the bureaucratic organization of society as a "mega-machine." A fully contemporary technics, however, must consider (as one of many problem areas) the underlying division or separation between "natural" and artificial intelligence reflected in the early agenda of cognitive science and AI. It would attend not only to the Heideggerian questioning of technics and the basis of the oppositions on which these sciences arise but also to how changes in technology constantly redefine the terms of both our conscious and unconscious engagements with it. The almost invisible interiorization of writing is a notable case in point, but only recently have scholars begun to attend to its immense consequences.[13] In cognitive science, an analysis based on technics would investigate how human cognitive functions are made possible by technical artifacts and how human cognition cannot be conceived outside of the shaping force and environment these artifacts constitute. Edwin Hutchins's study of how navigational instruments aboard a navy ship constitute a cognitive system is a rare and remarkable example of this much-needed approach.[14] Indeed, it is the general absence of this kind of analysis that explains—because it actually produces—the blurred boundary between AI and cognitive science. In the hiatus between wanting to *produce* artificial intelligence and wanting to *understand* natural intelligence, the constitutive role of technics unaccountably drops out of sight, even though both oppositions (produce/understand, artificial/natural) presuppose it.

In *La technique et le temps* Bernard Stiegler makes a decisive step toward establishing the centrality of technics in contemporary terms. Stiegler argues that the human (in contradistinction to the *anthropos*, which is only the human's genetic imprint), cannot be conceived outside of or apart from the advent and forces of *technē* and specifically of the tools that make the exteriorization of memory possible. With the invention of the human, that is, with what Stiegler calls the constitutive interplay of the "who and the what" in and through technics, the laws of biological evolution are in effect suspended. No longer does *Homo sapiens* evolve, but the milieu as shaped by technological artifacts, a theme Stiegler takes from Leroi-Gourhan. Characterizing our present cultural condition in terms of an "industrialization of memory," Stiegler offers a critique of cognitive science based on its denial of both the material supports necessary for writing in general and those temporal processes in which memory is externalized. In the cognitive sciences and for the first time in the history of Western thought, technics itself becomes a "heuristic vector," and a new explanatory paradigm finds its unity in a research project that begins with the idea that "we humans are ourselves computers."[15] Yet despite the fact that the cognitive sciences position "the technological artifact at the heart of their heuristic approach, they do not seem to accord any theoretical pertinence to the *technical fact* in the history of life":

> Thus, the process of corticalization [expansion of the cerebral cortex] at work at the same time that tools appear, inaugurating a new relationship of the living with its milieu, mediated by an artefactual layer which is also an artificial memory essential to the living human, is not grasped in its specificity: by the very force of the gesture that consists in effacing the traditional metaphysical oppositions between animal, human and machine, the sciences of cognition weaken themselves by simultaneously erasing the dynamic specificities that are engendered there, that is to say, the temporality of the process.
>
> That human knowledge is technological in its essence, that there are no possibilities of knowledge without the surfaces of inscription of an artificial memory, and that the concrete characteristics of these supports as organized inorganic matter constitute the reality of human cognitive operations, are thus ignored. Assuming a priori that a machinic simulation of thought, as production of a prosthesis of thought, is conceivable, the cognitivist model forgets the originary role of prosthesis in thought: what is not thought is the coupling of the who and the what as that which pre-dates the who and the what as such. (189)

If we assume that "prosthesis" is originary and not a supplementary technological extension or add-on, then not only the machinic simulation of thought but computation as well must be part of this primordial constitution of the "who and the what." But here, unfortunately, Stiegler becomes preoccupied with the Heideggerian/deconstructive problematic of language and fails to consider computation as an essential and increasingly constitutive aspect of technics. More important, by not pushing his critique deep enough historically he fails to locate the occlusion of technics at the very origin of cognitive science, that is, in its "repressed" relationship to cybernetics.[16] Like his fellow French poststructuralist philosophers, Stiegler never entertains the possibility that computation as a process may be part of poiesis, both in nature and its artificial simulation. Only Jacques Lacan, as a consequence of his encounter with cybernetics (as we saw in chapter 2), understood that a "symbolic order" can provide a basis for social organization precisely because it is a form of quasi-autonomous computation. This symbolic order, arising from counting practices and the emergence of self-organizing "laws" in primitive notation systems (Lacan remains vague about the origins), functions as a complex information-processing machine in-mixed in human behavior. As Lacan argues, cybernetics and the new digital information machines make these laws visible in and for themselves and thus establish the truism that "the world of the symbolic is the world of the machine." For Lacan, and contrary to contemporary misunderstandings, the symbolic order is not language but an abstract machine that only takes the exchanges that occur within language as its support.

Given the historical significance of cybernetics and the centrality of technics in human cognition, we turn with high expectation to Varela's *Invitation aux sciences cognitives*, since it foregrounds at the outset the importance of both. Indeed, it begins with the resounding declaration: "The sciences and technologies of cognition (which we will henceforth designate with the acronym STC) represent the most important conceptual and technological revolution since the advent of atomic physics, having a long-term impact on all levels of society" (21). Disappointingly, however, the precise nature of this impact is never explored. Instead, Varela asserts that cognitive science and technology are "confounded" and that "the distinction between applied science and fundamental research" (21) is hardly pertinent, since to consider either aspect alone would obscure the vitality of the movement and its conjunction of multiple perspectives. At the same time, he acknowledges that typical themes in cognitive science—perception, language, inference, and action—stand in parallel proximity to several "principle axes of technological development: image recognition, language understanding, syntheses in programming and robotics" (23). This amounts to saying that no matter which relays the other, technology and cognitive science are each implicated in the other's development.

Yet this relaying effect is precisely what we want to understand. In my view, it requires a different kind of conceptualization: we must first posit the technical and discursive assemblages that make the "the sciences and technologies of cognition" (STC) possible as such. Then, in view of the specific computational assemblages that began to take shape in the 1950s, we can chart the emergence of two series of reciprocally related technical/theoretical constructions: on one side, a succession of computational techniques concretized in the digital computer, artificial neural nets, and the dynamic interactions of body-brain-environment; on the other, a succession of computational models of the mind, ideas about the "language" of thought, mental process, cognition, and consciousness. In short, we find a series of distinct machines, each with its correlative discourse that explains how cognition occurs. In each case, the reciprocal relations between a specific machine and its specific, correlated discourse define a computational assemblage, which is understood to work precisely in and through this singular coupling of the two. The priority of technics is thereby maintained, and the brute materiality of the machine is not reduced to effects of language or conceptualization.[17]

This mutating computational assemblage originates in an "abstract machine," first imported from mathematics in Alan Turing's conception of the Turing machine. In postulating a formal equivalence between a set of effective procedures, or algorithms, and their instantiation in a mechanistic sequence of machine-state transitions, Turing made possible not only the modern computer but the very idea of cognition as a mechanically reproducible computational process. When Warren McCulloch and Walter Pitts proposed that the activity of networks of neurons in the brain could be treated as calculations in propositional logic and that these neural nets thus constituted a biological but computationally equivalent Turing machine, the seed of an alternative conception of computation was planted.[18] Materialized in devices like Oliver Selfridge's Pandemonium and Frank Rosenblatt's Perceptron, this alternative gave rise to a different model of cognition, based on the parallel processing of information in a dynamic network. Revived and further developed in the mid-1980s under the banners of connectionism and parallel distributed processing, this alternative finds its most recent incarnation in emergent computation and complex systems theory.[19]

In Turing's model of computation, states of a machine (a finite automaton with infinite memory) and stages in a performed calculation are set in a relationship of strict mechanical determination. In the recent alternative approaches, by contrast, computation occurs as the result of a self-organizing process among large numbers of elements or agents uncontrolled by any central processing unit or hierarchical command system. In these terms a seemingly unbridgeable gap opens up between the two models of how "thought" is understood to operate: the one is logical and procedural, the other dynamic and statistical. Yet both are guided and comprehended by computational, information-processing objectives made possible by this new kind of abstract machine. Because this machine transcends any particular material instantiation, it raises philosophical questions similar to those raised by the putatively transcendent status of representation in language, an issue (taken up in chapter 7) essential to behavior-based robotics.

Miraculous Birth, Symbolic Murder: The Beginnings of Classic AI

In his analyses of complex cultural events like the birth of tragedy and the genealogy of morality, the philosopher Friedrich Nietzsche suggests that the historian's search for the pristine, self-identical point of origin is doomed to be baffled by the discovery of a dissension and disparity at the very heart of things and that the supposed "origin" always and only serves to cover over a profusion of entangled and contradictory events.[20] The history of AI proves to be no exception. At its origin, operating side by side, we find two distinct research initiatives—one seeking to mechanize or automate thought using logic and algorithmic computation, the other seeking to model and reproduce mental processes by simulating the capacity of the cells in the brain to link together in self-organizing networks. Within a few years, however, these two approaches would be set into conflict by an altogether different kind of agency: DARPA, the section of the Defense Department then responsible for funding most AI research. Seymour Papert's fairy-tale version of this division and rivalry is worth quoting at length:

> Once upon a time two daughter sciences were born to the new science of cybernetics. One sister was natural, with features inherited from the study of the brain, from the way nature does things. The other was artificial, related from the beginning to the use of computers. Each of these sister sciences tried to build models of intelligence, but from very different materials. The natural sister built models (later called neural networks) out of mathematically purified neurons. The artificial sister built her models out of computer programs.
>
> In the first bloom of youth the two were equally successful and equally pursued by suitors from other fields of knowledge. They got on very well together. Their relationship changed in the early sixties when a new monarch appeared, one with the largest coffers ever seen in the kingdom of the sciences: Lord DARPA, the Defense Department's Advanced Research Projects Agency. The artificial sister grew jealous and was determined to keep for herself the access to Lord DARPA's research funds. The natural sister would have to be slain.
>
> The bloody work was attempted by two staunch followers of the artificial sister, Marvin Minsky and Seymour Papert, cast in the role of the huntsmen sent to slay Snow White and bring back her heart as proof of the deed. Their weapon was not the dagger but the mightier pen, from which came a book—*Perceptrons*—purporting to prove that neural nets could never fill their promise of building models of mind: *only computer programs could do this.* Victory seemed assured for the artificial sister.... But Snow White was not dead.[21]

Papert's dark conceit was prompted by a deep rupture within the history of AI. Officially emerging with Simon and Newell's Logic Theorist and later consolidated in their physical symbol system hypothesis, the "artificial sister" rose to dominance in the 1960s and '70s, only to be effectively challenged in the 1980s with the resurrection of biologically inspired neural net research—the "natural sister," Snow White in Papert's tale—now conducted under the broader terms of connectionism and parallel distributed processing. While Papert wonders whether this revival is really a happy ending after all, onlookers today may be inclined to interpret his allegory somewhat differently: since symbolic AI denied the importance of the material substrate or body and could only flourish with the murder of the "natural sister" (the first form of embodied AI), the recent valorization of "embodiment" in the new AI would seem to be just and inevitable revenge. In any event, the torch has clearly passed to the new AI. Stressing the concrete situatedness of autonomous mobile robots and influenced by the bottom-up, distributed approach of ALife, the new AI inaugurated by Brooks and others has positioned itself on the other side of this rupture, as we'll see in chapter 7.

To recount the history of AI as several stages in a progressive history, as Varela does, is misleading, however, inasmuch as the basic problems—what is intelligence? how can it be artificially simulated?—are not so much resolved or abandoned as successively reformulated from different vantage points and within different research agendas.[22] In this sense AI is constantly beginning again, while never altogether abandoning the accomplishments of preceding phases. Even though the "image of thought" neural net theory proposes differs radically from that of its predecessor, symbolic AI, it must account nonetheless for the syntax and logical structures on which the latter is built. Similarly, the new AI cannot really ignore neural net theory, so it mostly relegates it to the status of a tool or function, its instantiation in the brain only one instance of a biological implementation. In what follows I propose to map these various relationships as a series of decodings and recodings of different computational assemblages. I focus first on that conflicting inaugural moment in AI history summarized by Papert, although less in relation to the institutions brought into play than in terms of the two images of thought brought into conflict. Instead of resolving the conflict between symbolic and connectionist AI, the new AI shifts the conceptual field to include the relationship with an environmental "outside," cognition being redefined in terms of dynamical systems theory. As a consequence, its image of thought becomes unequivocally antirepresentational. The semantic difficulty posed by the nature of intelligence is resolved by defining intelligence as an emergent, adaptive response—a behavior that arises in and further enables the interactions between an agent or collectivity of agents and the larger environment.

As far as the institutional history of AI is concerned, this may be the place to mention Paul Edwards's *The Closed World: Computers and the Politics of Discourse in Cold War America*, which admirably counters the idealizing and progressive history-of-ideas approach.[23] Resituating analysis of the development of computers and the early history of AI and cognitive science in relation to America's Cold War politics and military agenda, Edwards shows how they participate in the construction of a "cyborg discourse." Understanding both human minds and artificial intelligences as information machines, cyborg discourse serves to integrate humans into the new techno-military complex of the "closed world." However, and unfortunately for my purposes, Edwards does not extend his critical history beyond the initial moments of symbolic AI, with scant attention given to its rivalry with neural net theory. In certain respects the subsequent history of AI—the return to neural net theory and the later development of autonomous agent theory—constitutes a break with and escape from "the closed world." Or rather, in Deleuze and Guattari's useful terms, it is a history of decodings and recodings, in the sense that cognitive functions are abstracted and deterritorialized (decoded) and then reinscribed (recoded) in a transformed context. We saw above how the cognitivist theory of cognition radically decontextualizes, or decodes, intelligence and cognitive capacity by understanding them as abstract computational functions, and we consider below how they are recoded in a symbol system. In D&G's theory, the functioning of an abstract machine in a particular assemblage ensures both the coding of closure ("the closed world") and the decoding that allows a fruitful break or line of escape. In these terms we can say that Edwards describes only those aspects of the militarized, Cold War computational assemblage that recodes and reterritorializes. However, by focusing exclusively on the political and ideological *capture* of forces unleashed by the new technology, we can also fail to recognize or downplay the conceptual innovations of the technics of computation, which have not only been harnessed to a wide variety of creative and more positive ends but also have ushered in changes that outstrip the Cold War ideological field.[24] To see only the negative side of this Cold War assemblage was precisely

Heidegger's mistake in his celebrated critique of technology, and this perspective continues to bias influential work in contemporary science studies.[25]

The first two chapters touched on many of the new ideas that provided the conditions for AI to emerge as a field of research, most notably Alan Turing's essay on computable numbers (1936) and his later essay on machine intelligence (1950) as well as the work of diverse individual talents in the 1940s and early '50s such as McCulloch and Pitts, John von Neumann, Claude Shannon, Norbert Wiener, and Ross Ashby. The first public discussions of cybernetics and information theory at the Macy Conferences brought together and disseminated much of this work, which consequently spurred various research projects that circulated it in wider orbits. But the development and ensuing use of the electronic, stored-program computer was AI's most essential precondition, and that is true of cognitive science as well. The magic year of 1956 saw the official birth of both: AI from a conference at Dartmouth that summer organized by John McCarthy; cognitive science from a symposium at MIT in September organized by George A. Miller. As the Dartmouth Conference organizers explained, the conference was predicated "on the conjecture that every aspect of learning or any other feature of intelligence can in principle be so precisely described that a machine can be made to simulate it."[26] One immediate effect of the conference was to ratify the term *artificial intelligence*, suggested by McCarthy. The MIT symposium had similar repercussions. Ostensibly devoted to information theory, it actually focused on information-processing psychology, as cognitivism was often called during its first decades. Miller later reflected: "I went away from the Symposium with a strong conviction ... that human experimental psychology, theoretical linguistics, and computer simulation of cognitive processes were all pieces of a larger whole, and that the future would see the progressive elaboration and coordination of their shared concerns."[27] In retrospect, the striking temporal proximity of the two conferences is superseded in importance by the fact that Allen Newell and Herbert Simon, two researchers from RAND, presented their work on the Logic Theorist at both. This was a catalyzing event, for the presentation was singularly instrumental in valorizing the new computational paradigm within which both artificial intelligence and cognitive science would first develop. But before considering the Logic Theorist, let us briefly assess the significance of the MIT symposium in the history of cognitive science.

In the most general terms, the symposium and the work that flows out of it mark the end of the behaviorist and stimulus-response model for understanding human behavior and the beginning of the information-processing model. In addition to Simon and Newell's paper (discussed below), two other seminal papers presented at the symposium can be regarded in this light. The first, Noam Chomsky's "Three Models of Language," described how Shannon's theory of information can be modified to produce the more complex and powerful formal grammar necessary for describing the syntax of natural language. This work would lead Chomsky to a new theory of transformational grammar and a devastating critique of B. F. Skinner's behaviorist account of language. As we saw in chapter 2, Chomsky theorized that a specific type of information-processing machine or automaton underlies our linguistic capacity, that is, our ability to recognize and generate grammatically correct sentences, and believed that this machine is an essential part of our innate cognitive faculties. The second paper, George Miller's "The Magical Number Seven, Plus or Minus Two: Some Limits on Our Capacity for Processing Information," demonstrated the empirical limits of human short-term memory in a variety of contexts.[28] Miller also showed how these limits can sometimes be surpassed by a method of chunking he called recoding. Within a few years, with coauthors Eugene Galanter and Karl Pribram, he would publish *Plans and the Structure of Behavior*, an immensely influential book that replaced the behaviorist notion of the reflex arc with the mediating function of internal representations:

> [Cognitivists] believe that the effect an event will have upon behavior depends on how the event is represented in the organism's picture of itself and the universe. They are quite sure that any correlations between stimulation and response must be mediated by an organized representation of the environment, a system of concepts and relations within which the organism is located. A human being—and probably other animals as well—builds up an internal representation, a model of the universe, a schema, a simulacrum, a cognitive map, an image.[29]

These internal representations are built up in ways that resemble or that can be modeled or simulated by computer programs. In fact, at the heart of all the symposium papers was the new idea that human beings are *natural* information-processing systems that take information from the environment (perception), process it (cognition), and act or make decisions on the basis of its output (behavior).[30]

At both the Dartmouth and MIT conferences, Newell and Simon simply stole the show by presenting a working computer program that instantiated "thinking," or what they would later call intelligent action. Specifically, they described the construction, operation, and first results of Logic Theorist—the first computer program to perform a mathematical proof.[31] It was written in Information Processing Language (IPL), a computer programming language that Newell, Simon, and Cliff Shaw had invented specifically for the program. This new higher-level programming language enabled Logic Theorist to use a list processing scheme that could be applied to both numerical and symbolic expressions.[32] Two innovative features account for the program's dramatic effect. First, no specific proofs for theorems were programmed into Logic Theorist's memory; and second, in addition to the axioms and rules of logic the program consisted of heuristics and search procedures for solving formal problems.[33] The inclusion of heuristics proved to be fundamental. Whereas algorithms are exact procedures for arriving infallibly at a predetermined result, heuristics are only search strategies that *may* work in solving a specified problem. Starting from an initial hypothesis (the root node in what became known as a search tree), Logic Theorist would move toward a determined goal by searching possible branches from the root node. In this manner it eventually succeeded in discovering proofs that were more elegant or succinct than the standard ones without the exact outcome being programmed or even known in advance.[34]

Logic Theorist thus solves a problem (how to prove a theorem in symbolic logic) by establishing a goal and the means to achieve it in algorithmic terms. For early AI, the problem-solving approach and the search strategy—and their implementation in a working computer program—were what was most essential. Search and heuristics clearly go together, since an exponential explosion of the number of possible solutions to a problem is common. This problem was anticipated by Claude Shannon in his paper "A Chess-Playing Machine," published in 1950, in which he emphasizes the importance of an automated search and the necessity of reducing its scope. The highest priority, Shannon reasoned, was the problem of "searching the space [of possible chess moves] for an acceptable solution."[35] It didn't have to be the best possible solution—only a workable one. Not coincidentally, Herbert Simon reports that he had entertained "thoughts about chess programming" in a RAND summer seminar in 1952 and had learned to program the IBM 701 in 1954. It was while Simon was working with Newell and Shaw at RAND on heuristics and problem-solving strategies that they realized that "we could use the computer to simulate all sorts of information processes and use computer languages as formal descriptions of those processes" (*Models of My Life*, 201). Their earliest efforts were directed toward writing a computer chess program, but the project was superseded by Logic Theorist. However, they did publish a revealing essay in which they assert that the essence of human thinking is problem solving, which is most clearly exemplified in the playing of chess.[36]

Alan Turing had also been very interested in chess as a "model for mechanical thought," and at Bletchley Park during the war he also spoke "about the possibility of computing machines solving problems by means of searching through the space of possible solutions, guided by rules of thumb."[37] In his unpublished essay "Intelligent Machinery" (1948) and more directly in "Computing Machinery and Intelligence" (1950), Turing confronted the question of whether "machines can think," arguing that there was no compelling reason why not. But it was still only an idea. That the Logic Theorist could prove theorems in logic without external help from its programmers thus seemed to make the notion of a thinking machine a reality. Certainly Simon and Newell thought so. Describing Logic Theorist, Simon flatly asserted, "We invented a computer program capable of thinking non-numerically" (190). And two years after Logic Theorist: "I don't mean to shock you. But the simplest way I can summarize is to say that there are now machines that think, that learn and that create. Moreover, their ability to do these things is going to increase rapidly until—in a visible future—the range of problems they can handle will be coextensive with the range to which the human mind has been applied."[38]

Considered philosophically, the question of what it means to think is certainly a vexed one; but no one could deny that proving a theorem in logic requires some kind of thinking. There was (and still is) the common sense objection that the machine itself does not really think—it only follows the instructions that its designers have programmed into it and therefore lacks the flexibility and creativity of human thinking. Yet this objection does not take into account what computers actually do and may be capable of doing, and it relies on unexamined assumptions about what makes possible and constitutes "human thinking." Thus there are really two questions here. First, what does it mean to say that a machine can think? Second, what does a human being do when he or she thinks? Simon and Newell answer both questions with a general theory of information-processing psychology based on the creation and manipulation of symbols. Both of these activities—whether they take place in a human or a machine—are forms of intelligent action made possible by what they call the physical symbol system hypothesis.

Simon and Newell's theory of the physical symbol system provided the cornerstone for the computational theory of mind, which was the underlying paradigm of both cognitivism and symbolic AI. The theory postulates that thinking, or cognition, is accomplished by the manipulation of symbols or tokens that represent some entity, pattern, or process in the world.[39] The symbols exist on two levels, that of their concrete embodiment (the physical) and that of the relations that allow them to be components of another type of entity called an expression, or symbol structure. The rules of their manipulation constitute a syntax, but because the symbols designate processes or entities beyond themselves, they include semantic constraints as well. Two types of symbolic action, designation and interpretation, are essential in this respect, since they evoke and perform processes by means of the expressions that designate them. Comprised of symbols, symbol structures or expressions are operated on by "processes of creation, modification, reproduction and destruction." The physical symbol system is thus "a machine that produces through time an evolving collection of symbol structures" while also existing "in a world of objects wider than just these symbolic expressions themselves" (40). This semantic dimension means that the symbol structures cannot be viewed as mere tokens in a formal system manipulated by rules independently of what the tokens represent. The heuristic search hypothesis, or principle means by which a physical symbol system solves problems, also goes beyond the constraints of a merely formal system: "The solutions to problems are represented as symbol structures. A physical-symbol system exercises its intelligence in problem-solving by search—that is, by generating and progressively modifying symbol structures until it finds a solution structure" (51). Whether these constraints actually anchor the symbol structures in external reality will be contested by the new AI.

The core hypothesis, in any case, is that such a physical symbol system

> has the necessary and sufficient means for general intelligent action.... By "necessary" we mean that any system that exhibits general intelligence will prove upon analysis to be a physical symbol system. By "sufficient" we mean that any physical symbol system of sufficient size can be organized further to exhibit general intelligence. By "general intelligent action" we wish to indicate the same scope of intelligence we see in human action: that in any real situation behavior appropriate to the ends of the system and adaptive to the demands of the environment can occur, within some limits of speed and complexity. (40–41)

This hypothesis, they emphasize, is to be verified or disproved by further empirical research. As with any hypothesis, it spells out certain assumptions while leaving others tacit or hidden. In particular, their hypothesis draws together and synthesizes various research efforts, beginning with Turing's concept of the universal machine and work on computability, subsequent developments of automated formal systems, especially the stored-program concept for the computer and the list-processing computer language, later formalized as LISP by John McCarthy in 1959–1960. Altogether, these precursor accomplishments make possible "the total concept," which is "the join of computability, physical realizability (and by multiple technologies), universality, the symbolic representation of processes (i.e., interpretability), and, finally, symbolic structure and designation" (46). This "total concept" yields a realizable recipe for the construction of an intelligent system.

Let us consider the physical symbol hypothesis more simply. A computer performs calculations by submitting symbols to a predefined repertory of possible operations. For example, it carries out mathematical operations on physical representations of numbers according to a rigorously mechanical procedure. To add 2 and 3, for example, they are first converted to binary numbers (0s and 1s), which are represented by voltage differences in an electrical current. These differences are brought together and "combined" in an array of switching circuits (the central processor). The resulting current (which would represent their sum) is then converted back from a binary number to the natural number 5. However, because this example involves only numerical calculation, it doesn't convey the full power of Newell and Simon's proposal. Since symbols can designate not only numbers but logical propositions, data structures, instructions, and so forth, the physical symbol system hypothesis applies to the manipulation of any type of symbolic expression, as long as the steps that constitute it can be written as a combination of algorithmic and heuristic procedures. As such, that is, *as* software, these procedures must be implemented in some form of hardware, either a digital computer or a biological brain. But it is the process—the manipulation of symbol structures—that defines thinking, not the medium or material substrate of the symbols.

In these terms the agenda for AI becomes clear: one only has to simulate a thought process by recasting it as a series of symbolic operators and expressions. This is similar to what Turing theorized in his essay "Computing Machinery and Intelligence"—find a mechanical equivalent for a thought process such that when the physical operation is carried out, the thought process is simulated—but in Newell and Simon's version it is accomplished at a higher level of abstraction. Yet the physical symbol system hypothesis implies something more: that "thought" does not operate directly in and on the world, but independently in representations composed of symbolic expressions and structures. At bottom, "intelligent action" consists of the manipulation and transformation of these expressions and structures, which can be expanded, broken down, reassembled, destroyed, and created. The efficacy of any particular expression or symbol structure can only be determined by comparing it with other expressions and structures. In this sense, the physical symbol system hypothesis locates the intelligent action within a closed realm of representations, in contrast to the behavior of the earlier cybernetic machines, which always took place in response to the larger physical environment. That a computer architecture provides the best image of this realm is directly affirmed when Allen Newell later rewrites the hypothesis as a universal, general-purpose computer program.[40]

Compared to the machines built or proposed by the cyberneticists, Simon and Newell's program/model not only operated at much higher, purely symbolic level of abstraction but claimed an independence from material embodiment itself. Significantly, their programs did not need to access the computer's machine language. Similarly, since in theory symbolic logic and the manipulation of symbol structures worked the same whether instantiated in a computer or biological brain, the physical implementation or substrate of intelligent action was no longer a matter of direct interest.[41] But Logic Theorist and General Problem Solver were not just theory—they produced tangible results. And they worked because of the software, that is, the underlying programming language, IPL, which was machine independent. (IPL went through five iterations before it was replaced by LISP.) Paul Edwards, in *The Closed World*, rightly emphasizes the multiple layerings of languages and systems that software requires, in order that conceptual independence can be achieved, asserting that "this insight into the possibility of a symbolic, machine-independent level of description in computing was the conceptual foundation of artificial intelligence" (246). Newell himself certainly understood that software was what AI was all about: "AI as a whole is founded on some striking methodological innovations, namely, using programs, program designs, and programming languages as experimental vehicles."[42] However, the machine independence of software or symbol manipulation allows and justifies a disconnection from the material brain, as the twin and coupled discourse of AI and cognitive science makes clear. In fact, Newell and Simon highlight this very separation: "Our theory is a theory of the information processes involved in problem-solving and *not a theory of neural or electronic mechanisms* for information processing" (cited in Edwards, *The Closed World*, 252). The brain—whether biological or mechanical—is replaced by the "psychology" of information processing.[43]

What we witness here is simply the displacement of one kind of computational assemblage by another.[44] The various cybernetic machines constructed or theorized by Shannon, von Neumann, Ashby, and Walter —and later the learning machines of Rosenblatt and Selfridge—are primarily dynamic systems physically situated in the environment and function with a relatively low-level information-processing capacity. The emphasis is always on the concrete behavior of these machines, but this behavior is sufficiently complex to support a discourse constituted by notions of feedback, adaptation, self-reproduction, and self-organization. Consequently these material automata evoke and to a certain degree can sustain an analogy with organic life. In the computational assemblage of early AI, on the other hand, the machine is displaced by the computer, but almost entirely in the latter's formal and conceptual aspect. Not only does the physical hardware so important to the cyberneticists drop out of sight, but so does the environment. Like the user himself, these physical actualities are shadowy and only implied. Indeed, the disembodiment of the subject and his or her reinscription in the psychology simulated by the symbol processing is rather striking. Though much attention is given to problem solving, decision making, searches, goals, logical operations, languages, and representation, these are processes without an identifiable subject. In this sense AI is truly a simulation of abstract thought.

The Cognitivist Paradigm

The importance of Logic Theorist was not lost on the participants at either conference, and its successful simulation of a specific kind of thinking went a long way toward establishing not only the viability of AI but a new research paradigm for cognitive science. Generally known as the cognitivist paradigm, or "the mind as computer" model, it quickly gained prominent adherents in AI and cognitive science as well as in the neighboring fields of linguistics and philosophy. Two examples, Noam Chomsky and Jerry Fodor, will be considered below. Nevertheless, *within* this paradigm there are degrees of variation and subtlety. One advocate, John Haugeland, states unequivocally that "AI ... is based on a theoretical conception as deep as it is daring: namely, we are, at root, *computers ourselves*."[45] But Newell and Simon's position is actually more nuanced. As Simon summarizes it in *The Sciences of the Artificial*, their founding hypothesis states that intelligence is the work of a physical symbol system; or, more strongly, that a physical symbol system is both necessary and sufficient to account for intelligent behavior. Physical symbol systems, furthermore, constitute a "family of artifacts," of which the computer and the human mind/brain are the most important. Thus while Haugeland starts with the idea of a computer as an automated formal system and then suggests that humans are more or less competent computers, Newell and Simon start with the idea of the physical symbol system and then find that it can be instantiated in different ways, as in human brains and computers. In other words, whereas Haugeland gives us an idealized model to which concrete instances seem more or less adequate, Newell and Simon provide a functional diagram for understanding what must take place in order for there to be cognition, or "intelligent action." While Haugeland essentializes and idealizes intellectual functions, Newell and Simon abstract, or deterritorialize, these functions and then reterritorialize, or recode, them in a symbolic, operational language.

The subsequent history of AI until the mid-1980s is precisely the record of the development and limits of the physical system hypothesis. According to Herbert and Stuart Dreyfus, the research program launched by Newell and Simon has gone through three ten-year stages.[46] The first stage (1955–1965) was dominated by problems of representation and search, specifically by the heuristics approach first deployed in Logic Theorist and further developed in Newell and Simon's more ambitious General Problem Solver. The second stage (1965–1975) continued to investigate how data and rules could be implemented in a systematic program but deliberately reduced its ambition to the construction of simple and isolated "micro-worlds," with the hope that once success was achieved with these restricted worlds, efforts could be directed toward more realistic, real-world problems. The best-known effort in this vein was Terry Winograd's program SHRDLU, constructed around 1970 at MIT, which could obey commands given in a subset of a natural language about a simplified "blocks-world." The third stage (1975–1985) focused primarily on what came to be known as the commonsense knowledge problem, that is, how to incorporate not only commonsense but the immense background of assumptions that humans take for granted when solving just about any practical problem. Here the problem of representation was addressed in terms of various kinds of frames and scripts. However, despite some limited success with certain "expert systems" that succeeded in narrowly constricted domains such as medical diagnosis, the intractability of the commonsense knowledge problem eventually led to the widespread feeling that AI research had reached an impasse.

Certainly by the mid-1980s many believed that classic AI research founded on symbol processing and a computational model based on algorithms and heuristics had run its course. Two projects in particular, notable for their grandiose ambition and signal failure, should be mentioned. The first, Japan's very expensive "Fifth Generation Project," was to be a computer distinguished by the ability to understand its users' needs (as communicated in a natural language) and to design programs that would meet those needs. However, it gradually became obvious to all concerned that the programming languages required to realize these ambitions were either too far in the future or inherently untenable. In either case, the ambition itself was flawed, and by the end of the decade the project was abandoned as an acknowledged failure.[47] The other project, called CYC and launched by Douglas Lenant and his group, faired only slightly better. After considerable success in the 1970s with programs for machine learning, particularly one called Automated Mathematician, in the early 1980s Lenant and his group embarked on an attempt to encode literally millions of assertions in a vast data base of common knowledge that, with the proper programming, would lead to "semantic convergence" and thus to something like the artificial (re)creation of common sense. The group assumed that if they encoded a complex hierarchy of interlocking frames into the computer's programming, it would gradually acquire a repertory of concepts and categories large enough to enable it to encode new statements. For example, consider the statements, "Napoleon died on St. Helena. Wellington was saddened." This relatively simple pair of declarations contains a number of embedded assumptions (that Napoleon and Wellington are human individuals, that all such individuals eventually die, that their deaths affect still living individuals, etc.), which are understood immediately by natural-language users everywhere, even if they know nothing about the specific context (here, European history). Even so, it took Lenant's team three months to encode the two statements. Again, the group assumed that the coding would eventually go much faster, one day reaching the point when it could be completely automated. Critics, however, remained skeptical, pointing out that "every sentence required the definition of a new and arbitrarily long chain of related categories, and reality seemed to branch out into an infinitely large number of unrelated concepts."[48] Indeed, this is the very problem that classic AI failed to resolve in the 1970s. Far from being a theoretical breakthrough, CYC simply continues to build up (and on) ad hoc representational schemes. It is the old AI writ large, and founders on the impossibility of recoding in machine language what appears to be coded in natural language.[49]

As already noted, the value of the physical symbol hypothesis was not limited to its conceptualization of human intelligence (and therefore of AI). During the 1960s it also came to provide the unity and theoretical grounding for cognitive science as a new discipline. No doubt one of the most important consolidations of the symbolic computational approach outside of AI came from Noam Chomsky's theory of generative grammar, the importance of which was immediately recognized with the publication of *Syntactic Structures* in 1957. Chomsky demonstrated that in order for a grammar to adequately account for natural language, it would have to be generative, that is, it would have to embed recursion rules in order to generate an infinite number of new sentences; but it would also have to be transformational, that is, it would have to embed rewrite rules that would allow one type of phrase structure to be transformed into another. Developed in order to understand the grammar of natural language, these kinds of rules could be easily applied to symbol strings and their concatenation and thus gave additional strength to what was considered to be the standard computational paradigm.

Chomsky, however, actually faced a double problem: to discover the rules that allow a speaker to understand and produce an infinite number of sentences, and to understand how these rules are acquired. The latter was a particularly vexing problem, given the observable fact that children in a specific linguistic environment are able to instantiate these rules in speech relatively quickly and on the basis of imperfect and fragmentary perceptions. As Chomsky himself has pointed out, no empirical theory—that is, no theory based on accumulation, association, and generalization—could possibly explain this fact.[50] The difficulty, simply put, was to ascertain where (outside of linguistic theory) these rules exist and how they are implemented by human speakers. Chomsky's solution was to distinguish between "competence" and "performance." Competence refers to the knowledge internalized by a speaker that unconsciously enables him or her to understand and produce an infinite number of new sentences. Generative grammar is the explicit account of this knowledge. Performance, on the other hand, refers to the production and comprehension of language by actual speakers in real time, in which other systems (memory, vocalization, etc.) intervene. While performance is obviously influenced by a number of individual contingencies, Chomsky believes that competence is guaranteed by an innate, hardwired feature of the human brain. It is this innate capacity, moreover, that enables a child to acquire a human language; *which* language depends only on the particular linguistic environment in which the child is brought up.

Although Chomsky makes no mention of the Newell-Simon hypothesis, there can be no doubt that his theory of generative grammar lends it both weight and credibility. This is made explicit in the philosophy of Jerry Fodor, where we find a direct linkage between the two. Building on Chomsky's work, Fodor argues for the necessity of what he calls a "language of thought" composed of bedrock linguistic-like mechanisms ensuring that a capacity for symbol manipulation is inherently part of our native cognitive apparatus.[51] Specific examples will be considered in the next section. What must be emphasized here—and it is reflected in Fodor's work—is that the development, relay, and support of the cognitivist computational hypothesis forms a single block extending from AI to linguistics to cognitive science and philosophy. Yet even while the symbolic computational paradigm functioned as the core hypothesis for cognitive psychology from the early 1960s well into the 1980s, completely displacing behaviorism, unease with the innateness hypothesis persisted among more empirically minded psychologists and philosophers. For one thing, its double functionality was troubling: by postulating the symbolic function as a biological given, the singularity of the human was also assured as a "natural" capacity; and by establishing a biological, predetermined basis for mental functioning, the rigid hierarchy of formal and abstract cognitive functions valorized by the cognitivist paradigm was firmly anchored in nature. Yet these assurances were advanced in a context outside of any dialogue with evolutionary theory or palpable interest in how "the language of thought" could be explained in Darwinian terms.[52] Here we see a clear instance of the structure-change dichotomy mentioned at the chapter's outset.

However, if we consider cognitive science and symbolic AI as two particular discourses, we can understand how this double functionality answers to the specific and dynamic logic of a decoding and a recoding. On the one hand, these new sciences both effectuate and valorize an abstraction and deterritorialization of cognitive functions, which are then reinscribed in the highly abstract, formal languages of symbolic logic and computer programming. On the other hand, this decontextualization (or decoding, in Deleuze and Guattari's sense) seems to call inevitably for a recoding at another level; specifically, the abstracted cognitive functions find their guarantee and necessary grounding in the innateness hypothesis of biological hardwiring. Yet this recoding doesn't simply reflect an ideological bias, since there seemed to be no conceptual alternative, at least until the rebirth of neural net research in the 1980s. However, with the resurgence of interest in the brain model, as neural net theory was sometimes called, an entirely new conception of how coding occurs becomes available, one that makes the language of thought look as if it were based on subsymbolic, statistical connections well beneath the threshold of consciousness. These connections are not given as the components of an innate structure but are the consequence of temporal material processes. Thought thus finds its condition of possibility not in an underlying but theoretically transparent set of logical and syntactic structures, but in the changes and continual self-modifications of a dynamic network.

A striking anticipation of this cataclysmic shift in the modeling of thought appears in Douglas Hofstadter's essay "Waking Up from the Boolean Dream, or, Subcognition as Computation."[53] Without abandoning the computational perspective, Hofstadter espouses his own version of subsymbolic cognition and sketches the essentials of a multiagent system. The freshness of Hofstadter's approach appears at the outset. To the question, what is AI? he responds with the question, what is the letter *a*? In other words, how are all *a*'s alike, and what is the intelligence already at work in the process of recognizing letters, which can be rendered in a wide diversity of scripts, typefaces, and degrees of legibility. From letters, of course, it is only a short step to symbols, but for Hofstadter the problem with symbols, at least as Newell and Simon define them in the physical symbol system hypothesis, is that they are too passive, and their manipulation according to rules doesn't correspond to the processes of actual thinking. Hofstadter argues, consequently, that it is not cognition but *subcognition* that is computational, and it involves what he calls active symbols, which "activate or trigger or awaken other symbols in a brain." These active symbols result from "team" activity in the brain:

> The brain itself does not "manipulate symbols"; the brain is the medium in which the symbols are floating and in which they trigger each other. There is no central manipulator, no central program. There is simply a vast collection of "teams"—patterns of neural firings that, like teams of ants, trigger other patterns of neural firings. (648)

The upshot of this formulation is that we are not symbol manipulators; rather, "we are manipulated by our symbols" (648).[54] For Hofstadter, these active symbols are what connect vast collections, or "clouds," of neural firings to semantic categories; hence "symbol triggering patterns are the roots of meanings" (650). Far from being a formal, rule-bound activity, thought springs from an evolutionary need to survive. Humans thus use these semantic categories in a makeshift way in order "to imagine a cluster of approximations of what may happen and anticipating some plausible consequences of them" (650). Hofstadter concludes that high-level, global, cognitive events are not in themselves computational, but epiphenomenal. They are constituted out of and driven by many smaller computational events at a substrate level. Given this perspective, the goal of AI is to bridge the gap between cognition and subcognition, which can only be done by giving up the Boolean dream for a "statistically emergent mentality" (654).

Neural Networks: Statistical Coding and the Recoding of the Symbolic

Having staked its claim that thinking is symbol processing and thus in principle capable of being emulated by software, classic AI research soon found itself increasingly dependent on advances in computer hardware. As Rodney Brooks observes, for a long time rapid increases in computing power kept AI researchers from having to reexamine their founding assumptions.[55] Even today the claim is often repeated: when machines attain a certain level of processing speed and memory, artificial intelligence will become a reality.[56] In contrast, for cognitive science the issue of "hardware implementation" posed a problem from the outset, since there is nothing in the brain that corresponds to physical symbols and their rules of operation. McCulloch and Pitts, of course, argued that because the brain's networks of neurons ("neural nets") functioned as "all-or-none processes," that is, as arrays of on-off switches, their operations could be treated in terms of propositional logic (i.e., as performing computations). But these networks were nothing like the architecture developed for the modern digital computer. In fact, as a computing device the brain doesn't resemble the computer at all: it contains no separable central processor or storage unit for memory; instead, it consists of a spongy mass of some hundred billion elaborately interconnected neurons, some with as many as ten thousand connections to other neurons. Moreover, there are few visible hints as to how this tangle of "wetware" could implement the kind of logical thinking instantiated by Newell and Simon's Logic Theorist, much less the range of mental activities exhibited by a normally functioning human being. It is hardly surprising, then, that the physical structure of the human brain would give rise to a completely different paradigm for computation, one that has challenged not only the dominance of classic AI but the cognitivist or symbolic computational hypothesis more generally.

There are several ways to tell this story, and even Papert's fairy-tale version still holds a certain interest. In 1949, following McCulloch and Pitts's research on neural nets, Donald Hebb postulated that learning could be explained as the result of progressive modifications of connections among neurons; specifically, when neurons fired together in response to the same stimulus, their connections were strengthened; if not, their connections weakened.[57] Building on this work, Frank Rosenblatt constructed an artificial device called the Perceptron, in which the links between connected nodes would imitate the behavior of neurons.[58] The Perceptron, Rosenblatt explains, "was designed to illustrate some of the fundamental properties of intelligent systems ... without becoming too deeply enmeshed in the special conditions which hold for particular biological organisms" (180).[59] An early version, the Photoperceptron, consisted of three units. The first, or sensor unit, was an array of photoelectric cells meant to function somewhat like the retina of the eye when exposed to a pattern of light and dark. The cells fed randomly into a network of artificial neurons consisting of variable switches that composed the second, or associator unit, the output of which fed into an activator unit. Since the objective was to train the Photoperceptron to recognize letters, the activator unit consisted of a bank of labeled lights. Meant to replicate one kind of brain activity, the idea behind the Photopercepton's functioning was simple: when presented with an input pattern from the photoelectric cells, the neural network unit would organize itself into a global state that would be the specific output pattern for that input.

Initially, this might not seem like much. However, because the network's capacity to organize itself depends on a combination of states in dynamic interaction rather than any single neuron's state, the network as a whole is capable of complex relationships between input and output. In fact, it can not only recognize old patterns but learn new ones. Although we now know that biological neurons behave much more complexly, artificial neurons capture an essential feature: each one is a "weighted" node in a highly interconnected network of similar nodes. In order to fire, or pass electrical current, it must attain a threshold level, which is determined by two quantities: its own internal weight and the combined weights of the inputs from its neighbors, which, like its own effect on them, can be either positive (excitatory) or negative (inhibitory). For any given input stimulus, consequently, each neuron is affected not simply by the stimulus itself but by how all of its neighbors in the network are also affected. If the sum total of all these weights is positive and exceeds the neuron's firing threshold, it fires. Since every neuron in the neighborhood affects every other neuron in the same manner, the overall effect is a complex and dynamic range of possibilities.[60] However, for the neural net as a whole to function, it has to reach a stable state, such that for a specific given input the corresponding or associated output always occurs, and in artificial neural networks this is accomplished by adjusting the weights of each individual neuron. In natural neural nets in the brain and central nervous system this learning process occurs spontaneously, through interactive feedback processes with the environment, as the creature discovers and repeats functional behavior. In artificial neural nets the weight adjustments can be either supervised and directed or allowed, like natural nets, to self-organize into stable patterns autonomously. However—and this is what is most important—in both natural and artificial nets the information that defines each input pattern is stored as a particular distribution of weights in the network's connections. It is not stored in any specific location, but as a pattern distributed throughout the network.

For the Perceptron, the weights had to be adjusted by Rosenblatt himself, a process that came to be known as training the net. By adjusting these weights Rosenblatt could teach the net to recognize a particular input pattern, one that might be given by the shape of the letter *S*, for example. It would not have to be a perfect *S*, for the Perceptron could compensate for noise, that is, missing elements or distortions. What's more, if some of the individual neurons malfunctioned, the net as a whole usually remained operational or, over time, would exhibit what would later be known as "graceful degradation." Like the human brain, which loses thousands of neurons daily, artificial neural nets like the Perceptron are highly fault tolerant.

McCulloch and Pitts had claimed that neural nets could perform several sets of computational functions in a propositional calculus and were thus equivalent to Turing machines. But Rosenblatt was interested in pursuing a different line of research, namely, discovering how information is stored and remembered in biological systems and how it influences recognition and behavior. Two alternative explanations had already been formulated. According to the first, sensory information could be stored "in the form of coded representations or images, with some sort of one-to-one mapping between the sensory stimulus and the stored pattern." According to the second, "images of stimuli may never be recorded at all, and the central nervous system simply acts as an intricate switching network, where retention takes the form of new connections, or pathways, between centers of activity" (179). The first hypothesis, that is, a coded, representational memory, means "that recognition of any stimulus involves the matching or systematic comparison of the contents of storage with incoming sensory patterns" (180). In the second hypothesis, contrastingly, "there is never any simple mapping of the stimulus into memory, according to some code which permits its later reconstruction. Whatever information is retained must somehow be stored as a *preference for a particular response*; i.e., the information is contained *in connections or associations* rather than topographic representations" (179). The second hypothesis thus does not distinguish between memory and activation: "Since the stored information takes the form of new connections, or transmission channels in the nervous system ... it follows that the new stimuli will make use of these new pathways which have been created, automatically activating the appropriate response without requiring any separate process for their recognition or identification" (180). To his credit, Rosenblatt recognized that the second hypothesis explains how neural networks actually work.

The connectionist hypothesis led Rosenblatt to another highly fruitful discovery. With the application of symbolic logic to switching theory and the development of digital computers, many researchers simply assumed the functional equivalence between neurons and the on-of switches that make up the circuits of these computers. But neural nets containing many random connections actually behave very differently from the logic arrays of a digital computer. Rosenblatt realized that this behavior does not readily lend itself to description in the language of symbolic logic and Boolean algebra. He decided therefore to formulate his own descriptive model in terms of probability theory, which is much better suited to describing the behavior of a cluster of connected neurons than symbolic logic. On this basis, he was able to demonstrate that "the fundamental phenomena of *learning, perceptual discrimination and generalization can be predicted entirely from six physical parameters*" (195). These six parameters are variables defined by the number of excitatory and inhibitory connections in the network, their threshold values, and the proportions among types of connections. These parameters essentially define the weights and the types of nodes that comprise the network. In sum, neural net devices like the Perceptron, with an acentric, distributed architecture and functionality, promised to lead to a major advance in understanding how information is stored in biological systems and to an alternative to symbolic computation.

In the early 1960s Rosenblatt was not the only one doing research on artificial neural nets. Oliver Selfridge built another pattern recognition machine worth mentioning. Called Pandemonium, it differed from the

Perceptron in that each of the units in the network was assigned to a particular feature.[61] If a sufficient number of these "feature demons" were activated, specialized "cognitive demons" would then "shout out" the results to a "decision demon"—the greater the evidence, the louder the shout. In this way a decision was reached that summarized the findings of many parallel and independent computations.[62] Devices like the Perceptron and Pandemonium, at this early stage in the development of the AI community, were generally viewed as worthy and valuable pursuits. Neural net theory was deemed neither a rival nor an alternative to symbolic AI, but simply a different avenue of exploration. This was due in part to the fact that while artificial neural nets could recognize and classify patterns of stimuli, it was not clear whether or not they could represent such acts. The evidence, on all accounts, suggested that the act of perception/recognition itself could not be symbolized and referred to, but had to be repeated again and again.[63] In the 1960s this and other questions instigated efforts to assess more thoroughly the potential and limits of neural nets, both in theory and in their implementation in constructed devices.

The most consequential examination was made by Marvin Minsky and Seymour Papert. An original participant in the Dartmouth Conference, Minsky had published one of the earliest and most influential papers on symbolic AI,[64] and he later would become one of its most distinguished advocates and researchers. But before that, in 1954, he had written his PhD dissertation at Princeton on "Neural Nets and the Brain-Model Problem," and had long been interested in neural net research. A gifted mathematician, in 1967 he published *Computation: Finite and Infinite Machines*, which became a classic of computational theory. Throughout this period Minsky had experimented with neural nets and even engaged in spirited public debates with Rosenblatt. In 1969 he and Papert, a colleague at MIT, published *Perceptrons*, which grew out of their mostly disappointing efforts to establish the scientific value of Rosenblatt's work and of neural nets in general. According to canonical accounts, the book literally delivered a deathblow to neural net research, which would not be resumed for some fifteen years. As Papert acknowledges in the fairy-tale version of these events quoted above, there had come a point when symbolic AI and neural net research were seen as rivals for Defense Department funding, and he and Minsky were the self-appointed assassins.

Curiously, a somewhat different picture emerges from the pages of Minsky and Papert's book, especially the revised edition of 1988, which includes a new prologue and epilogue written after the revival of neural net research in the mid-1980s.[65] Although they readily admit in the prologue that by the end of the 1950s "battle lines began to form along such conceptual fronts as parallel versus serial processing, learning versus programming, and emergence versus analytic description" (xi), their own purpose had been to make an objective, scientific assessment of the strengths and weaknesses of neural nets as learning machines and computational devices. In the period following the publication in 1962 of Rosenblatt's own book, *Principles of Neurodynamics*, which summarized neural net research up to that point, "progress had come to a virtual halt because of the lack of adequate basic theories" (xii). In particular, there was no rigorous theoretical account of why nets could recognize certain patterns but not others. In attempting to answer this question, they discovered other limitations as well, most importantly, an inability to perform the XOR (exclusive or) computation. Minsky and Papert also pointed to the mathematical intractability of learning rules for neural nets above a certain level of complexity, essentially arguing that there could be no reliable algorithm for training these nets to solve any but the most elementary problems. Meanwhile, important advances in AI were being made "through the use of new kinds of models based on serial processing of symbolic expressions" (xi). On balance, then, the symbolic approach "suddenly seemed more satisfactory" (xi).

This same tone, at once rigorous and friendly, also animates their epilogue, which attempts to address the claims of "the new connectionism," as the mid-1980s revival of interest in neural networks led by David Rumelhart, James McClelland, and fourteen collaborators who form the PDP Research Group is usually called.[66] On the one hand, Minsky and Papert readily acknowledge that distributed parallel processing can do many things and that connectionist networks must be included in a full understanding of how the brain actually works. It seems certain, they assert, that the brain is composed of many small, highly specialized networks that are interconnected in multiple ways. On the other hand, the strength of these networks is also their weakness, inasmuch as the widely distributed representations that emerge from them cannot be made the object of symbolic representation, and they lack the power of extension and generalization that characterize systems built on symbol processing. (I return to this issue below.) Thus, while refusing the role of the "enemy" of connectionism, in the revised edition the authors remain consistent with their original intentions as well as their overall negative judgments as set forth in the 1969 edition.

The revival of neural net theory and the advent of the new connectionism is an event of considerable interest and importance. Before we consider particulars, a brief look at the wider cultural context may be in order. In *Invitation aux sciences cognitives*, Varela points to the renewal of interest in self-organizing systems and nonlinear mathematics, citing in particular Ilya Prigogine and Isabelle Stengers's *Order out of Chaos* (published in France in 1979 under the title *La nouvelle alliance*), as well as the new availability of powerful computers (55). Similarly, in *Connectionism and the Mind*, William Bechtel and Adele Abrahamsen emphasize the development of powerful new approaches to network modeling that gave rise to new architectures and new training techniques as well as to new mathematical models of description. They also point to a growing interest among cognitive scientists, who were spurred by an increasingly acute sense of the limits of symbol-processing models. The rule systems of the latter seemed too complex, brittle, and ad hoc compared with actual human cognition, which "seemed to be relatively free of such limitations."[67] Yet to a certain extent these explanations of the sudden and widespread interest in neural nets and nonlinear mathematics and decentralized models presuppose one another, and therefore what they are meant to explain. Perhaps the renewed interest in neural networks forms part of a larger cultural shift, a movement of decentralization visible in social and economic organization, technologies, scientific models, theories of self and mind, and theories of knowledge.[68] Whatever the reasons, the shift in the basic model of AI and cognitive science is striking. Whereas the "mind as computer model" postulates intelligence as the manipulation of symbols according to a syntax of well-defined rules, the neural net brain model understands intelligence as a coherent global state emerging from the dynamic and only statistically knowable interaction of numerous nonmeaningful components. Not surprisingly, the merits of the two conflicting models have been extensively discussed, and their differences have incited serious philosophical debate. Perhaps the best place to begin is with the two most salient points in Minsky and Papert's critique.

First, Minsky and Papert showed that a neural net could not perform the XOR computation, which yields a positive output if and only if its two inputs are different; otherwise, it returns a negative output. Since its negation, NOT(XOR), is a computationally universal function—meaning that all other functions can be derived from it, and thus a computer can be built from this function alone—this appears to be a serious limitation indeed. For this very reason, in fact, Minsky and Papert argued that neural nets are not universally computational. In the mid-1980s, however, Rumelhart and McClelland's group were able to show that simply by adding another layer of neurons between input and output layers (usually referred to as a hidden layer) neural nets gained the capacity to perform this computation.[69] Second, Minsky and Papert insisted on the mathematical intractability of learning rules for neural nets above a certain level of complexity, essentially arguing that there could be no reliable algorithm for training these nets to solve any but the most elementary problems. Yet this "inherent" limit was also surpassed, mainly through the development of what are called gradient descent algorithms, in which a learning rule (basically, a formula for adjusting the weights) is applied repeatedly until the difference between the desired and actual output is eliminated (see Rumelhart, Hinton, and Williams, "Learning Internal Representations," 321–328). The most effective algorithm of this type, known as "back propagation" and discovered simultaneously by several researchers in the mid-1980s, made it possible to train a multilayered, "feedforward" network in real time. About the same time, other types of network architecture were also developed, most notably autoassociative networks, in which all the nodes are connected to one another, and recurrent networks, in which layers have both feedforward and feedback connections. Since these architectures produce highly nonlinear effects, new efforts were directed toward determining what configuration of weights and paths is responsible for the output. In sum, the mid-1980s saw connectionist research advance simultaneously on two fronts, with researchers discovering both new kinds of architectures and new learning algorithms.[70]

One of the most important of these advances occurred in 1982, when physicist John J. Hopfield published a remarkable essay, "Neural Networks and Physical Systems with Emergent Collective Computational Abilities," that greatly stimulated interest in neural net research.[71] Hopfield was interested in the peculiar dynamics of spin glass, a certain class of alloys whose magnetic properties depend on how their atoms are arrayed in lattices with up or down spin orientations. The spin orientation of each atom, moreover, would influence that of its neighbors, leading these atoms to evolve collectively to global states in a manner that could be described by the mathematics of the rapidly developing field of nonlinear dynamical systems theory. Specifically, the states could be represented by attractors in a phase space mapping of all the possible states that the spin glass system could assume. The attractor states would indicate points of equilibrium or stability between the system's random heat loss and the interacting organizational forces of the spin glass atoms. Hopfield realized that there was a deep analogy between the dynamics of this system and that of a certain kind of autoassociative neural net (now known as a Hopfield net) in which all nodes are connected to one another. Given an initial input, the network will evolve toward and then self-organize around one of these attractors, which thus corresponds to a stable pattern. In fact, the network can reliably store as many patterns of information as there are attractors. Not only does the Hopfield net *not* have to be trained, but because it seeks configurations of equilibrium spontaneously it can be effectively deployed to solve computationally difficult problems like "constrained optimization"—the traveling salesman who must visit a cluster of cities and has to figure out the order that will result in the least repetitive sequence is the best-known example. In settling into patterns of stability, however, a Hopfield net only goes "downhill," as it seeks the equivalence of local energy minima. To overcome this limitation, Geoffrey E. Hinton and Terence J. Sejnowski added a hidden layer to a Hopfield net, creating what they called a Boltzmann machine (after the great scientist of statistical mechanics), capable not only of modifying its own connectivity and thus of unsupervised learning but also of moving "uphill."[72] This made it possible to solve a variety of constrained optimization problems. As Cowan and Sharp state, it "provides a way in which distributed representations of abstract symbols can be formed and therefore permits the investigation by means of adaptive neural nets of symbolic reasoning" ("Neural Nets," 102).

Summarily, then, by the mid-1980s all of Minsky and Papert's objections to neural nets had been refuted or overcome, leading to the rebirth of neural net research and a new beginning in the quest to model intelligence and simulate the workings of the brain. This rebirth essentially amounted to a series of spectacular successes in the theory and application of connectionist models and parallel distributed processing, successes all the more impressive in that they were perceived against a background of failure and dead ends in symbolic AI. Two accomplishments in particular should be mentioned. In a striking application of back propagation, Sejnowski and Charles Rosenberg trained a neural net machine called NETtalk to "read" and pronounce printed English words.[73] Perhaps more impressively, D. E. Rumelhart and J. L. McClelland were able to teach a neural net how to form the correct past tenses of both regular and irregular English verbs.[74] Remarkably, they discovered that in the process of learning the neural net actually made the same kinds of mistakes (and at the same rate) that children make in learning the past tense.

Since then neural net theory has made great strides in understanding how the brain and central nervous system work, particularly in regard to sensory perception and motor control, memory, and learning. Beyond any doubt, these accomplishments demonstrate that neural nets are capable of pattern recognition, categorization, and (to a limited degree) generalization. Of particular importance, neural nets can encode content addressable memory, a capability that brings a whole new dimension to theories of learning and cognition not available to computer-inspired models of the brain based on symbolic computation. The literature on these developments is extensive, but even a brief indication of how the brain represents input from the sensory world will indicate the richness of this system.[75]

When we bite into a ripe peach or take a sip of good-quality wine, we are immediately assailed by a profusion of tastes and smells. The same is true of course for our other senses when we dive into a fresh pool of water or listen to a complex piece of music. How does our brain process any one of these highly distinctive onslaughts of fresh sensory information? And how is it represented in the brain? Let's consider the taste of a peach. Four basic types of cell receptors (corresponding to sweet, sour, salty, and bitter) are found on the tongue. When the juice from the peach hits these receptor cells, they are excited and produce a unique activation pattern determined by the excitation levels of all four cell types. The pattern is not a mix but a unique combination, like a signature or footprint. And of course there are always multiple patterns, produced in temporal waves. These activation patterns are passed to successive layers of neurons where they are transformed into other activation patterns and passed to other neuron groups, and so on throughout parts of the brain. What happens to this series of activation patterns will partly depend on whether this is a first time experience or a repetition, in which case there is a very similar activation pattern already stored in memory. If not, it will be passed upward to higher cognitive levels. These activation patterns can be represented or coded as vectors in a state space. The peach activation pattern would be coded by a vector defined by the four values of the excitation levels of each of the four receptor cell types. This vector, in turn, can be mapped as a point located in a four-dimensional space, with each receptor value determining the length along one of the four axes. Though difficult to visualize, the vector points in this 4-D space would represent all possible tastes, with differences and similarities in taste reflected by the distances between individual vectors. Peaches and apricots, for example, would be found in one region in fairly close proximity, while black olives and green olives would be found close together but in another region. The combinatorics of vector coding thus constitute a simple but rich system that can represent the extraordinary range of "sensory subtleties [the nervous system] encounters" (Churchland, *The Engine of Reason*, 21).[76] In the brain itself, of course, the activation patterns are stored in the variously weighted synaptic connections among the trillions of networks of neurons.

The Philosophers Clash

Even with the eye-opening breakthroughs achieved by neural net research in the 1980s, the research agenda of classic AI did not simply fade away, nor was it dispatched to the historical dustbin. Indeed, in its own way symbolic AI continued to set the agenda, at least for research on "higher" cognitive and intellectual functions, though often indirectly from the precincts of math and logic within philosophy, while relegating connectionism to neuroscience. Following Jerry Fodor's *The Language of Thought*, several philosophers have taken up the attack against connectionism, arguing that, unlike computational or symbolic AI, it cannot provide a fully adequate model of cognitive functions. The ensuing philosophical debates over the relative merits of symbolic AI and connectionism dramatize the fundamental differences between symbol-processing and neural net approaches to computation and how each understands what intelligence is and how it operates. The first valorizes the formalized procedures and relations of syntax and logic; the second, the perception of pattern and the dynamics of self-organization. While the first must postulate a series of homuncular agents that implement or carry out cognitive tasks at lower levels, the second must assume that cognitive functions somehow emerge from the bottom-up dynamics of a self-organizing neural network.

These debates may call to mind how the two approaches reach back to and recapitulate differences within the history of Western philosophy. For Plato, intelligence entailed the anamnesis of innate Ideas imprinted on the soul, whereas for Aristotle it involved the perception of analogies in the actual world. Later, Hobbes, Descartes, and Leibniz would understand thinking as basically a form of reckoning (i.e., calculation), whereas for Locke, Hume, and the British empiricists it was a process of mental association. Symbolic AI and connectionism manifestly draw on and extend these familiar oppositions into new realms, with something new emerging on both sides: symbolic AI does not simply repeat the long familiar tenets of rationalism but brings precision and detail to our knowledge of the mechanics of symbol processing; similarly, connectionism is not merely associationism reborn, but rather a different kind of computational architecture that may reveal how symbolic structures are implemented at subsymbolic levels, both in natural and artificial computational devices. The clash between the cognitivist philosophers Jerry Fodor and Zenon Pylyshyn and the connectionist researcher Paul Smolensky goes to the heart of these differences, which are explored below in relation to a dynamical systems theory of language. A second encounter—between the philosophers Daniel Dennett and Paul Churchland—over the nature of consciousness, will in turn implicate these differences in the phenomenology of conscious states and the possibility of conscious machines.

In "Connectionism and Cognitive Architecture: A Critical Analysis," Fodor and Pylyshyn try to pinpoint the inherent weaknesses of connectionism and parallel distributed processing for a cognitive theory of mind.[77] Detailed and rigorous (at least in its own terms), the essay argues that there are certain things that an adequate theory of cognition or intelligent behavior must be able to account for. The classic theory of cognitive architecture (i.e., Newell and Simon's physical symbol hypothesis, Fodor's language of thought) meets these conditions, whereas connectionism fails. The authors do not deny that connectionism may explain how this cognitive architecture is implemented in a subsymbolic domain, but in itself, they argue, it is *not* an adequate theory of cognition.

At the outset Fodor and Pylyshyn acknowledge that connectionist models are representational, that is, that a connectionist network can represent a thought or mental state. The problem is that connectionist representations, unlike the representations of classic cognitive architecture, lack a combinatorial syntax and semantics. In other words, a connectionist network has no inherent *structure* that would allow for complex mental representations. They give a simple example. In the classic model, if we are given the proposition **A & B**, we can logically infer **A** with a preservation of truth value. However, although a connectionist network can represent this same relationship, the relationship among constituent parts is fundamentally different. In the connectionist network the relationship between **A & B** and **A** is only a causal or associative one. In other words, since the connectionist network effectuates a mapping of an input onto a statistical distribution of weights among nodes, the relationship of the particular distribution that yields **A & B** to the one that yields only **A** involves no structural relation like part/whole or logical entailment. In short, for any array of symbols modeled by connectionism the relationship among the symbols can only be "atomic," whereas for the classic architecture it can be either atomic or syntactic and semantic.

Put another way, representations in connectionist networks are distributed over microfeatures. For the proposition "Mary drinks a cup of coffee," the connectionist representation will in effect consist of a concatenation of nodes that give something like "+Mary-subject; +drink-verb; +cup (of coffee)-object." But simple changes in the distributions of weights in the nodes might yield "A coffee of cup drinks Mary." This is because the connectionist architecture lacks the kind of inherent, systematic constraints that define language. It should be noted, however, that instead of the term "language," Fodor and Pylyshyn speak of "thought or mental processes." As they later explain, "linguistic capacity is a paradigm of systematic cognition, but it's wildly unlikely that it's the only example" (329). Leaving this distinction aside for the moment, the crucial point for Fodor and Pylyshyn is that while connectionist networks can represent linguistic propositions, they do not instantiate the kinds of relationships (i.e., syntactic and semantic constraints) that are required to account for human cognition.

In contrast, because the classic cognitive architecture of symbol systems entails "not just causal relations among representational states but also relations of syntactic and semantic constituency," it thereby provides an adequate cognitive model. Fodor and Pylyshyn discuss this adequacy specifically in terms of *productivity*, *systematicity*, and *inferential coherence*. By productivity they mean that unbounded expressive power is achieved by finite means, mainly by means of recursive structures. (Here again, the influence of the language model is clearly evident.) By systematicity (sometimes called compositionality) they mean that "the ability to produce/understand some sentences is intrinsically connected to the ability to produce/understand certain others" (330). Anyone capable of thinking the thought "Tim loves Mary" will also be capable of thinking the thought "Mary loves Tim." In short, thought has a composite structure. What guarantees systematicity, of course, is syntax, and Fodor and Pylyshyn contrast the mastery of syntax with the phrase book approach to speaking a foreign language. While a phrase book can provide an additive, one-to-one correspondence between utterances and meanings (i.e., a lexicon), it cannot convey the relational and systematic understanding that is absolutely necessary to form and comprehend thoughts that are not in the phrase book, even though some or all of the constituent parts might be found there. Inferential coherence, finally, restates and expands on the first example given above. By virtue of syntax, logical relations of entailment and inference are intrinsic to thought. As Fodor and Pylyshyn put it, the mental representations that express thought are "structure-sensitive." Although logical and syntactical relationships can be represented by connectionist networks, there will always be gaps and inconsistencies that the structure of classic architecture does not allow. Hence the positive features that distinguish connectionist networks and parallel distributed processing are irrelevant for their value as cognitive models. At best, connectionist networks can only serve as a theory of implementation for the classic model of symbol processing.

In "Connectionism, Constituency, and the Language of Thought," Paul Smolensky meets this critique head on and offers a powerful rebuttal, arguing that connectionism should be viewed as a refinement, not an implementation, of the classic symbolic approach.[78] More precisely, connectionist models offer "a truly different cognitive architecture, to which the Classical architecture is a scientifically important approximation" (287). This reversal of Fodor and Pylyshyn's argument stands on two separate claims: first, that Fodor and Pylyshyn's critique does not take into account fully distributed representations and therefore only applies to locally structured networks in which units or nodes represent a single feature instead of being distributed over the entire network; and second, that fully distributed networks can in fact embody compositionality and structure-sensitive processing. Smolensky's countercritique thus amounts to the charge that Fodor and Pylyshyn have focused their attack on an overly simplified version of connectionism, in effect reducing it to a form of neoassociationism. Yet Smolensky does not directly contest the language-of-thought model that underlies classic cognitive architecture. Instead, he argues that "distributed representations provide a description of mental states with semantically interpretable constituents" (298).

According to Smolensky, "There is no complete, precise, formal account of the construction of composites or of mental processes in general that can be stated solely in terms of context-independent semantically interpretable constituents." More simply, there can be no completely adequate formal (syntactic and semantic) account of mental processes independent of context. This amounts to saying that the language-of-thought hypothesis cannot possibly do what it claims to do. As Smolensky puts it, "There is a language of thought—but only approximately; the language of thought by itself does not provide a basis for an exact formal account of mental structure or processes—it cannot by itself support a precise formal account of the cognitive architecture" (298). One might have already suspected that this was the case and that the classic model only gives a highly abstract and idealized picture of thought, not an account of how thought actually works. Yet the connectionist perspective gives this critique an unexpected precision. Summarily, then, Smolensky's rebuttal of the cognitivist position amounts to a double claim: not only can connectionism in fact do the things that Fodor and Pylyshyn say it cannot, but the classic model to which connectionism is held up as the standard is itself inadequate as a model of thought or cognition.

Before entering into the details of Smolensky's countercritique, more must be said about how he characterizes connectionism. In contrast to the classic architecture, in which mental representations are made up of symbols manipulated according to clearly defined rules, mental representations in connectionist networks correspond to "vectors partially specifying the state of a dynamical system (the activities of units in a connectionist network)" (287). Connectionist networks thus give us what Smolensky calls a split-level cognitive architecture: "The syntax or processing algorithm strictly resides at the lower level, while the semantics strictly resides at the upper level" (288). At the lower level of nodes and links there is only the dynamic interaction of states described by mathematical relationships; only at the upper level do large-scale patterns emerge that can be interpreted or assigned a meaning. The strength of this split-level architecture appeared earlier with the introduction of the Perceptron: since the actual processing is carried out by many simple elements working in parallel, flexibility, speed, and graceful degradation are inherent features.

As we have seen, cognitivism's strongest criticism is directed at the way connectionist networks represent a mental state. Since a representation is essentially a pattern of logically and semantically connected microfeatures (as in the example, "Mary drinks a cup of coffee"), connectionist networks have no unequivocal way of representing either a context-independent entity or the constraints that syntax imposes. To represent "coffee" one would have to produce the pattern for "cup of coffee" and then subtract the pattern for "cup," or some such operation. In an extended demonstration that Smolensky refers to as "the coffee story," he explains how the problem can be resolved using the mathematics of vector coding; thus, he states, "The representation of *coffee* is a collection of vectors knit together by family resemblance" (293). But this solution, he acknowledges, is "too weak to serve all the uses of constituent structure—in particular, too weak to support formal inference—because the vector representing *cup* cannot fill multiple structural roles" (294). He then offers a strong solution to the problem of constituent structure by means of what he calls tensor product representations. Basically, he shows how the latter can be combined and superimposed on one another. The technical details of this part of Smolensky's argument are difficult to summarize, however, and need not detain us here (see 294–297).

Having resolved this problem, Smolensky argues that the "agenda for connectionism should not be to develop a connectionist implementation of the symbolic language of thought, but rather to develop formal analysis of vectorial representation of complex structures and operations on those structures that are sufficiently structure-sensitive to do the required work." More generally, he believes that "when powerful connectionist computational systems are appropriately analyzed at higher levels, elements of symbolic computation appear as emergent properties" (298). The central hypotheses of classic cognition, which concern principles of memory, inference, compositionality, and constituent structure, are to be understood therefore as approximations of complex mental processes. Connectionism's agenda for these approximations is to find

> new ways to instantiate them in formal principles based on the continuous mathematics of dynamical systems. . . . The concept of memory retrieval is reformalized in terms of the continuous evolution of a dynamical system toward a point attractor whose position in the state space is the memory; we naturally get content-addressed memory instead of location-addressed memory. (Memory storage becomes modification of the dynamics of the system so that its attractors are located where the memories are supposed to be; thus the principles of memory storage are even more unlike their symbolic counterparts than those of memory retrieval.) When reformalizing inference principles, the continuous formalism leads naturally to principles of statistical inference rather than logical inference. (298)

Yet there is more at stake here than a redirection of research, even though this redirection has been astonishingly successful. While it was clear from the start that connectionism would expose "the hidden microstructure in . . . large-scale, discrete symbolic operations" (299), it has still not been able to completely bridge the gap between the micro- and macro-levels of cognitive functioning. More important, even if there is now wide agreement that the "mind is a statistics-sensitive engine operating on structure-sensitive (numerical) representations" (293), as Smolensky puts it, in certain essential respects the classic cognitive model remains intact. Far from being dismantled, it has simply been demoted to the status of an approximation.

Another way to get at the issue here is to suggest that while Smolensky's approach provides the terms for a fruitful exchange between the two models, the truly radical implications of the connectionist model remain mostly in abeyance. That radicality stems from connectionism's decoupling of syntactic and semantic coding—its split-level cognitive architecture, in short. This architecture displaces computational information processing to a subsymbolic domain, while on an upper level it attempts to account for semantic coding (including memory) and rule-bound relationships in terms of attractors in the state space of a dynamical system. Ultimately its success would seem to hinge on its ability to model not only abstract and logical thinking (as Smolensky shows it can), but the complex operations of natural language as well.

Language and Attractor Syntax

One of the most exciting prospects for the neural net modeling of language can be found in Jeffrey Elman's research. Elman boldly proposes that human natural language processing is actually the instantiation of a dynamical system and works as such; it is not, therefore, the result of a computational machine that works the way most linguists have assumed since Chomsky's groundbreaking work of the 1950s. Deploying the terms and concepts of dynamical systems theory, Elman argues that the internal representations of words are not symbols but locations in state space, that the lexicon or dictionary is the structure in this space, and that the processing rules are not symbolic manipulations but dynamic attractors that pull the system state in specific directions. According to this understanding, when we form sentences that are guided and shaped by grammatical rules, we are not processing thoughts by accessing fixed, coded blocks of information and following tracks and pathways preimprinted or hardwired in our computational brains; instead, we are being pulled and twisted by invisible attractors in a dynamic force field that is constantly changing as we ourselves traverse it, in part because we change and shape it according to how we pass through it. As Elman himself puts it: "Objects of mental representation are better thought of as trajectories through mental space than things constructed" (199).[79] The material substrate of this process is a "recurrent neural network," that is, a neural net with both feedforward and feedback (or ascending and descending) pathways. These networks have incredibly complex activation patterns, mainly because the feedback or descending pathways make information about past activity available for current processing.[80]

Among cognitive scientists, Elman is widely known for his work with a specific type of recurrent neural net, which produces a temporally structured activation pattern. As he often emphasizes, time is an essential aspect of language use: we hear, make sense of, and in turn utter words in a temporal process; our speaking, hearing, and understanding of language all depend on patterns of expectation and anticipation, different kinds of memory as well as many contextual parameters that classic language theory mostly ignores. In this light, the assumptions of classic theory make our use of language seem very mechanical and remote from actual experience. According to this theory, words are conceived of as lexical items in a passive data structure that the speaker processes. First, words are subjected to acoustic and phonetic analysis; then their internal representations must be accessed, recognized, and retrieved from permanent storage. Following this, the internal representations have to be inserted into a grammatical structure defined by a syntax, and so forth. Our own experience, contrarily, is that speaking is more like following a path we make as we go along. For this reason Elman's theory that time is essential to language and that language is the unfolding of a dynamical system obviously holds great promise.

In "Language as a Dynamical System" Elman explains how his theory has been developed and tested using a simple recurrent neural network. Recall that neural nets can be trained to encode representations (both verbal and visual) as unique activation patterns. To begin, Elman trained a recurrent network to encode simple sentences formed from a small lexicon of twenty-nine nouns and verbs, using a learning set of ten thousand sentences. Each word thus trained represented a vector in the network's total state space. To understand what this means we must keep in mind that what the network "learned" had nothing to do with "how the form of the word was correlated with its properties" (i.e., its sound or meaning), but only with its distributional behavior in relation to other words. Yet this single relationship turned out to be very rich and revealing. Elman then tested to see if the network could predict successive missing words from among sentences in the training set. Since the network had not memorized these sentences, his purpose was to determine the degree to which the network would supply the correct type of word—most often a verb or noun as subject or direct object—and thus had made a classification of word types based solely on their distributional behavior and co-occurrence associations. When Elman determined that the network could indeed make such predictions with a high degree of accuracy, he then did tests and calculations based on hierarchical clustering of vectors (for an explanation, see "Language as a Dynamical System," 205) to determine more precisely the network's category structure of classifications. The results were mapped in a tree diagram (fig. 6.1).

The diagram makes it clear that the network has learned to distinguish verbs from nouns; verbs, furthermore, are distinguished according to whether a direct object is necessary, optional, or required. Nouns, on the other hand, are subdivided into animates and inanimates, the former being further subdivided into humans and animals (which then breaks down into large and small), while the category of inanimates subdivides into breakables, edibles, and miscellaneous.

Having shown that the network has structured the lexicon as a highly differentiated state space, Elman turns to the notoriously difficult problem of syntax, specifically in order to show how embedded relative clauses can be modeled, especially those in which the verb endings are different over several successive phrases. (For example: "The girls who chase the boy who follows the cat run noisily.") Essentially, Elman is able to demonstrate that a grammatical sequence of word vectors delineates a vector space trajectory, and that grammatically similar sentences have similar trajectories. Most crucially, successively nested relative clauses are coded as similar but spatially distinct cycles within the activation space. (See Elman's fig. 8.7, reproduced below as fig. 6.2.) In short, small grammatical differences are encoded as slight but dynamically relevant differences in a vector space trajectory. Grammatical structures can thus be said to have "signature trajectories"; furthermore, the network can distinguish between the same words in different grammatical contexts.

While cognitive scientists in the last twenty years have been able to show how neural nets in the brain code sensory perceptions,[81] language has been the last holdout. But Elman's work suggests that a combination of neural net and dynamical systems theory can eventually yield a more complete understanding of human language use in all its varied aspects. Once it can be shown that higher cognitive functions like language can be understood in the same conceptual terms and underlying framework as sensorimotor and perceptual functions, then we are well on our way to a bottom-up approach that is concordant with our intuitive understanding of the physically integrated human body, instead of a mélange of theories that apply to different parts and functions, as if the body were an assemblage of mechanically related ad hoc attachments and the mind (or brain) its magical controller.

To be sure, higher cognitive functions like language and high-order perception are distinctly more complex than sensorimotor movement,

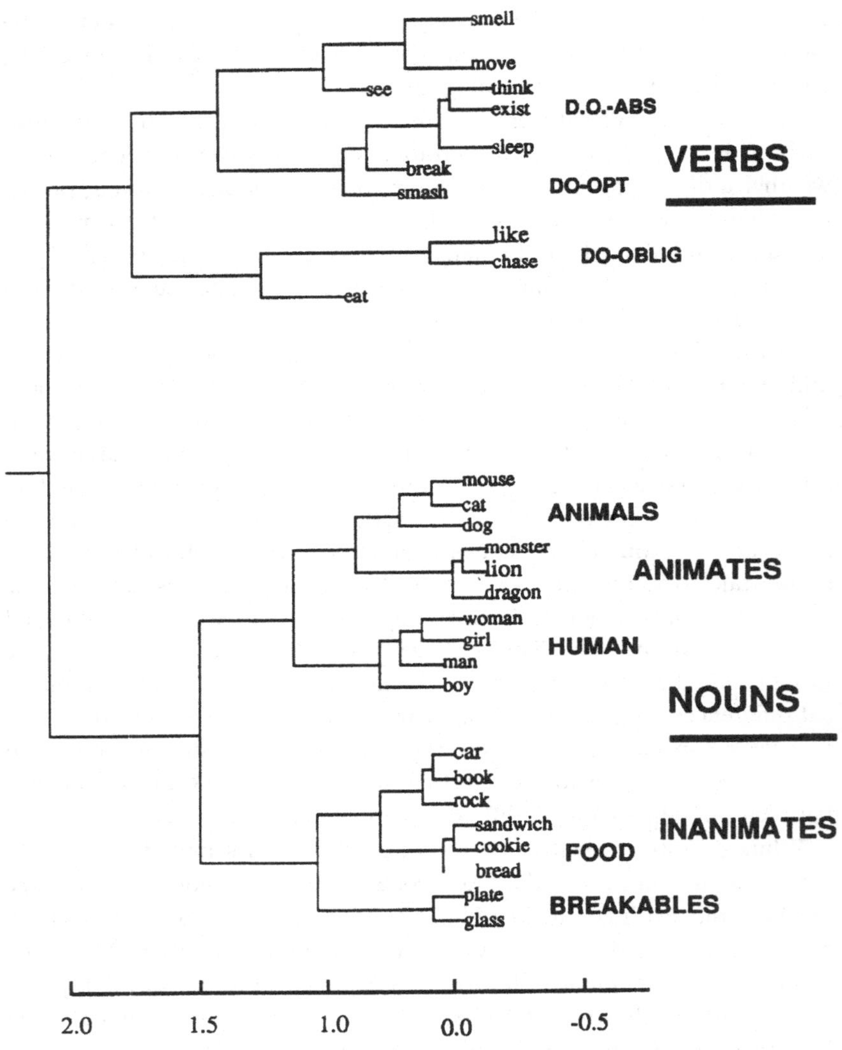

Figure 6.1
Elman's clustering diagram. Jeffrey Elman, "Language as a Dynamical System," in *Mind as Motion: Explorations in the Dynamics of Cognition*, ed. Robert F. Port and Timothy van Gelder (Cambridge, Mass.: MIT Press, 1995), 206.

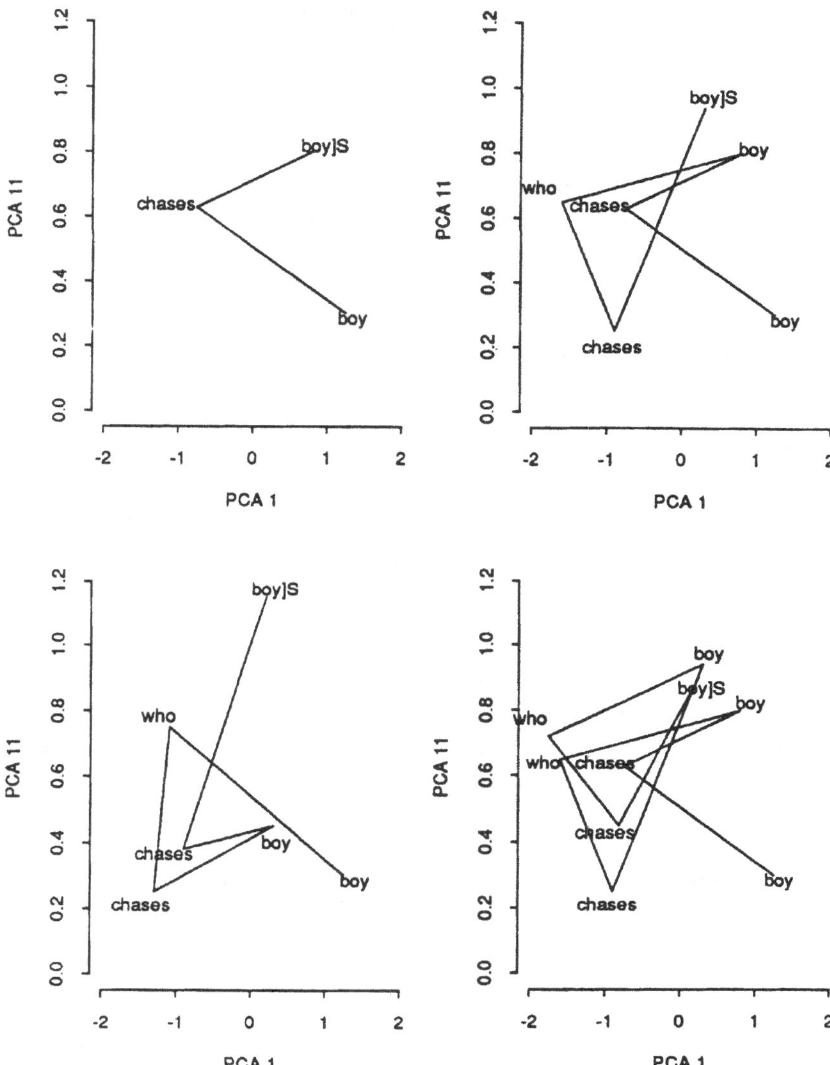

Figure 6.2
Elman's trajectories through state space. Jeffrey Elman, "Language as a Dynamical System," in *Mind as Motion: Explorations in the Dynamics of Cognition*, ed. Robert F. Port and Timothy van Gelder (Cambridge, Mass.: MIT Press, 1995), 214.

and their operations and functions are constantly being extended and reshaped by the techniques and technologies that constitute our physical and cognitive environment. Nevertheless, the combination of neural nets and dynamical systems makes for a powerful model, which is likely to be part of the conceptual toolkit that will enable us to understand how we interact with and are shaped cognitively by these technologies. The NN/DS model also tends to suggest that there is no clear or inherent hierarchy between perception and language use, as the computational cognitivists earlier posited. In fact, there is compelling research that understands these two functions as very closely allied and even codependent. The work of Douglas Hofstadter and Melanie Mitchell on analogy making as a form of high-level perception is one example; another is Jean Petitot's studies of both language and perception from the perspective of what he calls morphodynamics. Let us briefly consider the latter.

Petitot's morphodynamics is a further development of René Thom's catastrophe theory.[82] According to Thom (who was Petitot's teacher), there are seven simple types of catastrophe, or ways that one kind of topological structure can be transformed into another. For obvious reasons Thom's work has been especially useful to embryologists interested in morphological change, that is, in how a body changes shape and form as it grows. In general terms Petitot can be said to extend Thom's theory to other kinds of morphological change by applying dynamical systems theory more fully. Given a state space in which there are several attractors, what kinds of changes can bring about the disappearance of these attractors and the appearance of new ones? Technically, this is the provenance of bifurcation theory, which is concerned with the critical changes that cause attractors to appear and disappear. From this viewpoint, morphodynamics studies how complex structures can emerge from bifurcations.

Like Elman, Petitot seeks to understand how linguistic syntax results from particular attractors in a state space (the underlying dynamical system is again a neural network), but Petitot puts greater emphasis on transformation, or catastrophes in Thom's sense. To this heady theoretical mix he also brings ideas from the field of cognitive grammar, as represented primarily by the work of Ron Langacker. One way to explain cognitive grammar is to go back to Aristotle's idea that language structurally mirrors the world we perceive through our senses (i.e., that agents and actions are symbolized in nouns and verbs, etc.). Founded on a similar idea, cognitive grammar finds (or assumes) such mirrorings at a much deeper level. Specifically, it elaborates what is known as the iconicity thesis, which "asserts that there exists a homology (though not, of course, an isomorphism) between the structure of the [visual] scene as an organized Gestalt and the structure of the sentences which describe it" ("Morphodynamics," 249). All of which suggests that there is a deep-level connection between verbal meaning, visual perception, and mental images. Herbert Simon reflects the iconicity thesis in a passage that Petitot quotes:

> We experience as mental images our encodings of words we have read or of memories we have recovered.... The most common explanation of mental images in cognitive science today ... is that such images, whether generated from sensations or memories, make use of the same neuronal equipment that is used for displaying or representing the images of perceptually recorded scenes.... On this hypothesis, a mental picture formed by retrieving some information from memory or by visualizing the meaning of a spoken or written paragraph is stored in the same brain tissue and acted on by the same mental processes as the picture recorded by the eyes. (250)

Petitot seeks to understand these mental processes as transformations in a state space of what cognitive grammar understands as cognitive archetypes, or Gestalten. These transformations of Gestalten would suggest that both verbal and visual syntax are markers in the unfolding of a dynamic process. Following Thom's geometric-topological conception of syntax, Petitot believes that "there exist syntactic Gestalts constituting a perceptively rooted *iconic protosyntax*" (249). Indeed, if one accepts the idea that "many linguistic structures (conceptual, semantic, and syntactic structures) are organized in essentially the same way as visual Gestalts are," in short, if one accepts the idea of an "*iconicity* of syntax" (and many studies in cognitive linguistics find evidence for it), then a way to address the problem of the missing link between language and perception in classic linguistic theory becomes possible. What interests Petitot, therefore, is "an *intermediary* representational level where perceptual scenes are organized by *cognitive organizing Gestalts* and *image schemas* which are *still* of a perceptive nature but *already* of a linguistic nature" (249). Given this interest, in his article on morphodynamics Petitot focuses specifically on how a form of visual syntax that first involves the scanning of boundaries and then the extracting of singularities from instances of "contour diffusion" can be accounted for precisely using these conceptual tools.

Petitot's work is highly technical and draws on multiple disciplines, but even this brief sketch reveals that it no longer assumes that linguistic syntax is a unique and intrinsic feature of language; instead, syntax is conceived as a transformation of deeper primitive visual structures, or "scenes." As a consequence, there is no basis for language's assumed superiority over visual modes of apprehension or expression, since both linguistic and perceptual syntax are understood to emerge from underlying dynamical transformations of activation patterns in the brain that encode these primitive scenes as cognitive archetypes. This contrast between verbal and visual modes of attention, nevertheless, continues to be a particularly vexed problem in theories of consciousness.

Consciousness, Human and Otherwise

For a variety of reasons, the 1990s saw an energetic resurgence of interest in the study of consciousness, both among philosophers and scientists working in the fields of cognitive science, neuroscience, and AI. New studies of higher primate behavior and advances in brain imaging technology no doubt contributed to this interest, but don't fully explain it. Consciousness, of course, is a very complex phenomenon, and there has been surprisingly little agreement on how it should be defined and explained, or even how important it is in many human activities. Two perspectives on consciousness, offered in Daniel Dennett's *Consciousness Explained* and Paul Churchland's critique of this work,[83] highlight (though at a different level) several of the issues in cognitive science and neural net theory discussed above; more important, they lead, by way of the question of whether consciousness can be simulated by a machine, to the best-known critique of classic AI.

According to Dennett, what is presented to consciousness is a confused mass of "content-discriminations" that

> yield, over the course of time, something rather like a narrative stream or sequence, which can be thought of as subject to continual editing by many processes distributed around in the brain, and continuing indefinitely into the future. This stream of contents is only rather like a narrative because of its multiplicity; at any point in time there are multiple "drafts" of narrative fragments at various stages of editing in various places in the brain. (113)

Since consciousness is precisely what is produced by these highly distributed editings of multiple drafts, there can be no single "Witness" or "Central Meaner" at the center of a Cartesian theater directing its operations. As Dennett readily concedes, this latter image is part of a whole family of metaphors that he wants to replace with another, "trading in the Theater, the Witness, the Central Meaner, the Figment for Software, Virtual Machines, Multiple Drafts, [and] a Pandemonium of Homunculi" (455). However, since metaphors are the "tools of thought," it is not a matter of merely substituting one set for another, but of using the new set to build a theory of consciousness that is more consistent with scientific and phenomenological fact.

Essential to his theory is the hardware/software distinction. At the outset, Dennett acknowledges the neural net theory of the brain and that the brain is a massively parallel computing device bearing no visible relationship to the now classic structure of a serial, discrete-state, programmed digital computer. For Dennett, however, the billions of dynamically interacting neurons in our brains are only the underlying material "hardware." When humans learn language, this hardware is restructured to run as a virtual serial computer, thus giving humans the capacity "to represent and process information in a structured sequence of rule-governed representations unfolding in time" (264–265). Thus, Dennett assumes, our brain as a parallel processing device works as a virtual machine that can emulate (or run as software) a serial machine, here instantiated as (or in) language. But what allows a human mind to emerge from the operations of this brain machine is not simply language. Appropriating Richard Dawkins's idea of the "meme" as cultural replicator, Dennett argues that once "our brains have built the entrance and exit pathways for the vehicles of language, they swiftly become parasitized . . . by entities that have evolved to thrive in just such a niche: memes" (200).[84] Through the infestation of memes and their constant reprogramming by the brain, human consciousness and the larger field of human culture become the sites of natural selection and evolutionary process.

In his critique, Churchland argues that Dennett has perverted a genuine scientific advance (distributed parallel processing and neural net theory) in order to return to an earlier model of symbolic computation. But this perverse maneuver is based on a deeply confused analogy. While it is true that computational machines can generally emulate other computational machines, it is essentially a matter of replicating input-output functions, in which the fundamental differences between the two modes of computation are bypassed and ignored. Dennett wants to ignore these differences because he is ultimately committed to a language-centered theory of consciousness. As a consequence, Churchland charges, Dennett's theory is rooted in a historical and cultural prejudice that no longer has any scientific basis:

> The prototype of language-like activity has exercised an iron grip over all theoretical attempts to account for human cognition since Aristotle. But it is a false prototype for cognitive activity, even in humans. Here in the closing decades of the twentieth century, we have finally unearthed and have begun to explore the power

of a very different prototype: distributed vectorial processing in a massively parallel recurrent neural network. It is now beyond serious doubt that this is the principal form of computational activity in all biological brains, and we have begun to see how to explain the familiar forms of cognitive activity with the strikingly novel and fertile resources that this new prototype provides. (265–266)

For Churchland, furthermore, Dennett's theory is not only scientifically wrong but ethically offensive: having made consciousness essentially linguistic, Dennett must deny any form of consciousness to animals. A few pages later, Churchland summarizes his critique:

Finally, Dennett's account of consciousness is skewed in favor of a tiny subset of the contents of consciousness: those that are broadly language-like. Human consciousness, however, also contains visual sequences, musical sequences, tactile sequences, motor sequences, visceral sequences, social sequences, and so on. A virtual serial machine has no especially promising explanatory resources for any of these things. A recurrent parallel network does. (269)

In sum, Churchland believes that Dennett has missed the fundamental scientific truth that an array of connected recurrent parallel networks offers the most promising model for understanding how the brain accomplishes higher cognitive functions, including consciousness.[85]

Despite these fundamental differences, there is one issue on which Dennett and Churchland completely agree: the feasibility of machine consciousness. With little or no reticence, in fact, the question of an artificial consciousness was often raised in the 1990s. Churchland, for example, devotes an entire chapter to the question: "Could an electronic machine be conscious?" His answer is straightforward: not only is it possible in principle but rather likely, once it becomes possible "to construct electronic implementations of the sorts of networks that in us are implemented biologically and neurochemically" (236). In fact, research like Carver Mead's development of a silicon retina (to which Churchland devotes several pages) demonstrates that it is already under way. Dennett takes the same position but arrives at it by a different route. Since, according to his theory, we are machines and human consciousness is itself machinic, it is not much of a leap to reason that machines can be conscious too.[86] Dennett takes up some of the implications of this position in a section called "Imagining a Conscious Robot" (431–440) as well as in his later essay "The Practical Requirements for Making a Conscious Robot."[87] In the latter he reveals that he is working with a team led by Rodney Brooks at MIT to construct a robot, named Cog, with higher cognitive functions. "Many of the details of Cog's 'neural organization'," he declares, "will parallel what is known (or presumed known) about their counterparts in the human brain."

Although the details of this neural organization remain unavailable, Cog and its fellow humanoid robot, Kismet, merit at least a brief description. The two robots were designed and built at MIT's Artificial Intelligence Laboratory, which is directed by Brooks, and reflect the central importance accorded to perception and social attention, respectively, in theories of consciousness. As Brooks describes them in *Flesh and Machines*, each robot has achieved remarkably humanlike behavior in these two restricted domains.[88] The development of Cog began in 1993, as an attempt to make a vision system that "works like that of people, and with eyes that can saccade and verge, and that look like human eyes" (87). Physically, Cog has a head, two arms and a torso, which is mounted on a heavy platform (fig. 6.3). In photographs, Cog often

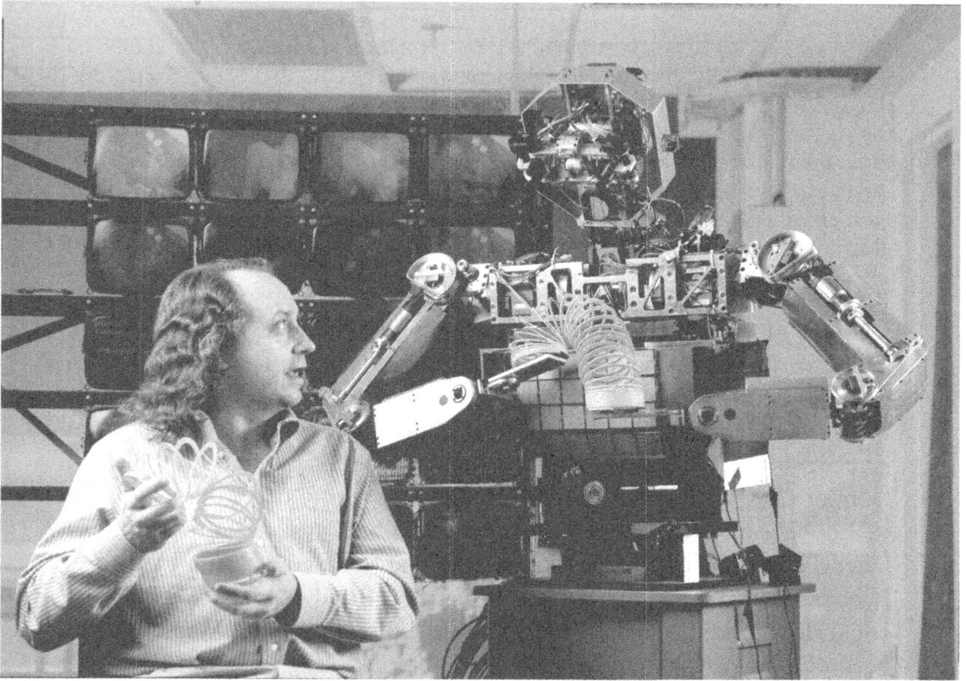

Figure 6.3
Rodney Brooks with Cog. Peter Menzel and Faith D'Aluisio, *Robo sapiens: Evolution of a New Species* (Cambridge, Mass.: MIT Press, 2000), 62–63. © Peter Menzel/menzelphoto.com.

appears in front of a bank of computer monitors that display in grainy images what Cog is actually "seeing." Notably, each one of Cog's two eyes has two camera eyes that give it both peripheral and foveal vision. The eyes are mounted on gimbals that can pan and tilt; additionally, head and neck motors give it more freedom of movement. It also has a gyroscope so that, like the human inner ear, it can sense its own head motion. With Cog, Brooks asserts, "We have been able to duplicate the mechanical aspects of the human visual system" (88).

As Brooks acknowledges, "The difficult thing is how to process the images" (88). Though Cog can do things like recognize human faces and track moving objects, there are many things—mostly involving recognition and discrimination—that it cannot do very well at all. (Brooks lists what Cog can do well and not so well on page 90.) The chief obstacle is that scientists do not yet have the algorithms that can convert a visual pattern of intensities into a rich description of what humans can perceive in the world.

> What appears to be a sharp boundary to our eyes and vision system combined does not necessarily show up in an obvious way in the data that comes from a digital camera. If we see a pen lying on a desk, we can see a sharp boundary between the pen and the desk. But often when we look at the intensities of light from each little square pixel in a digital image, there is no clear boundary. Pixels corresponding to parts of the pen, and parts of the desk, just two or three pixels apart, may have exactly the same intensity values. Somehow our brain is getting a much more global understanding of what is going on, and it then perceives the boundary. (89)

Brooks concludes from this that we must be "missing something fundamental" in our understanding of how human vision is organized (91). What he doesn't say, but it seems implied, is that Cog is ideally suited to research that one day may fill this gap.

Cynthia Breazeal, who helped Brooks design and build Cog, has spent an enormous amount of time doing simple things like holding an object in front of Cog, who would then saccade toward it and reach out and touch it. Looking at videotapes of these interactions, Breazeal noticed that in her behavior she always assumed more than Cog could actually do. As Brooks describes it, "She had picked up on the dynamics of what Cog could do and embedded them in a more elaborate setting, and Cog had been able to perform at a higher level than its design so far called for" (92). This led to her thinking about social interaction, particularly at an unconscious level, and about those basic learning situations that occur between mother or primary caregiver and child. This led in turn to the construction of Kismet, something of an offshoot of Cog but really more of

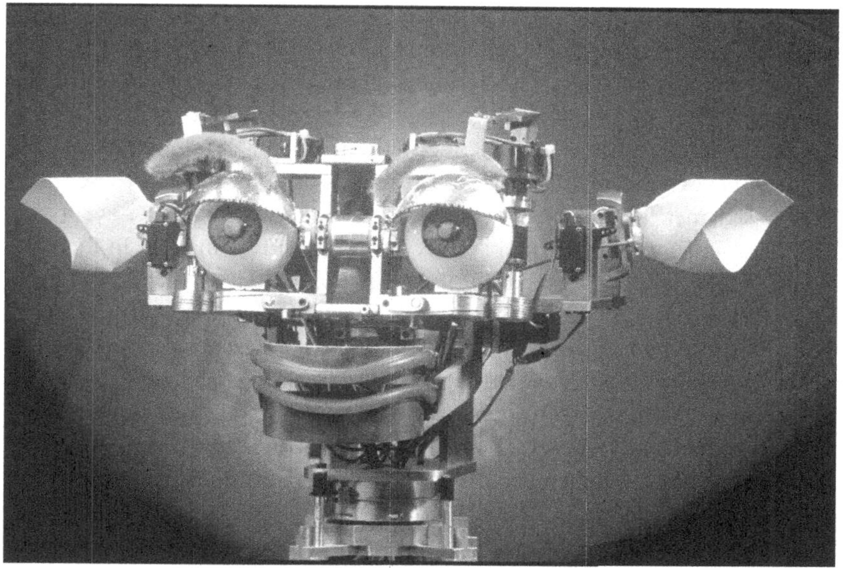

Figure 6.4
Kismet showing happiness. Peter Menzel and Faith D'Aluisio, *Robo sapiens: Evolution of a New Species* (Cambridge, Mass.: MIT Press, 2000), 68. © Peter Menzel/menzelphoto.com.

an abstractly expressive "face." Physically, Kismet has two eyes and, like Cog, the capacity to move them as well as its neck and head; but it also has microphone ears, eyelids, eyebrows, and a mouth—all of which gives it the capacity to make facial expressions. The expression in figure 6.4 is one of happiness.

For Kismet to make facial expressions and to respond appropriately to human expressions obviously requires a very elaborate visual attention system, which was designed by Breazeal and another of Brooks's students, Brian Scassellati. Like Cog, Kismet saccades toward whatever catches its attention. Three kinds of things will work: things in motion, things with saturated colors, and things with skin color. However, Kismet does not simply observe these things passively but acts in accordance with its own set of "drives," or basic needs: it has a need to be stimulated by humans (a social drive) and to play with toys or objects (a stimulation drive); but it also has a "fatigue syndrome," which makes it tired over time.[89] In addition to being affected by these internal drives, Kismet is emotionally affected by the prosodic signals it receives through its ears and auditory system.[90] More specifically, Kismet's mood or emotional state is conditioned by three variables: *valence*, a measure of its happiness; *arousal*, a measure of its stimulation versus how tired it is; and *stance*, which is how open it is to new stimuli. Basically, Kismet tries to keep its drives in balance by getting people to interact with it. In this sense it is a very sophisticated cybernetic machine, though curiously the word is never used in Brooks's or Breazeal's descriptions. In these interactions Kismet displays its emotional state by moving its ears, eyebrows, and lips, and by altering the prosody of the sounds it makes. In fact, the photographs in Menzel and D'Aluisio's *Robo sapiens* (68–71) register quite a range: calmness, happiness, sadness, anger, surprise, disgust, tiredness, and the state of sleep. Needless to add, some people are much better at interacting with Kismet than others, children in particular being more apt to engage with sustained interest. In their interactions with Kismet both Breazeal and Scassellati comfortably assume the role of caregiver to infant, so it is not surprising that their published research is framed by theories of child learning and psychology.[91]

Witnesses who observe Cog and Kismet interacting with humans often feel that there is more going on than there actually is. Kismet in particular seems to elicit the feeling that it has intentions and awareness and therefore some nascent form of consciousness. Brooks attributes these feelings to our tendency to "over-anthropomorphize," even in purely human activities. "We attribute too much to what people are doing," he states (*Robo sapiens*, 58), making the category error of mistaking the appearance of something and a description of the appearance for a description of the mechanism that explains it. Like Dennett, he believes that human beings are machines—highly complex machines, but machines nonetheless. There is no reason in principle, therefore, why life, intelligence, and even consciousness cannot be constructed artificially. But if so, why haven't there been more dramatic breakthroughs, especially given the great leaps in computational power that new computers make readily available? Brooks believes that there is still "something missing" from our models, particularly of life and consciousness. It is not anything mysterious and ineffable, as in the old biological theory of vitalism; rather, it comes from a lack of complexity in the models themselves, like some deeper principle that is staring us in the face but has yet to be recognized. Just when the needed discoveries might come remains unpredictable. In the meantime, he intends to seek what is missing—the "missing stuff" he calls it—by building machines that approach the human from the bottom up.

Increasingly, the position that Dennett, Churchland, and Brooks assume toward machine consciousness is gaining ground, while earlier counterarguments like John Searle's "Chinese room" polemic against strong AI (and implicitly machine consciousness) seem less convincing.[92] Searle argues that a mechanical or algorithmic implementation of a thought process, even when performed by a man emulating a Turing machine (or its read/write head), fails to exhibit or instantiate real intelligence because it lacks understanding or intentionality. In other words, it is not conscious of, nor does it understand, what it is doing. Searle illustrates his claim with the following thought experiment. A man in an isolated room receives through a slot in the door sheets of paper covered with unintelligible "swiggles." However, with the aid of a code book of instructions (as in: if two swiggles, write two swoggles), he is able to reply in kind, and send back similarly marked sheets of paper. Outside the room it is known that the messages are written in Chinese, and it is assumed therefore that the man inside understands the Chinese language. But clearly, Searle proclaims, the man doesn't understand Chinese, no more than a computer understands the computations it performs. At best one could say that he *simulates* such an understanding. From this thought experiment Searle concludes that no computer, no matter how fast and powerful, will ever be able to think. This is because its formal operations are defined syntactically, as a sequence of purely symbolic manipulations, and lack the necessary semantic dimension—the *intentionality*—necessary for understanding and authentic intelligence.

Searle's argument has provoked a number of important critical responses by Douglas Hofstadter, Daniel Dennett, and David Chalmers, to name only three. The reason, I believe, is not simply because of the issues involved but also because of Searle's deceptive clarity, which obscures the extent to which his argument against strong AI is entangled with and even displaced by a host of other important issues and questions, such as: What is the relationship between intentionality and consciousness, or between understanding and consciousness? And what does it mean to understand a language? Accordingly, one might interpret the Chinese room as a simple parable about meaning, that is, a machine can't possibly understand Chinese (or any other language) if it doesn't understand what its words or written symbols *mean*. The best known reply to Searle, designated the "systems reply," argues that while the man in the room does not understand Chinese, the room itself does. Hofstadter and Dennett both take this position.[93] Basically, it asserts that all the things in the room—the man, the pieces of paper, the code and instruction books, and so forth—constitute a distributed system that exhibits intelligent behavior. But Searle rejects this argument out of hand, since for him understanding (which he silently substitutes for intelligent behavior) requires conscious mental states, which cannot reasonably be attributed to a room. However, in *The Conscious Mind* David Chalmers retorts with a more elaborate counterargument, involving the step-by-step substitution of the man's neurons with computational demons and then with a single demon in which the neuron-level organization of the human brain is completely duplicated.[94] With these substitutions, the room becomes a dynamical system in which neuronal states are determined by the rules and manipulations of the symbols, and the system as a whole is sufficiently complex to have "the conscious experiences [including qualia] of the original system" (325), that is, of a human being.

This rendering of the room as a dynamical system is necessary because, as Chalmers argues, the slips of paper are not merely a pile of formal symbols but "a concrete dynamical system with a causal organization that corresponds directly to the original brain," and it is "the concrete dynamics among the pieces of paper that gives rise to conscious experience" (325). With his fiction of neuronal substitution and demons, Chalmers thus makes fully visible what was submerged in Searle's version: the full systemic complexity of the computational room. Actually, Hofstadter anticipates this very aspect of Chalmers's argument when he writes that "the program on those 'bits of paper' embodies the entire mind and character of something as complex in its ability to respond to written material as a human being is, by virtue of being able to pass the Turing test" (375). But Chalmers also attacks Searle precisely on the issue of how the computer in the room (or the computational process that *is* the room) is implemented. He agrees with Searle's assertion that a computer program is purely syntactical but points out that the program must be implemented physically in a system with "causal dynamics." In more fully accounting for the causal dynamics of the activities in the Chinese room, albeit by means of a simulation accomplished with artificial neurons organized like the brain, Chalmers makes a compelling argument that the system would be capable of having conscious states. He thereby joins the ranks of contemporary philosophers who share the view that "the outlook for machine consciousness [and strong AI] is good in principle" (331).

As a postscript to this argument, it is worth noting that Searle himself never denies the possibility that "a machine could think." To the contrary, we *are* machines and we can certainly think, he asserts. But we are biological machines, and intentionality (or consciousness) is a biological phenomenon. Thus his argument really falls into two parts. The first part, illustrated by the Chinese room thought experiment, asserts a negative: that no program running on a digital computer is capable of intentionality (i.e., consciousness or thought). This also means that "the computational properties of the brain are simply not enough to explain its functioning to produce mental states" (40). The second part, which is unduly developed and therefore usually ignored, argues that thinking, or consciousness, is essentially biological in nature and therefore cannot be reproduced without a causal, material system equivalent in complexity to our own biochemical system. It means that thinking requires a body located within—and which would be part of—the world. While the first part of Searle's argument was (correctly) understood to be a hostile critique of the operational and functionalist approach of early AI, the second now finds wide agreement among contemporary neuroscientists and those with a biologically inspired approach to the building of intelligent machines.[95]

It may be surprising, nevertheless, that the original biologically inspired neural net approach should give new life to the quest to build intelligent machines, given that the symbol-processing cognitivist approach it displaced was so closely allied to it. Moreover, while important differences among the cognitive scientists and philosophers who have taken up these debates continue to exist, increasingly the central philosophical issues have shifted away from classic cognitivist versus neural net architecture, strong AI versus weak, and so forth, toward questions about representation, particularly as a consequence of how dynamical systems theory and theories of emergence effect causal explanation.[96] Indeed, with Brooks's inauguration of behavior-based robotics, which is discussed in the next chapter, the antirepresentational bias of the new AI becomes an explicit issue.

In closing, I mention one further complicating twist concerning the way in which coding and decoding—both in symbol processing and neural networks—destabilize the commonsense notion of representation. Though not considered directly, this destabilization has been an undercurrent throughout this chapter. For example, when Smolensky demonstrates that connectionism can provide an adequate version of classic cognitive architecture by a *re*coding—specifically, a vector coding in terms that are functionally equivalent to the coding accomplished by symbol and rule manipulation—he argues that connectionism can provide an adequate account of representation. Yet there is a strong sense in which connectionism also brings about a *de*coding. Whereas the classic architecture can only allow *either* truthful propositions *or* instances of incoherence and contradiction, a connectionist architecture opens onto (just as it reposes on) chaos and dynamic, unpredictable change. In contrast to the closed (but generative) classic cognitive model, defined by abstract and idealized syntactic and semantic structures, the neural net model of "mind"—whether instantiated in a biological brain or an artificial neural network—is understood to be a statistical engine operating in a stochastic field. At the same time, this fundamental decoding of the classic cognitive model is presented as the outcome of a contained dynamical process, with clear limits and boundaries. In other words, the dynamic remains internal to the mind/brain. However, Maturana and Varela's theory of autopoiesis suggests a more radical alternative: through their notion of perturbation, a system is "open" to the outside while its autonomy and closure is still maintained. As we'll see in the next chapter, the new AI goes even further and brings about a more open and active alignment with the outside, which is no longer conceivable as simply a source of input or perturbation but becomes an active field in its own right. In this sense, by shifting the dynamic of the mind/brain/machine to its coupling with the (or an) external environment, the new AI performs a decoding of connectionism itself. This decoding occurs *not* as an internal counterdynamic among or across levels within a bounded neural network—as in connectionism's decoding of classic cognitivism—but as a shift of the dynamic itself to a fully immanent articulation with the outside. And what this new opening attempts to internalize and set to work in the process of constructing robotic life is the power of evolution itself.

7 The New AI: Behavior-Based Robotics, Autonomous Agents, and Artificial Evolution

Humans can't build a robot as smart as themselves. But, logically speaking, it is possible for such robots to exist.

How? Cobb had asked himself throughout the 1970s, How can we bring into existence the robots which we can't design? In 1980 he had the bare bones of an answer. One of his colleagues had written the paper up for Speculations in Science and Technology. "Towards robot consciousness," he'd called it. The idea had all been there. Let the robots evolve.

—Rudy Rucker, *Software* (1982)

Until the late 1980s, two distinct theoretical orientations had shaped the history of artificial intelligence. The first, that of classic AI, was to construct intelligent systems from the top down, on the basis of rule-bound symbol systems. Herbert Simon and Allen Newell's Logic Theorist, presented in 1956, was officially recognized as the first working example. In their "physical symbol system hypothesis," which provided a theoretical basis for this approach, they defined the symbol system independently of its material substrate. Ideally, therefore, it could be instantiated in any medium, for example, a biological brain or a digital computer. Such a system, comprised of an accretionary set of symbol structures whose operations reflected the underlying rules of logic, syntax, and conceptualization, was deemed both necessary and sufficient for all intelligent action. In the corresponding cognitivist version assumed by cognitive science, the structures underlying this capacity were understood to be somehow hardwired in the brain.

The second—and alternative—theoretical orientation was based on the brain's actual dynamic behavior—its parallel processing of information in neural networks. It led to the construction of machines like the Perceptron and Pandemonium, but apparently intractable limitations made this approach less viable. In the mid-1980s, however, the addition of another, so-called hidden layer of neurons produced an upgraded, much more versatile version of neural net computation in which earlier weaknesses were overcome. Renamed connectionism, it could provide a basic explanation of pattern recognition, learning, and memory in terms that were consistent with the brain's inherent capacity for self-organization and adaptability.

Given that each approach has well defined and complementary strengths, it would seem that the terms were set for a larger synthesis. Yet this is not what happened.[1] Instead, an altogether new AI emerged, and with it a renewed argument for a dynamical systems approach indifferent to or even opposed to connectionism. In the 1990s, spearheaded by Rodney Brooks's development of a behavior-based (as opposed to knowledge-based), bottom-up approach to the construction of autonomous mobile robots, the new AI made great strides. Drawing on the lessons of ALife research, Brooks's colleagues and followers Luc Steels, Pattie Maes, Maja Mataric, and Randal Beer developed notions of emergent functionality, autonomous agent theory, collective intelligence in multiagent systems, and a rigorous dynamical systems approach. Having rejected symbolic computation as part of the baggage of the old AI, the new robotics allied itself with Francisco Varela's theory of enaction, a new theory of cognitive science based on embodiment and concrete situatedness. In Varela's view, whether cognition takes place by means of a rule-based manipulation of symbols (the computer model) or by means of self-organizing neural networks (the biological model), the act of cognition itself is still envisioned as occurring "in the head." Cognition, however, must be relocated "in the world," where it actually occurs. Indeed, for Varela and his teacher and colleague Humberto Maturana the act of cognition is precisely what brings about a structural coupling that enables the very possibility of the (or a) world's emergence. This view quickly became the more or less official position of behavior-based robotics and the new AI. Meanwhile, it also became evident that progress in robotics increasingly depended on the application of programming techniques that could evolve not only neural net controllers but new robotic morphologies as well. However, to be maximally effective, evolutionary programming is usually combined with computer simulations, with which the new robotics has always been uneasy. What results is a strong exigency to bring computation back into the mix—a form of emergent computation perhaps, but computation nonetheless.

Viewed historically, this trajectory promises to bring to fruition the original ambition of cybernetics to fashion a complete theory of the machine, according to the theoretical perspectives first elaborated by

W. Ross Ashby and John von Neumann in the early 1950s, as discussed in chapter 1. Specifically, Ross Ashby had demonstrated how a machine possessed of a "requisite variety" of internal states would inevitably optimize itself through self-organization, and von Neumann had shown theoretically how it would be possible to build self-reproducing automata that would evolve generations of increasing complexity. Now, some fifty years later, research in evolutionary robotics seems to be readying itself to leap over the "complexity barrier," as von Neumann called it. When it comes, this leap will not only initiate a new phase in the evolution of technology but will mark the advent of a new form of machinic life.

One visible sign of the new AI's ascendance and conceptual hegemony is the widespread contemporary use of the term *agent*. In the structuralism of the 1960s the term designated a type of function that initiated, sustained, and/or relayed a sequence of actions. Its reappearance in the discourse of AI, robotics, and cognitive science trades on the same kind of abstractness and ambiguity that explains its earlier appeal: Does the term denote a class of subjects, objects, or functions?[2] From its use in phrases like "modeling cognitive agents," "designing autonomous agents," or simply "embodied agents," we can infer that an agent can be a person, animal, insect, robotic machine, or even a software program (a body of code). It seems hardly fortuitous, moreover, that the term's reappearance coincides with various breaks with representation in contemporary philosophy, which makes perfect sense, given that a representation always implies a subject. Presumably, then, actions without representations—or beyond the possibility of representation—are undertaken by agents. Agents thus appear when the concept of a subject defined in relation to a world of representations is no longer useful or necessary.

A world of technical immanence defined by interacting machines constitutes one such world, although *world* cannot be the proper term here but serve only as a placeholder. That such a world is posited and even constructed in certain sectors of advanced AI is certainly a noteworthy event, for it brings into focus a tendency at work from AI's very inception. As noted in the previous chapter, from its beginnings AI has posited a model of intelligence in which (a rather abstract) subject is tacitly dependent on technics—not only the computer but a whole computational assemblage. Yet it never takes into account the consequences of this unacknowledged dependence. Consider a simple example. The mere presence of a sharp bone or heavy stone in an anthropoid's grip opens a field of possible acts, thus instantiating a form of intelligence that can be assumed again and again but that cannot be attributed retroactively to the anthropoid. Only with repeated acts and their inscription in memory—that is, the reorganization of neurons in the cortex and the expansion of the cortex itself—does intelligence become an inherent attribute.[3] Up to that point, it would be more accurate to say that the anthropoid is the operator, or agent, of the intelligence implicit in this primitive tool. Thus the subject's "intelligence" always lags one step behind the technology that makes it possible; the agent, on the other hand, is at one with the intelligence of the system of which it is a part. As we shall see, with Rodney Brooks's behavior-based robotics we (re)enter this realm of pure technological immanence.

Ants: A Model of Collective Information Processing

Before considering this realm, we can get a glimpse of what is to come by returning briefly to Douglas Hofstadter's critique of classic, or symbolic, AI. In "Waking Up from the Boolean Dream, or, Subcognition as Computation," Hofstadter argues against the assumption that information processing occurs through the manipulation of fixed and static symbols, proposing as an alternative the notion of "active symbols" that function like a colony of ants.[4] What distinguishes Hofstadter's critique from others, including those of Varela and Brooks, is that he invokes an understanding of the cognitive enterprise in collective and distributive terms. Though not explicitly, he thus gestures toward a view beyond and no longer defined by the putatively unified individual subject. With the new AI, this collective dimension of cognition will come into sharp relief in the study of multiagent systems, distributed AI, and swarm intelligence.[5] Indeed, the collective approach to intelligence has become one of the most exciting new paths in current AI research.

In proposing the ant colony as alternative model of cognition, Hofstadter provides one of the earliest and most fruitful descriptions of a multiagent system. He first suggested the ant colony metaphor in his earlier book, *Gödel, Escher, Bach*, in order "to set up an extended metaphor for brain activity—a framework in which to discuss the relationship between 'holistic,' or collective, phenomena, and the microscopic events that make them up" (646). The metaphor was inspired by a passage in E. O. Wilson's *The Insect Societies*, where mass communication is defined "as the transfer, among groups, of information that a single individual could not pass to another" (quoted by Hofstadter, 646). For Hofstadter, the implications for an alternative to the cognitivist understanding of information processing are clear:

One has to imagine teams of ants cooperating on tasks, and information passing from team to team that no ant is aware of (if ants are "aware" of information at all—but that is another question). One can carry this up a few levels and imagine hyperhyperteams carrying and passing information that no hyperteam, not to mention team or solitary ant, ever dreamt of.

I feel it is critical to focus on collective phenomena, particularly on the idea that some information or knowledge or ideas can exist at the level of collective activities, while being totally absent at the lowest level. In fact, one can even go so far as to say that *no* information exists at the lowest level. (646)

The lowest level, of course, would be a subcognitive level. Since the formation of information at a higher level results directly from the ants' collective activities rather than their individual activities, the information is emergent.

In Hofstadter's collective model of cognitive processes, no central program, programmer, or processing unit is given or implied. Whereas in a standard computer program "you can account for every single operation at the bit level," tracing high-level functions downward to subroutines and finding a global reason for the manipulation of every single bit, in an ant colony

a particular ant's foray is not the carrying-out of some global purpose. It has no interpretation in terms of the overall colony's goals; only when many such actions are considered at once does their statistical quality then emerge as purposeful, or interpretable. Ant actions are not the "translation into machine language" of some "colony-level program." No one ant is essential; even large numbers of ants are dispensable. All that matters is the statistics: thanks to it, the information moves around at a level far above that of the ants. Ditto for neural firings in the brain. Not ditto for most current AI programs' architecture. (653)

Hofstadter's interest in the ant colony serves primarily as a metaphor for what he calls a "statistically emergent mentality," that is, an explanation for how "subcognition at the bottom will drive cognition at the top" (654). His remarks, however, have become what John Holland has referred to as "the classic description of agent-based emergence."[6] In emergent systems, as we have seen in previous chapters, persistent global patterns and properties like self-organization arise from the independent, nonlinear interactions of many simple lower-level rules, elements, or agents. We find emergence, for example, in cellular automata, neural networks, the immune system, and other physico-chemical systems, but also in biological ecologies and human economies as well. In the ant colony, information is not possessed or communicated by any single ant or given in advance in a genetically inscribed program; rather, it arises solely through the interactions of a large number of ants and is instantiated only in the behavior of the collectivity as a whole, which is more than the sum total of individual behaviors. Because of this emergence, the colony as a whole must be understood as a multiagent, information-processing system.

To be sure, more needs to be said about the "computations" that the ant colony as a multiagent system actually performs. However, this would require that we know much more about the different classes of ants and their limited repertoire of genetically programmed behavior, how this behavior is regulated for the welfare of the colony, and how a primitive sign system of chemical traces (pheromones) is instituted to provide a collective cartography and external memory. In this regard, the recent work of Eric Bonabeau, Marco Dorigo, and Guy Théraulaz on swarm intelligence has been groundbreaking.[7] Drawing together research from a range of diverse fields, these authors show how swarms of social insects—ants as well as bees, termites, and wasps—actually constitute powerful problem-solving systems that can be said to exhibit very sophisticated collective intelligence. These systems are built on principles with which we are already familiar: relative autonomy of many individual agents, distributed functioning, and emergence. These principles give natural computational systems a robustness, flexibility, and adaptability that systems based on centralized control and sequential (rather than parallel) programming noticeably lack. Moreover, the emulation of such systems has been instrumental to the development of what is distinctively new in many areas of contemporary science. Recall that Swarm is the name that Christopher Langton gave to the software system he began to develop in 1995 that would enable scientists to simulate behavior in highly distributed, multiagent systems.[8] As Bonabeau and colleagues note, the term *swarm intelligence* was first used in describing cellular robotic systems (7); today, they add, the field of "swarm-based robotics" is growing so rapidly, it is difficult to keep up with (19). Indeed, swarm robotics holds great promise, perhaps most spectacularly in its convergence with nanotechnology.

The Robotic *Merkwelten*

With his invention of subsumption architecture for mobile robots, Rodney Brooks initiated a movement that was continued in the work of his colleagues Luc Steels, Pattie Maes, and Maja Mataric. This movement will now be sketched in broad outline. Like Brooks himself, these colleagues worked in allegiance and solidarity with a version of cognitive science based on Varela's notions of enaction and embodiment while also drawing explicitly on ALife research. In 1995 Randall Beer, working within the same movement, introduced a more rigorous version of dynamical systems theory into the field. Equally important, the simulated insectlike robot Beer used to demonstrate the theory's cogency required a neural net controller that could only be programmed by evolutionary programming techniques. However, the use of evolutionary programming and simulation in the construction of mobile robots, which had actually begun in the early 1990s, posed new theoretical and practical problems. For one, in so far as these new forms of computation were actually required to make the robots functional, computation became a supplemental necessity, reentering by the back door, so to speak, from its peripheral position in behavior-based robotics. As it became evident that artificial evolution offered not only a viable but an inevitable path for future progress, the new subdiscipline of evolutionary robotics based on applications of evolutionary computation necessarily assumed a new prominence.

In 1989 Brooks and Anita Flynn attracted considerable attention with the publication of "Fast, Cheap and Out of Control: A Robot Invasion of the Solar System," where they argued that for space exploration it would be more feasible to use a large number of small, insectlike "mobots," each designed to accomplish a simple task, than a single large, complex, multipurpose robot.[9] While use of the latter would entail solving nearly insurmountable problems of communication and control, the mobots could easily be assembled using current off-the-shelf parts and launched immediately with expendable rockets. Before Brooks and Flynn's proposal, the reigning assumption had been that mobile robots needed an onboard computer programmed to plan and coordinate its activities within a three-dimensional model of the world. This assumption was based on the idea that cognition—understood as modeling and planning—mediates between perception and action in the world. However, years of working on vision-based perception and motion planning for robot manipulator arms had convinced Brooks "that the so-called central systems of intelligence—or *core AI*, as it has been referred to more recently—was perhaps an unnecessary illusion, and that all the power of intelligence arose from the coupling of perception and actuation systems."[10]

This reorientation led Brooks to what he calls subsumption architecture, which approaches robot construction from the bottom up. Instead of attempting to build a robot with cognitive skills roughly similar to

Figure 7.1
Brooks's MIT robots. Rodney A. Brooks, "Elephants Don't Play Chess," in *Cambrian Intelligence: The Early History of the New AI* (Cambridge, Mass.: MIT Press, 1999), 119.

those of a human being, he begins with small, mobile constructions that have a limited repertory of simple functions: the six-legged insectlike Genghis, for example, is built to traverse extremely rugged terrain. Tube-shaped Herbert, on the other hand, can avoid obstacles and locate and pick up soda cans (fig. 7.1).

The key to this architecture is that these simple functions are not integrated by a top-down, globally extensive controlling mechanism. Instead, each function interacts independently with a single aspect of the world through sensors and actuators only connected through a controller designed to adjudicate conflicts. In contrast to the traditional approach, where sensory information about the environment is fed to a modeling module, then a planning module, then an action module, and finally to actuators, in subsumption architecture functional layers are added incrementally, starting with the simple capacity to move around and building up to more complex tasks.

Brooks's diagram indicates this basic difference in approach. In subsumption architecture (fig. 7.2b), the idea is to build a robot that can first move about in the real world without colliding with things. New functions ("to explore," "build maps," etc.) are then added to this basic platform. As a consequence, and in contrast to the method of "traditional decomposition" (fig. 7.2a), there is no centralized model of the world rep-

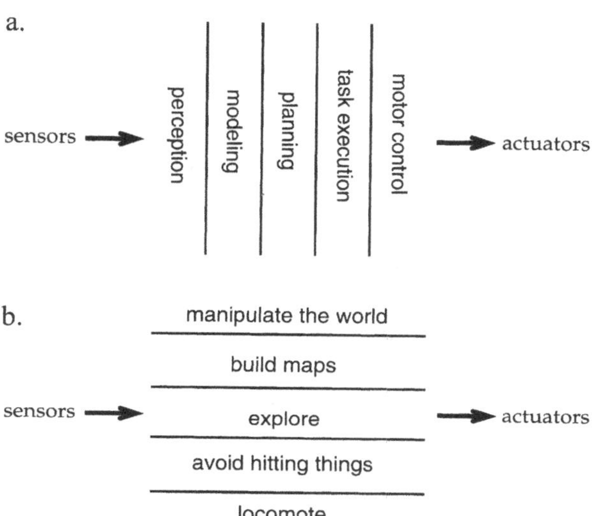

Figure 7.2
Brooks's subsumption architecture diagram. Rodney A. Brooks, "New Approaches to Robotics," in *Cambrian Intelligence: The Early History of the New AI* (Cambridge, Mass.: MIT Press, 1999), 67.

resented within the robot's various systems and no explicit separation of input data and computation. Instead, both are "distributed over the same network of elements" (67), and control is implemented through "networks of message-passing augmented finite state machines (AFSM)" (66). Because these distributed functions are not regulated by means of a centralizing representation of the world, as in earlier generations of robots like "Shakey" built at Stanford, Brooks's robots do not require long periods of number-crunching before they can negotiate even a simple space. In fact, Brooks believes that human beings evolved in a similar way, first becoming highly mobile creatures that interacted robustly with different aspects of the environment through distributed systems. Consciousness is a "cheap trick" that comes late in the developmental process; as an emergent property, it increases the system's functionality but is not essential to its architecture.

Robots built according to the principle of subsumption architecture exhibit two key aspects, situatedness and embodiment, which Brooks explains as follows:

Situatedness The robots are situated in the world—they do not deal with abstract descriptions, but with the "here" and "now" of the environment which directly influences the behavior of the system.

Embodiment The robots have bodies and experience the world directly—their actions are part of a dynamic with the world, and the actions have immediate feedback on the robots' own sensations. ("New Approaches to Robotics," 60)

By way of explanation, Brooks adds that an airline reservation system is situated but not embodied—since it interacts with the world only through the sending and receiving of messages, whereas an industrial spray-painting robot is embodied but not situated—since it doesn't perceive any aspect of the object it paints but only carries out preprogrammed instructions.

Situatedness and embodiment, more or less as Brooks defines them, soon became the basis for a reorientation in cognitive science as well, thanks in part to Varela's recognition of Brooks's success. In "Intelligence without Reason" (originally published in 1991), Brooks solidifies this connection with the new cognitive science, adding intelligence and emergence to situatedness and embodiment:

Intelligence They [behavior-based robots] are observed to be intelligent—but the source of intelligence is not limited to just the computational engine. It also comes from the situation in the world, the signal transformations within the sensors, and the physical coupling of the robot with the world.

Emergence The intelligence of the system emerges from the system's interactions with the world and from sometimes indirect interactions between its components—it is sometimes hard to point to one event or place within the system and say that is why some external action was manifested. (139)

Though these two added criteria are perhaps less well defined, taken together the four establish the basic parameters for behavior-based robotics research in the 1990s. Other themes—learning, adaptability, and collective or social behavior—are mentioned but remain peripheral. For Brooks's followers, however, they will play a more significant role.

Brooks's early essays constitute a sustained critique of the computational and cognitivist bent of classic AI. Indeed, this critique is fully explicit in his notion of subsumption architecture. While not denying that certain kinds of functional intelligence can be built or programmed into AI systems, Brooks believes that real-world intelligence can only emerge from agents or robots that are fully situated and embodied in the world. Like Varela, he sees abstraction as classic AI's most fundamental "sin," and in two key essays argues against the abstraction from perception that representation and symbol manipulation entail. In "Intelligence without Reason," he cites evidence from ethology and neuroscience for his claim that such abstraction is not necessary for intelligent behavior.

He implies that an autonomous robot (or agent) exhibits a form of immanent intelligence—immanent because it does not transcend the machinery in which and by which it acts. In "Intelligence without Representation," Brooks argues more specifically that in AI programs "abstraction reduces the input data so that the program experiences the same perceptual world (*Merkwelt* in Uexküll 1921) as humans" (84). The term *Merkwelt* comes from J. von Uexküll's pioneering work in animal ethology, where it conveys the sense in which the perceptual world is species-specific, since constrained by each animal's unique sensory apparatus, morphology, and capacity to move.[11] When applied to robotics it becomes obvious that the *Merkwelt* of humans and machines cannot possibly be the same. If a machine is to function in the world, it must have or define its own *Merkwelt*, not a representation of the human *Merkwelt* interposed between itself and its field of activity. Human representations necessarily transcend the concrete particulars of embodiment and situatedness. If these representations mediate the robotic *Merkwelt*, they will necessarily make robots slow, inflexible, and highly fault *in*tolerant. By designing robots to function in their own *Merkwelt*, using a layered, bottom-up approach, Brooks is able to get them up and doing things in the noisy real world.

Brooks's pragmatic achievement and the theoretical reorientation on which it reposes greatly stimulated the construction of autonomous mobile robots throughout the 1990s. One indication of the decade's accomplishment is strikingly displayed in Peter Menzel and Faith D'Aluisio's *Robo sapiens: Evolution of a New Species*, which offers color photographs and detailed descriptions of a whole range of robotic creatures.[12] Not surprisingly, both Cog and Kismet, two humanoid robots with humanlike cognitive capacities that originated in Brooks's lab (see chapter 6), figure prominently in its pages. Another eye-catching project is Honda's human-sized P3 robot, which is capable of walking and negotiating stairways. Whether we want to call these impressive developments "the evolution of a new species" or an expanding new branch of the machinic phylum, there can be no doubt about their tendency to blur any hard-and-fast distinction between the human and the machine.[13]

AI After ALife

Among the practitioners and theoreticians of the new AI associated with Brooks in the early 1990s, three in particular were quick to draw out the implications of recent research in ALife for behavior-based robotics. Luc Steels, who directs the AI research lab at the University of Brussels and who coedited with Brooks *The Artificial Life Route to Artificial Intelligence*, developed the concept of emergent functionality.[14] Pattie Maes, who worked in Steels's lab as a research scientist before moving to MIT, where she co-organized with Brooks the Artificial Life IV conference, shifted the focus to the construction of autonomous agents and systems. Meanwhile, both Steels and Maes published essays in Christopher Langton's journal *Artificial Life* and formed relays between ALife research and behavior-based robotics. At the same time, Maja Mataric, while pursuing a PhD with Brooks at MIT, began to take behavior-based robotics in the direction of collective intelligence, robotic learning, and the design of adaptive group behavior. Each of these researchers thus brought something new to the line of research that Brooks had initiated.

In "The Artificial Life Roots of Artificial Intelligence," Steels assesses the contributions of ALife for the new AI, by which he means a subgroup within the AI community that stresses embodied intelligence and the behavior-oriented approach, in contrast to the knowledge-oriented approach of classic AI.[15] In this essay, as well as in his contribution to *The Artificial Life Route to Artificial Intelligence*, Steels spells out the terms of a fundamental reorientation in AI research. What the new AI finds most valuable in ALife research is the bottom-up approach and its interest in emergent complexity. Taken together, these contributions point to the need for a thorough reconceptualization of intelligence—one that builds on but goes beyond what is implicit in subsumption architecture and behavior-oriented robotics.

Classic AI research equates intelligence with knowledge-based conceptual and cognitive performance. Operating by means of symbol systems that must be engineered and programmed from the top down, artificial intelligence basically amounts to an abstract, disembodied manipulation of software. This decoding and recoding of mental functions means that intelligence becomes portable and can be installed in various physical systems, both natural and artificial. Thus far, expert systems based on the old AI have been very effective in limited domains but remain highly specialized and completely dependent on human planning and design in the engineering of hardware and the programming of the software. In contrast, the new AI gives primary importance to the bottom-up processes by which intelligence emerges and evolves in biological life, particularly in interactions with the environment that enhance the agent's present situation and increase its chances for survival or in which new kinds of organization and cooperation among multiple agents emerge. Taking up this new orientation, the new AI assumes that it is far preferable to build up or evolve intelligence in artificial life forms, as embodied and situated agents, than to attempt to abstract formal methods, computational procedures, and search strategies from human cognitive activities and emulate them mechanically. In this shift the meaning of intelligence also changes, with the emphasis now falling on adaptation and learning. Intelligence, in short, becomes adaptive behavior. Top-down planning and design yield to strategies for implementing evolutionary processes.

Having established the importance of ALife research for the new AI, Steels develops one of his central themes: how generating complexity through emergent functionality may further enable the construction of artificial physical systems (or agents) that exhibit intelligent behavior. Regrettably, Steels eliminates from discussion multiagent systems as well as simulation—the former because of space limitations, the latter because simulation is associated with the standard computational paradigm and yields only virtual—not real world—behavior. This is regrettable because, far from being peripheral, simulation and multiagent systems are actually essential to the new AI. Even with these exclusions, however, Steels's definition of emergent functionality constitutes a significant conceptual advance.

First, his preliminary definition: "A behavior is emergent if it can only be defined using descriptive categories that are not necessary to describe the behavior of the constituent components. An emergent behavior leads to emergent functionality if the behavior contributes to the system's self-preservation and if the system can build further upon it" (78). Steels then considers three distinct types. The first and simplest occurs when a new functional behavior arises as a result of side effects. For example, two distinct behavior systems are built into a mobile robot (fig. 7.3), one producing wall-seeking and the other obstacle-avoidance. However, as a side effect of the interaction of these behaviors with the environment the robot acquires another functional behavior: wall following. Since this behavior is neither built in nor planned but definitely increases the robot's functionality, it can be called emergent. Steels notes that many adaptations in nature follow a similar pattern, nature being quick to take advantage of accidental effects if beneficial to its creatures. A second type of emergent behavior occurs when a temporal or spatial structure produced as a side effect by the activities of many agents turns out to be of use or beneficial. Steels's example (ironically, given the exclusion noted above) is the formation of a path in a multiagent system. When an ant discovers food in the environment and carries it back to the colony, traces of pheromone are released along the ant's route. The pheromone in turn attracts other

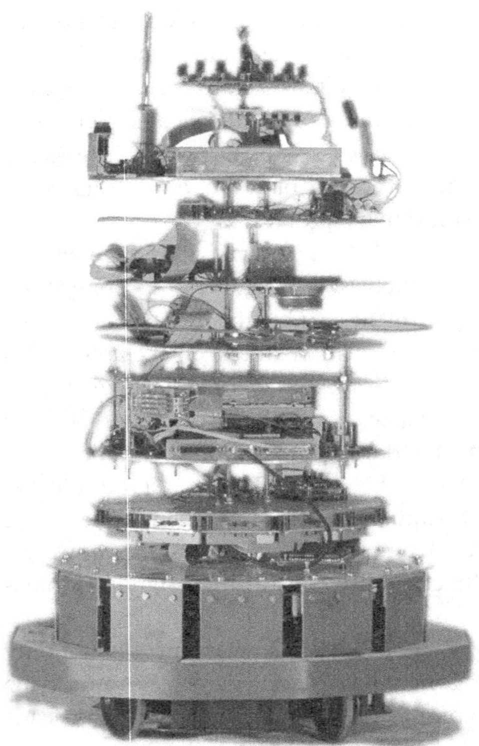

Figure 7.3
Steels's robot. © VUB Artificial Intelligence Lab. Permission to reprint granted by Luc Steels.

ants, which then discover the same food source and leave more pheromone as they traverse the route. Thus a temporal structure functioning as a path is rapidly marked out, which provides an added functionality for the colony even though it endures only as long as the food source remains viable. Finally, the third type of emergent behavior occurs when the formation of new behavior systems leads to the progressive buildup of more complexity. Since convincing examples do not yet exist, this type really represents a research objective: to engineer emergent complexity in robotic systems.

The problem boils down to one of applying the research with artificial neural networks, genetic algorithms, and the creation of ALife worlds like Tom Ray's Tierra to the construction of robotic systems that can evolve.

Following Ray's work in particular, it would seem that in order for robotic systems to evolve, two crucial conditions have to obtain:

1. There is enough initial complexity to make the agents viable, and there should be a diversity of agents. The buildup of complexity would be due as much to the competitive interactions among these agents as to their interactions with the world.

2. The ecological pressures on agents are real and partly come from other agents. There are no pre-defined fitness functions or rewards. (99)

From this we might infer that the third type of emergent functionality could only arise in an environment of robotic systems where there is a rather complete mimicking of a natural environment. In any case, Steels clearly understands that "emergent functionality is not due to one single mechanism but to a variety of factors, some of them related to internal structures of the agent, some of them related to the properties of certain sensors and actuators, and some of them related to the interaction dynamics with the environment" (101).[16] As discussed below, dynamical systems theory allows all three factors to be brought into play and to function together as aspects of a single system.

In a later essay, "The Homo Cyber Sapiens, the Robot Homonidus Intelligens, and the 'Artificial Life' Approach to Artificial Intelligence," Steels describes a robotic ecosystem containing a "growing repertoire of adaptive structural components (called behavior systems)" (12) out of which he hopes intelligent behavior will one day emerge.[17] An accompanying photograph shows a mobile agent, a charging station and a "parasite" lamp that drains energy from the charging station. Of equal interest is the evolutionary parallel suggested by the speculative twin figures Steels proposes: "Homo Cyber Sapiens," a future human whose biological brain capacity has been augmented through its interface with a computer, and "Robot Homonidus Intelligens," a future autonomous robotic agent with the complexity of humans (13). Steels argues that the fulfillment of the same research objective, that is, a full understanding of the origins of intelligence and the mechanisms of its evolution, will bring about both. Once again, this entails a grasp of the principles with which biological systems operate and their application to the "construction of artificial systems" (10). But as Steels adds immediately, the word "construction" is actually inappropriate, because "one of the main ideas is that intelligent autonomous agents cannot be built but should evolve in a process similar to the way that intelligence evolved in nature: using a combination of evolution by natural selection and adaptivity and development as in the development of the biological individual" (10). And indeed, this idea will turn out to be essential to the future development of robots.

In addition to the necessary inclusion of an evolutionary approach, Steels also stresses the importance of a complex dynamical systems perspective for current and future research. Specifically, he repeats the now familiar argument against the reductionist method—that behavior at a higher level cannot always be accounted for by identifying the components and the laws of their interaction operating at a lower level:

> There are properties at each level which cannot be reduced to the level below, but follow from the dynamics at that level, and from interactions (resonance) between the dynamics of the different levels. In the case of intelligence, this means that it will not be possible to understand intelligence by only focusing on the structures and processes causally determining observable behavior. Part of the explanation of intelligence will have to come from the internal dynamics, the interaction with the structures and processes in the environment, and the coupling between the different levels. (11)

From here it is an easy step to model behavior as a dynamical system in which the agents' "structures" are dynamically coupled with "processes" in the environment.

It should be noted that when Steels invokes dynamical systems, he really means Varela's specific notion of structural coupling. Like others of the new AI, he subscribes to Varela's theory of cognition as embodied action (or enaction), which is virtually equivalent to Brooks's twin notions of embodiment and situatedness. Translated into the practice of constructing physical robots, this entails giving priority to building embodied intelligence by implementing direct links (sensors and actuators) with the world, which is basic of course to subsumption architecture. Brooks himself stressed the importance of sensory organs and the ability to move in the evolution of human intelligence. However, in their attempts to build in greater complexity in relation to action selection, learning, and adaptability, Steels, Maes and others influenced by Brooks have placed more emphasis on autonomy. Yet this too reflects Varela's influence, conveyed specifically by his organization (with Paul Bourgine) of the first European conference on ALife, "Toward a Practice of Autonomous Systems," held in 1991. Steels himself testifies to this influence in "Building Agents out of Autonomous Behavior Systems" (his contribution to *The Artificial Life Route to Artificial Intelligence*), where the conceptual weight falls not so much on mobile robots as on autonomous agents or autonomous systems of agents. This displacement widens the scope of the new AI while also giving it a slightly different conceptual focus, for not all autonomous agents are physically embodied, situated agents, but can also exist as software agents and simulations.

In this shift toward autonomous agent research, Pattie Maes has played perhaps the most influential role. Her edited collection, *Designing Autonomous Agents: Theory and Practice from Biology to Engineering and Back*, provides a useful overview of important early work.[18] Its publication followed closely on the 1990 conference "Simulation of Adaptive Behavior: From Animals to Animats," to which she contributed "A Bottom-up Mechanism for Behavior Selection in an Artificial Creature."[19] As Maes notes, the "animat approach"—"animat" is short for artificial animal—is an essential part of the new wave of autonomous agent research. In fact, like the European conferences on ALife following Varela and Bourgine's "Toward a Practice of Autonomous Systems," the animat series of conferences has given considerable weight to themes and methods identified with the new AI. At the same time, the results of this new research—both the simulations and the new robotic machines—are generally understood to be part of a widening growth and development of ALife.

Like Steels, Maes demarcates the objectives of this new research orientation from those of traditional AI. In the preface to *Designing Autonomous Agents* she notes that its new architectures emphasize "a more direct coupling of perception to action, distributedness and decentralization, dynamic interaction with the environment and intrinsic mechanisms to cope with resource limitations and incomplete knowledge" (1). Although she closely echoes Steels, in "Modeling Adaptive Autonomous Agents" her emphasis on autonomy supersedes robotics per se.[20] Maes first proposes a working definition of an adaptive autonomous agent: "An agent is called *autonomous* if it operates completely autonomously, that is, if it decides itself how to relate its sensor data to motor commands in such a way that its goals are attended to successfully. An agent is said to be *adaptive* if it is able to improve over time, that is, if the agent becomes better at achieving its goals with experience" (136). She then distinguishes three types of agent: (1) robots that inhabit the physical world; (2) software agents that inhabit the cyberspace environment of computers and computer networks, like the "knowbots" that navigate through these networks in order to find data of a particular kind; and (3) agents that inhabit simulated physical environments, like the synthetic actors or animated virtual agents in computer-simulated worlds. This simple typology distinguishes Maes's position from those like Steels, who at the time considered only physical robots.

Seeking to identify the deepest problems impeding the construction of adaptive, robust, and effective autonomous agents, Maes acknowledges that problems of action selection and learning from experience seem to entail intractable "computational complexities." This does not mean that aspects of the old AI must be reintegrated with the new but that "more fundamental research" is needed. Maes observes that notable successes in designing autonomous agents have been pragmatic and task driven. But the objective now is to move beyond an approach that ends up looking like "a bag of hacks and tricks" and instead to develop one that embodies general laws and principles. The idea of emergent functionality or emergent complexity still appears to be the most promising, but its specific application to these problems remains on the horizon.

Here a simple analogy of my own may serve to indicate the kind of difficulty posed by this lack of general laws and principles. Solving the problem of action selection and learning in robotics may require a conceptual leap comparable to the leap from an animal or hominid call system (screams, cries, finger pointing, fist shaking, etc.) to fully articulate spoken language. Linguists generally agree that language did not develop from the former; rather, it seems to have emerged from a number of nearly simultaneous evolutionary changes (including increased brain size, enhanced vocalization, tool use, forms of socialization, etc.) that suddenly jelled together and produced a new emergent capacity. But exactly how this happened remains unknown. Similarly, roboticists may one day build or evolve great complexity—and greater intelligence—in robots and autonomous agents, but they may not be able to understand analytically how this leap occurred. Although research in three crucial areas—social learning in multiagent systems, understanding behavior and the environment as forming a single dynamical system, and mimicking nature's evolutionary strategies of development—has been very fruitful, it remains uncertain whether the behavior-based approach is adequate for dealing with higher-level, more properly cognitive, behavior. While complex systems theory offers the most viable theoretical framework thus far, it is not clear yet whether it can supply the general laws and principles that Maes calls for.[21]

Before proceeding further it may be instructive to consider a specific autonomous agent in some detail. One striking success is Pengi, an autonomous software agent designed by Philip Agre and David Chapman to play the commercial video game Pengo.[22] The game is played on a two-dimensional labyrinth of moveable ice blocks, along the paths of which bees roam randomly. The player manipulates a penguin through the labyrinth, trying to kill bees and avoid being killed by them. Both bees and penguins can kick ice blocks, causing them to slide; if a sliding block touches either one, it dies instantly. In the video game a joystick allows the player to move the penguin up, down, right or left, and a button allows it to kick an ice block. In Agre and Chapman's version, the video game is simulated and the player is the autonomous agent Pengi, who is composed of a "central system made of combinatorial logic, a moderately realistic visual system, and a trivial motor system" (274), all implemented in LISP, a high-level programming language used mainly for AI research.

Pengi's play is enabled by deictic representations, which are representations defined in terms of the agent's immediate circumstances or the activities in which it is currently engaged. Thanks to its visual system Pengi can see and track penguins, bees, and blocks of ice. Although Pengi possesses no objective representation of a bee, he (or she) can recognize and respond to the bee-that-is-chasing-me-now or the bee-that-I-can-kill-by-kicking-this-block-of-ice. These deictic representations are characterized by entity and aspect: the-ice-block-I-am-kicking is an entity; the-ice-block-I-am-kicking-is-moving-away is an aspect. Such entities and aspects do not add up to a representational model of the Pengo world but only to a set of routines for actions within it, as in: if a bee is moving toward me, I must move to avoid it. Albeit limited in number, such routines can be combined to fit new situations that develop as the environment changes. Since the blocks of ice are always shifting position in relation to the ceaseless movement of the penguins and bees, there could be no fixed map of this world in any case. To know where it is and what is happening, Pengi simply looks at the screen like a human video game player—rather than like a computer that plays by modeling and planning. Thus Pengi's choices are not based on detailed representations and elaborate reasoning but on immediate possible reactions to the situation *now*. But unlike either behaviorism's stimulus-response or the computational machinery of decision making, these deictic representations allow complex behavior to emerge by being keyed to dynamic situations—that is, to agent-environment interactions.

For Agre and Chapman, Pengi's behavior recalls Herbert Simon's description of an ant slowly making its way across the irregular surface of a beach.[23] As the ant traces an erratic and difficult-to-describe path, halting here and detouring there but nonetheless progressing toward its goal, its behavior appears to be quite complex in the eye of an observer. However, as Simon notes, the complexity lies not in the ant but in the ant's traversal of this irregular surface. Similarly, while Pengi's behavioral repertoire is not in itself complex, his behavior grows quite complex as he enters into a set of dynamic relationships with a constantly changing environment. Agre and Chapman note that Pengi plays the game better than either of them, but that is not the point; nor is their experiment an argument against representation. What Pengi demonstrates, rather, is how much can be accomplished with limited representations, improvisational strategies, and simple routines when keyed to dynamic interactions in specific situations. If Pengi's complex but mostly improvised behavior validates Agre and Chapman's argument against presituational planning, as they call it, it is precisely because the representations Pengi deploys are generated by specific situations and have validity or usefulness only within that context.[24]

Pengi typifies one type of autonomous agent among a whole new fauna of artificial creatures that have steadily increased in number and type since the late 1980s. Their behavior is complex enough to warrant the question of whether there is more there than can be predicted and accounted for by the details of their construction and programming. This is essentially the question of emergent functionality addressed by Steels. But Steels was also well aware that complex emergent behavior is most likely to arise in a field of interacting multiple agents; hence his more recent research with a robot ecosystem ("Artificial Life Roots"). Meanwhile, and from the very beginnings of her graduate research in the early 1990s, Maja Mataric has focused on group behavior in behavior-based multirobot systems.

In "Designing Emergent Behavior: From Local Interactions to Collective Intelligence," Mataric presents an overview of her research objectives and some of her experimental results.[25] Her goal is to understand how simple local interactions among a collection of artificial autonomous agents produce complex and purposive group behavior. The agents are physically identical mobile robots about twelve inches long, equipped with bump sensors and a forklift for picking up, carrying, and stacking pucks. Twenty in number, each robot carries a radio transceiver that enables it to broadcast its own state and to receive information about the states of other agents at the rate of one byte/second. Furthermore, their control systems are set up to make them capable of what Mataric calls "interaction primitives": collision avoidance, following, dispersion, aggregation, homing, and flocking. For example, using the robots' capacity to detect obstacles with infrared censors (IRs), she devised the following two formulae for avoidance behavior:

Avoiding Other Agents:
If another robot is on the right
turn left
otherwise turn right.
Avoiding Everything Else:
If an obstacle is on the right only
turn left, go.
If an obstacle is on the left only
turn right, go.
After three consecutive identical turns
backup and turn.
If an obstacle is on both sides
stop and wait.
If an obstacle persists on both sides
turn randomly and back up. (435–436)

Using these formulae, other behaviors can be built up. *Following*, for example, can be implemented using the inverse of the *collision avoidance* behavior. Behaviors can also be combined, although this sometimes entails adding new sensors.

Under conditions of sufficient density, *collision avoidance* and *following* can produce more complex global behaviors. For instance, chemotropotaxic ants exhibit emergence of unidirectional traffic lanes. The same lane-forming effect could be demonstrated with robots executing *following* and *avoiding* behaviors. However, more complex sensors than IRs must be used in order to determine which direction to follow. If using only IRs, the robots cannot distinguish between other robots heading toward and away from them, and are thus unable to select whom to follow. (436)

By adopting various combinations of basic behaviors (the "interaction primitives"), she has devised formulae for dispersion, aggregation, homing, and flocking. As the following formula suggests, *flocking* turns out to be the most complex:

Weight the inputs from
avoid, follow, aggregate, disperse
then compute a turning vector.
If in the front of the flock,
slow down.
If in the back of the flock,
speed up. (438)

Inspired by Craig Reynolds's ALife simulation of bird flocking with boids, she admits that implementing robot flocking involves more complex dynamics and therefore requires a more detailed approach.

Figure 7.4
Mataric's "Nerd herd." Maja Mataric, "Designing and Understanding Adaptive Group Behavior," *Adaptive Behavior* 4, no. 1 (1995): 6.

Mataric's ultimate objective is to produce more complex behaviors, such as foraging and puck sorting—and eventually learning. These more complex behaviors would "emerge as temporal sequences of basic interactions, each triggered by the appropriate conditions in the environment" (438). *Foraging*, for example, would be initiated as a dispersing behavior, as each agent begins to search for food (i.e., pucks). Once it obtains food, it starts for home (*homing*). Along the way, it may encounter another agent also carrying food (i.e., in the same state, which would be indicated by a simple radio signal). If so, it would then follow it, eventually forming a flock if enough agents find food in the same vicinity (fig. 7.4). Mataric emphasizes that this work establishes "the effectiveness of the behaviors in our basis set [the simple interactive behaviors] by showing necessity (they are not reducible to each other) and sufficiency (they can generate a large repertoire of more complex agent interactions)."[26] She concludes, however, that in order for basis behaviors "to be a truly effective substrate of adaptive behavior, they must serve as a substrate for efficient and general learning."

In subsequent work, Mataric explores strategies by means of which robots can learn adaptive group behavior from one another and thus learn to behave socially. As she notes, social learning is ubiquitous in nature, and its propensity appears to be innate: "Animals imprint, mimic and imitate adults of their own kind instinctively, often without obtaining direct rewards or even successfully achieving the goal of the behavior" (453). Animals also mimic their peers, especially if the behavior leads to visible rewards, and usually avoid behaviors that do not, like eating poisoned substances. Obviously this is a large field of study, with many different distinctions—between imitation and mimicry, for example. Mataric focuses specifically on learning social rules, or behaviors that do not produce immediate payoff for the agent but benefit the group as a whole. In contrast to related research in game theory, she wants to find out "what is required for learning social strategies in situated agent domains where, due to incomplete or nonexistent world models, inconsistent reinforcement, noise and uncertainty, the agents cannot be assumed to be rational" (454). To do this she sets up a test environment with three autonomous robots that will gather food (pucks) and bring it home during a specific time period ("day"), and rest during another specific time period ("night"). She then postulates three types of necessary reinforcement: direct reinforcement, observation of other agents' behavior, and observation of reinforcement received by other agents. These types are implemented by means of a set of known learning algorithms (like a small reward for a definite progress achieved toward a specific goal), which are programmed into the robots using Brooks's Behavior Language.

As in her previous experiments, the robots use radios to communicate their state to the others, as in "holding food," "finding home," and so forth. The learning algorithm is activated whenever an agent finds itself:

1. near a large amount of food away from home,
2. receiving an agent's message,
3. within observing range of another stopped agent,
4. within observing range of another moving agent,
5. within interference range of another stopped agent,
6. within interference range of another moving agent. (459)

In previous work utilizing this same setup Mataric had tested the learning algorithm, finding that within an average fifteen-minute trial run the robots were able to learn foraging by learning to select appropriate individual behaviors for each state (459). But here the objective was for the robots to learn to yield and share information in a foraging task, with this learning behavior being implemented by means of reinforcement functions added to the learning algorithms. Overall, the results were successful. Groups of robots using the social rules always outperformed groups with only "greedy individual strategies" (460). As one might expect, rules that produce the most immediate reward were learned the fastest; for example, the social rule of sharing information about food was by far the hardest to learn, as the benefit to the agent had the least direct payoff (460). This suggests that altruistic social rules are perhaps best learned "genetically," an intuition supported by biological data.

In these experiments Mataric observed that the speed at which learning occurred was often directly reduced by hardware error and noise. Yet these unavoidable factors did not disable the learning algorithm; they only slowed it down.[27] In future work she intends to apply the basis behavior idea to other social and cooperative tasks with multiple agents, using both homogeneous and heterogeneous groups of robots. She also intends to use the strategies of genetic programming developed by John Koza for automatic generation of basis behavior sets for specific domains.[28] This will take her research into the increasingly important realm where behavior-based robotics as strictly defined begins to blur into evolutionary robotics. But before considering this development, we must take up the coupling of behavior-based robotics with dynamical systems theory, for that is where evolutionary programming was first deployed in robot construction.

Autonomous Agents in/as Dynamical Systems

Although behavior-based robotics research assumes a dynamical systems perspective from the outset, only with Randall Beer's essay "A Dynamical Systems Perspective on Agent-Environment Interaction" does it become a fully analytic framework for understanding the interactions between agents and their environment.[29] Beer demonstrates how a dynamical systems perspective can be rigorously applied to the construction of autonomous robotic agents, while also aligning this approach with new research in cognitive science that uses dynamical systems to address problems in cognition.[30] Thus while the vocabulary of embodiment and situatedness is retained, Varela's centrality is eclipsed by the more analytic concepts of dynamical systems theory (attractors, phase space, bifurcations, etc.). For this reason alone, the essay marks a significant conceptual advance, even though Beer considers only the interactions between a single autonomous agent and the environment and omits multiagent systems.

Beer begins by rehearsing the argument made by Brooks and Steels that AI research must now concern itself with embodied agents, specifically with autonomous agents situated in changing real-world settings within which the agents must interact in versatile and robust ways, rather than continue with attempts to emulate high-level intellectual skills like language acquisition, problem solving and abstract reasoning. From an evolutionary perspective, these high-level skills appeared long after motor and perceptual skills and the "basic capacity for situated action that is universal among animals" (175). Beer then proposes dynamical systems theory as the descriptive/theoretical model that can best characterize the interactions between an agent and the environment. Considered separately, agent and environment each constitute a dynamical system with its own state variables and parameters and thus possible behaviors in a state space of trajectories indicating stable limit sets, or attractors, basins of attraction, and so forth. Beer's central idea is to couple the two dynamical systems, by making some of the parameters of each system functions of some of the state variables of the other. He assumes that the coupled agent-environment system "exhibits only convergent dynamics" (181), that is, that the values of the state variables converge to some limit set, as we usually observe in the "adaptive fit" between animals and the environment in nature, rather than diverging to infinity or chaos. Of course there are borderline cases, as when it is uncertain whether the agent's body should be treated as part of the environment or part of the agent.

Needless to say, the coupled system of agent-environment constitutes a feedback system. Acknowledging the influence of cybernetics and specifically of W. Ross Ashby, Beer points out that one dynamical system cannot "in general 'steer' the trajectory of another along the desired path. It is therefore perhaps most accurate to view the agent and its environment as mutual sources of perturbation, with each system continuously influencing the other's potential for subsequent interaction" (182). In figure 7.5, Beer represents schematically the coupling of agent A and environment E to form a dynamical system. Note that S represents a *sensory function* that couples environmental state variables to agent parameters, while M represents a *motor function* that couples agent state variables to environmental parameters.

In order to survive in nature, a living organism must maintain the integrity of the network of biochemical processes that keep it alive—what Maturana and Varela call an "autopoietic system." Again acknowledging the influence, Beer suggests that such a system serves as a crucial constraint on the organism's behavioral dynamics. In other words, this

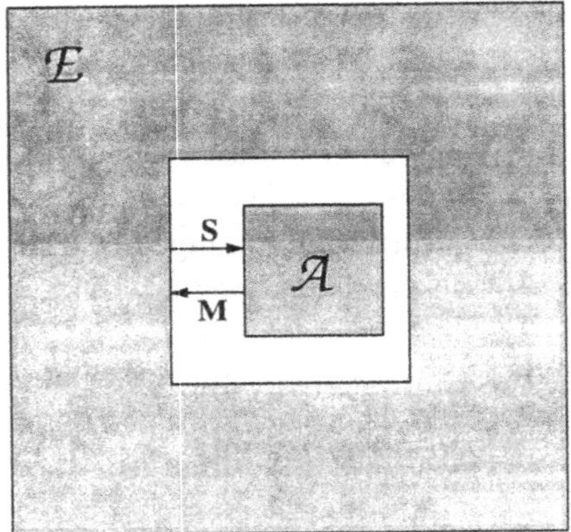

Figure 7.5
Coupled agent and its environment. Randall D. Beer, "Computational and Dynamical Languages for Autonomous Agents," in *Mind as Motion: Explorations in the Dynamics of Cognition*, ed. Robert F. Port and Timothy van Gelder (Cambridge, Mass.: MIT Press, 1995), 131.

constraint defines a trajectory or set of behavioral possibilities that, as long as the organism remains within it, allows it to survive despite perturbations and obstacles. For an artificial agent this constraint could be defined by particular performance criteria, such as the ability to carry out a variety of tasks. In both cases, only a subset among all the possible behavior trajectories of the agent-environment system is admissible. Beer proposes an illustration of such an "adaptive fit" (fig. 7.6).

Assuming, then, that the coupling of an agent with an environment establishes a unique dynamical system with its own specific constraints, what new factors does this framework put into play and what does it allow us to understand? Perhaps most important, it brings about a fundamental shift in perspective, since the agent's behavior is now seen to reside in the dynamics of the coupled system E and A and not in the individual dynamics of either one alone. As Beer points out, "This suggests that we must learn to think of an agent as containing only a latent potential to engage in appropriate patterns of interaction. It is only when coupled with a suitable environment that this potential is actually realized through the agent's behavior in that environment" (183). In these terms evolution in nature appears to be the trying out of many different agent dynamics, with only those that on average prove capable of satisfying autopoietic constraints long enough to allow the organism to reproduce being retained. As for the construction of artificial autonomous agents,

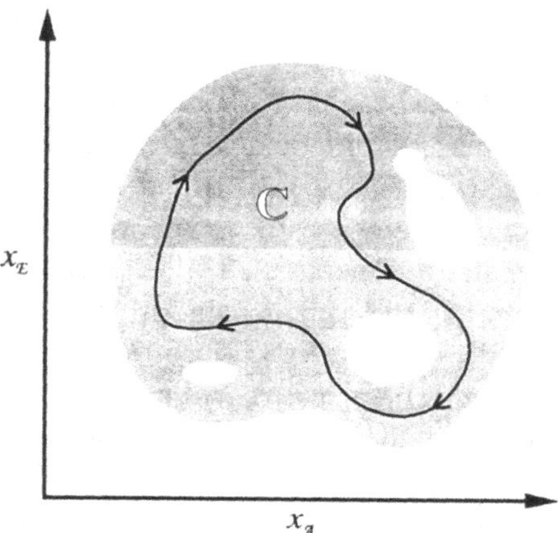

Figure 7.6
Beer's illustration of an adaptive fit. Randall D. Beer, "Computational and Dynamical Languages for Autonomous Agents," in *Mind as Motion: Explorations in the Dynamics of Cognition*, ed. Robert F. Port and Timothy van Gelder (Cambridge, Mass.: MIT Press, 1995), 133.

Beer believes that the dynamical systems framework suggests solutions to two kinds of problems: "the synthesis problem," or how to construct an agent that does what we want in a given environment; and "the analysis problem," or how to understand what an agent actually does in a given environment.

To illustrate, Beer considers the walking behavior of a simulated artificial insect from these two aspects. First, in relation to its synthesis or construction, the key to the creature's capacity to walk resides in the controllers that regulate the movements of each of its six legs. Each controller, which consists of a five-neuron neural network, is connected to an angle sensor (input) that gives the leg's angle in relation to the body, and to a signal (output) that moves a foot attached to a leg either up or down. When raised to the up position, the leg also swings. The problem is to coordinate the controllers so that the creature "walks." In other words, walking would be one of the possible behaviors of the dynamical system formed by coupling the creature's body E to the neural network A that controls it. To manually design the algorithm for setting the weights of the neural net controller would be a truly daunting task, so Beer resorted to genetic algorithms as a search technique. The space to be searched (defined by about fifty network parameters) was encoded as binary strings whose performances in training the network were then evaluated. By continually mating relatively successful strings, eleven different locomotion controllers were evolved.

Second, in his analysis of these performances Beer demonstrates that the neural net controllers settle on patterns of interaction with the environment that correspond to recognized gaits (like the "tripod gait") of insects in the world. More specifically, three classes of locomotion controllers were found: (1) "reflexive pattern generators," which evolved when the sensors were activated during the genetic algorithm (GA) search; (2) "central pattern generators," which emerged when the sensors were disabled during the GA search; and (3) "mixed pattern generators," which evolved when the sensors were sometimes available and sometimes not. Each class has its own strengths and weaknesses. The first makes maximum use of the dynamics already in the environment, but this advantage also makes the agent vulnerable to sensory loss. The second enables the agent to walk without sensory input, but it is not able to fine-tune its skill in response to environmental changes. The third class offers a mix of the first two: it works better with its sensors intact, but it is still able to function (i.e., to walk) without them. As Beer points out, such mixed organizations are the most typical among biological pattern generators. Analysis of the trajectory in the controller's state space portrait revealed two single-point attractors that corresponded to the two leg positions (up and down). However, the most interesting phenomenon that emerged was the systematic flipping back and forth between the two fixed-point attractors as the leg moved from the stance phase to the swing phase. The leg's swing thus corresponded to a bifurcation in the state space and thus to a branch point between two different basins of attraction and hence two different results from the specific dynamic of forces acting on the leg at this moment. What most merits emphasis, finally, is that this dynamic behavior was neither that of the neural net controller alone nor that of the agent's legs, but of the two acting together in a coupled dynamical system.

Evolutionary Robotics

In order to produce this coupled dynamical system, Beer had to resolve a formidable design problem: finding the algorithms that would enable the neural net leg controllers to perform in a manner resulting in a specific action (i.e., the artificial insect would be able to walk). The creature's basic design was given in advance, and an evolutionary programming technique was deployed to make that design physically workable. In nature, of course, there is no such split: an organism's morphology and nervous system evolve together as an interactive relay system. This simple fact raises some obvious questions: Was the controller all that could be evolved? That is, would it not be possible to apply evolutionary techniques more globally, to the robot's design and controllers together, especially since the design of control systems by hand could become exponentially more difficult as the robot's desired behavior increases in complexity? These questions suggest that the implementation of a workable neural net controller, as in Beer's example, is hardly a peripheral or subsidiary design issue. Indeed, in the formation of the new discipline of evolutionary robotics these questions become central.

In one sense, as Luc Steels had already suggested, evolutionary robotics was simply a continuation of the application of ALife techniques to the construction of mobile robots. This is the perspective adumbrated by Rodney Brooks in "Artificial Life and Real Robots," an early acknowledgment of the importance of evolutionary strategies in the building of autonomous robots.[31] After reviewing his previous work in behavior-based robotics, Brooks admits that progress with robots learning new behaviors has proven to be difficult, mainly because programming each new behavior must be done by hand. However, ALife "has developed techniques for evolving programs for controlling situated, but unembodied (i.e., simulated), robots," and thus may provide "techniques to evolve programs to control physically embodied mobile robots" (3). In fact, it was none other than Christopher Langton who suggested that Brooks program his physical robots "genetically."

Genetic programming is preferable to genetic algorithms because whole programs can be evolved rather than bit-string representations of solutions for particular predefined problems. In the most fruitful example of genetic programming, John R. Koza has developed successful strategies for evolving short LISP programs that have been applied to a range of programming tasks. Notably, Koza has applied genetic programming techniques to the "base behaviors" in Maya Mataric's behavior-based robot programs, which were written in Brooks's Behavior Language (BL). Although the results were only run in simulation, Brooks felt that they warranted further exploration. Consequently, he states, he plans to develop a high-level language called GEN that can be compiled into BL to facilitate the genetic programming of physical robots.

Yet Brooks also expresses an uneasiness concerning the "methodological dangers of using simulations as a testing medium in which to evolve programs which are intended eventually to run on physical robots" (9).

He admits, in fact, to having been very careful to avoid using simulations in his previous work. For him the dangers are clear: either a great deal of effort is wasted trying to solve problems that arise in simulations but "simply do not come up in the real world with a physical robot (or robots)," or programs that work well in simulations fail completely on real robots because of differences in real-world sensing and actuation. Real-world dynamics, moreover, can sometimes be exceedingly difficult to simulate. Nevertheless, despite these drawbacks, Brooks is enough of a pragmatist to accept the advantages that simulation can bring, as long as a "sufficient validation regime" is also developed.

In one respect Brooks's bias against simulation recalls Varela's objections to Langton's version of ALife.[32] Both assume that since the agents in the simulations are not physically embodied, they are somehow ontologically lesser beings. While understandable for Brooks, who has devoted his life to making physical robots, this view appears somewhat problematic. For one thing, it clearly diminishes the significance of much ALife research, which has been of such enormous benefit to contemporary robotics. In any case, the status and usefulness of simulation within robotics has been a matter of ongoing debate, despite several striking successes in evolving a morphology *and* controller together in a simulated world. In the most spectacular example, Karl Sims's "virtual creatures," blocklike entities are evolved that can swim, walk, jump, follow, and compete for a cube in a three-dimensional virtual world in which physical forces like gravity, friction, and fluid viscosity are also simulated.[33] Sims used evolutionary programming techniques to produce genotypes "structured as directed graphs of nodes and connections, and they can efficiently but flexibly describe instructions for the development of creatures' bodies and control systems with repeating or recursive components" (28) (see figs. 7.7 and 7.8, the still image from his computer animation). The phenotypes or individual creatures that result then compete in various contests in a virtual world, with the winners receiving a higher fitness rating that allows them to survive and reproduce.

By the mid-1990s, the successes of Beer, Mataric, Sims and others had ratified the primary objective in evolutionary robotics: to deploy evolutionary programming strategies and simulation (when necessary or pragmatic to do so) in the construction of more complex physical robots. Nevertheless, while success came rather quickly in the evolution of neural net controllers, particularly for movement and navigational tasks, the evolution of new adaptive morphologies has proved to be more difficult. Adaptive hardware configurations like field-programmable gate arrays

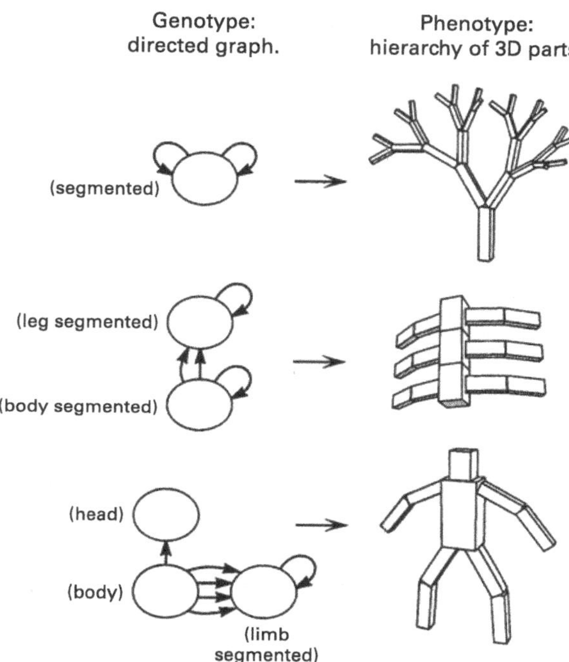

Figure 7.7
Sims's graphs and creature morphologies. Karl Sims, "Evolving 3D Morphology and Behavior by Competition," in *Artificial Life IV*, ed. Rodney Brooks and Pattie Maes (Cambridge, Mass.: MIT Press, 1994), 31.

Figure 7.8
Sims's Virtual Creatures. © 1994 Karl Sims.

(FPGAs) offer one promising path. One problem with adaptive or evolvable hardware, however, is that, compared with computer simulations, it requires tremendous expenditures of time to set up and test in multiple versions. But this is only one of a number of promising paths currently being explored. I conclude this brief survey with capsule reports on recent developments by two prominent research groups.

The Sussex group, centered at the University of Sussex since the early 1990s, is usually credited with first introducing the term *evolutionary robotics*.[34] In 1991 Inman Harvey introduced species adaptation genetic algorithms (SAGA), a conceptual framework for exploring the dynamics of genetic algorithms in instances where bit strings that define genotypes are allowed to increase in length. Harvey argued that while GA of fixed length are best employed for solving predetermined problems in optimization in a finite search space, for ALife applications a different kind of search was necessary—an "evolutionary search . . . around the current focus of a species for neighboring regions which are fitter . . . while being careful not to lose gains that were made in achieving the current *status quo*."[35] For this kind of search, that is, for the evolution of a species rather than a global search, GAs of different lengths were needed, hence SAGA. Harvey soon teamed up with Phil Husbands and David Cliff, also at Sussex, for the construction of actual robots.

One of the group's most impressive achievements to date has been with visual navigation systems for mobile robots. In one instance, they were able to "incrementally evolve . . . visually guided behaviors and sensory morphologies for a mobile robot expected to discriminate shapes and [to] navigate towards a rectangle while avoiding a triangle."[36] Of more theoretical interest, an earlier effort involved the coevolution of parts of a robot body with a control system. Here they also used a visually guided robot with a neural net controller programmed using genetic algorithms, but the robot's "chromosomes" were organized into two parts, one specifying the network architecture; the other, the morphology of the visual system. Having built into the robot a number of visual receptors scattered around target points on its "body," each genetically determined version of the controller would use a different combination of these receptors. Thus the group could experiment with different controller/visual receptor configurations without constantly changing the hardware.

This kind of strategy leads toward what Stefano Nolfi and Dario Floreano in *Evolutionary Robotics* have called self-organizing machines. As leaders of a second prominent group of roboticists (who are physically divided between the Swiss Federal Institute of Technology at Lausanne and the Research Group of Artificial Life in Rome), Nolfi and Floreano have made robotic self-organization a primary theoretical goal. On the experimental level the group's research is distinguished by their use of a small, two-wheeled mobile robot called the Khepera, which offers three notable advantages: (1) miniaturization (since it is only 55 mm in diameter, experiments require a small working surface); (2) a modular, open architecture that allows variation and expandability (for example, it can be equipped with a variety of light sensors and/or a gripper module); and (3) easy interface and compatibility with larger robots. In their book, Nolfi and Floreano describe many experiments using Khepera; one series, for example, is devoted to coevolving predator-prey robots and begins by equipping the predator with a superior vision module and the prey with twice the capacity for speed.

On the basis of these and other experiments, Nolfi and Floreano claim that their work marks a break with behavior-based robotics and the layered, incremental approach to decomposition initiated by Brooks. This approach, they argue, puts the entire burden of deciding how a desired behavior should be decomposed, or broken down into simple basic behaviors, on the designer. (To be fair, Brooks himself has expressed an awareness of the limits of his approach, as Nolfi and Floreano acknowledge.) Increasingly, however, contemporary robotics seeks to produce complex behavior that emerges unexpectedly from local interactions among its various elements and in relation to a changing environment. Since such global or higher-level complex behavior cannot be designed "by hand," evolutionary strategies become inevitable. For Nolfi and Floreano, therefore, it is no longer a matter of simply deploying genetic algorithms and other strategies of genetic programming to evolve neural net controllers and, in the best case, controller and new body morphologies simultaneously. (In point of fact, the authors give too little attention to the latter.) Rather, for them the key issue is getting the designer further outside of the construction loop. Referring to the Sussex achievement in evolving a controller and visual receptor morphology together, they state: "A more desirable solution ... would be a self-organized process capable of producing incremental evolution that does not require any human supervision. This ideal solution spontaneously arises in competing coevolving populations with coupled fitness, such as predator and prey scenarios" (15). Overall, then, artificial evolution appears to be by far the most effective method for creating conditions in which machines can self-organize and reconfigure themselves in order to adapt to changing environmental conditions.

However, in order for evolutionary robotics to advance, not only must "bodies and brains" (i.e., morphologies and controllers) be evolved together, but the design and fabrication processes must also be included. Only then will robots as a species of artificial life be able "to bootstrap and sustain their own evolution."[37] This, at least, has been the view advocated and in surprising ways partly realized by Hod Lipson, who has pushed the edge of evolutionary robotics in a growing body of innovative research.[38] Lipson and Jordan Pollack first contributed to this view of robotics with their GOLEM Project (GOLEM is an acronym for genetically organized lifelike electro mechanics). The project consists of three coupled parts, with each accomplishing a separate function in an almost fully automated three-stage process: first, a computer program designs a population of robots using evolutionary computation; second, the robots are run in a simulation program to determine fitness based on their ability to move horizontally on a flat plane; and third, the most successful are then materialized using a commercial rapid prototyping machine (in effect, a 3-D printer). In the final stage of the experiment, the locomotion ability of the physical robots is tested against their ability as measured in the simulation.

Needless to say, these are very simple robots, made from the most elementary building blocks. Short linear bar/actuators are connected together with ball joints to form a body (a morphology), while artificial neurons are connected to one another and to the actuators to form a brain (a controller). There are no sensors. The extreme simplicity of the building blocks is necessary in order that body and brain can coevolve simultaneously. Mutations can modify any aspect of the robots, by adding, removing, and modifying bars, neurons, and actuators. Initially a population of 200–1,000 machines in which these components are randomly combined is produced by the program, and then their locomotion ability is tested in the simulation. In an iterative process, fitter machines are selected, and offspring are created "by adding, modifying and removing building blocks, and replacing them in the population" (283). The process continues for several hundred generations. Typically, Lipson and Jordan note, "several tens of generations passed before the first movement occurred," for movement requires that, "at a minimum, a neural network generating varying output must assemble and connect to an actuator for any motion" (284). The best performing robots "were then *automatically* replicated into reality" (284) in the third stage, which they describe as follows:

> The solidifying stage was automatic but used a hand-coded procedure describing a generic bar, joint and actuator. The virtual solid bodies were then materialized using commercial rapid prototyping technology.... This machine uses a temperature-controlled head to extrude thermoplastic material layer by layer, so that the arbitrarily evolved morphology emerges as a solid three-dimensional structure ... without tooling or human intervention. The entire preassembled machine is printed as a single unit, with fine plastic supports connecting between moving parts; these supports break away at first motion.
>
> The resulting structures contained complex joints that would be difficult to design or manufacture using traditional methods.... Standard motors are then snapped in, and the evolved neural network is executed to activate the motors. (284)

Human intervention over the entire process is thus kept to a bare minimum (snapping in the motors and setting the parameters in the simulation). Remarkably, the resulting physical robots performed as well in reality as their virtual ancestors did in simulation. Also notable is that for the particular fitness criterion selected (locomotion on a flat plane), the evolutionary dynamics converged on three different solutions, or "species": a tetrahedron, an arrow and a pusher, which moved by ratcheting, antiphase synchronization, and dragging, respectively. (See fig. 7.9, where they appear from top to bottom. The images on the left are the physical robots; those on the right their simulations.) Overall, the GOLEM project takes a giant step forward in coevolutionary robotics. Not only are robotic body and brain coevolved, but the results extend "from virtual simulations ... to the reality of computer designed and constructed ... machines that can adapt to real environments."[39]

Lipson's next major step was to build a physical robot that could reproduce itself.[40] Again, of necessity, the building blocks are extremely simple. In this case they are modular cubes containing electromagnets, the strength of which can increase or decrease, thereby allowing connections to be easily formed and broken. Each cube, furthermore, is split into half along a slanted plane, with each half fitted with swivels allowing it to move in increments of 120 degrees (fig. 7.10). Each module also has its own power source and microcontroller, and modules communicate both power and data through their faces. The microcontroller is programmed to execute "a motion schedule governed by time and contact events" (163). These features enable an assortment of modules to form and change into any number of arbitrary arrangements.

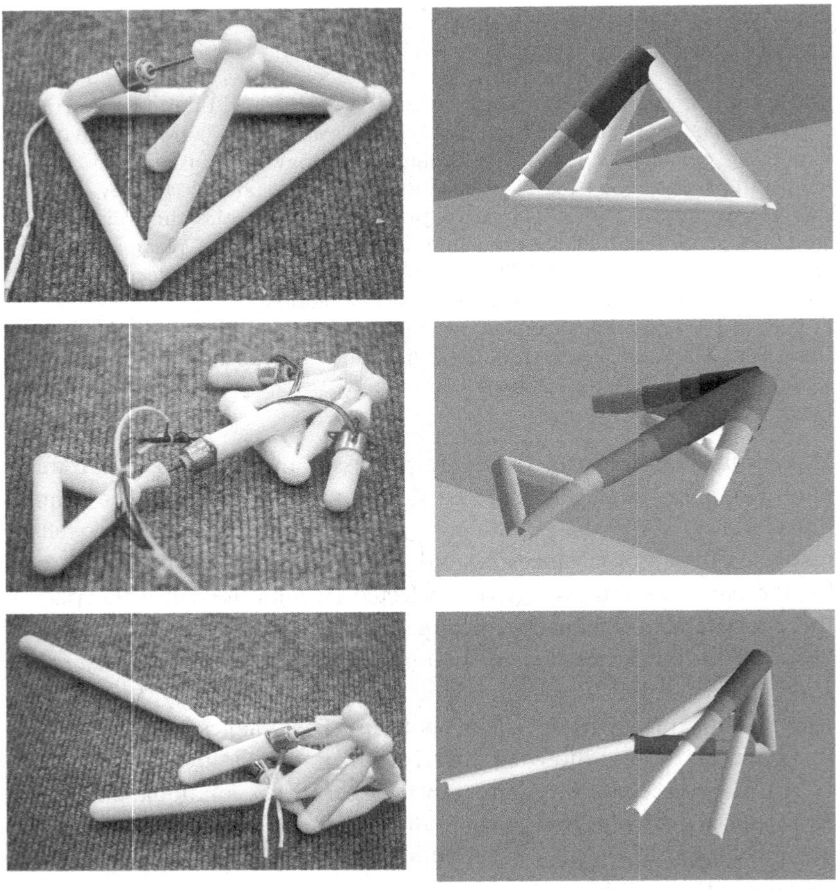

Figure 7.9
Lipson's Tetrahedron, Arrow, and Pusher. Jordon B. Pollack, Hod Lipson, Gregory Hornby, and Pablo Funes, "Three Generations of Automatically Designed Robots," *Artificial Life* 7, no. 3 (2001): 216.

In the experiments Lipson describes, three-cube and four-cube columns of modules were able to reproduce themselves. In order for self-reproduction to occur, of course, new material has to be produced or provided. Here additional modules were supplied manually at two specific "feeding locations." Rather than being passively built, the "replica reconfigures itself to assist in its own construction" (163), a striking feature of the process of self-reproduction that Lipson does not explain. The process is also surprisingly fast: it took 2.5 minutes for the four-cube

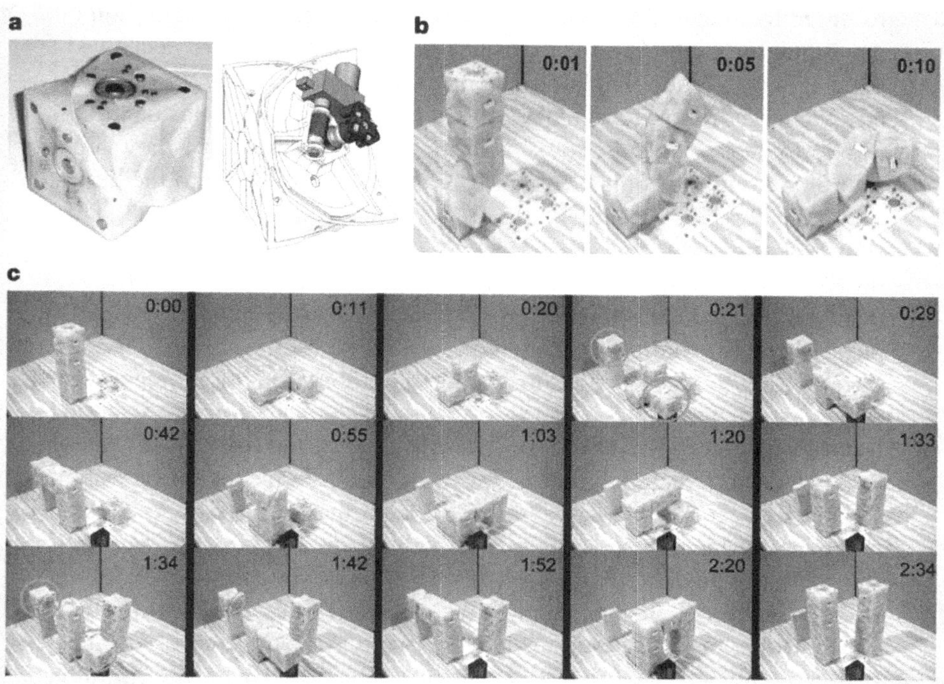

Figure 7.10
Lipson's self-reproducing robot. Victor Zykov, Efstathios Mytilinaios, Bryant Adams, and Hod Lipson, "Self-Reproducing Machines," *Nature*, 12 May 2005, 163.

column to make a replica of itself, and only 1 minute for the three-cube column. Lipson does not say whether these columns could in turn have reproduced themselves (presumably they could have) but does say that other reproducing forms are possible with these modules.

In "From Virtual to Physical Artificial Life," his keynote address at the ALife X conference (2006), Lipson described a number of fertile research paths currently being explored by his team at Cornell.[41] For coevolutionary robotics, the most important are new or better ways "to cross the reality gap," that is, to move from robot simulation (especially the task of evolving robot controllers) to the materialization of physical robots and new opportunities to discover and develop a physical medium in which physical ALife can evolve by changing its own morphology autonomously. In terms of the first, the most innovative proposal involves reversing the usual approach: instead of evolving a controller in simulation and then transferring it to the target physical robot, sensory data recorded by the physical robot is used as fitness criteria in the evolution of the simulator itself. In short, a robot and its simulator are coevolved. Starting with a crude simulator, a controller is produced and downloaded into the target robot. Its actual observed behavior in turn yields data that is used to evolve a new set of simulators. The best simulator is now used to produce another controller, which is again downloaded into the target robot. The cycle is repeated until an acceptable controller is found. This convergent method considerably reduces both the amount of noise produced or encountered in the usual method (evolve a simulated controller, then transfer it to the physical robot), as well as the number of costly physical tests and adjustments that are always necessary.

Perhaps even more important, the "exploration-estimation" algorithm developed for this method can also be used for much more than simply transferring controllers to robots: it can be used by the robot itself "to estimate its own structure"; that is, it can be used as a means for the robot to infer its own morphology and thus create a model of itself. As Lipson points out in his lecture, using an internalized simulator as a way of letting the machine discover for itself how its own morphology is interacting with the environment (i.e., its own dynamics) is similar to "identification for control," an engineering method for creating robust control systems. It is also similar to a primitive form of learning and may be fruitfully explored in those terms as well. But however it is viewed, it should be clear that the problems associated with using simulation in robotics have here been laid to rest. For in Hobson's internalized simulator the differences between simulation and actual robot behavior have been introduced into the productive evolutionary dynamic itself.

As for the problem of developing new, changeable, and evolvable morphologies, a whole new spectrum of possibilities now presents itself. Lipson is avowedly most passionate about the potential of rapid prototyping machines, especially if they can be produced at a cost that would make them available for use in the home.[42] If a fabber (digital fabricator) could be built that could indeed "print out" an entire robotic machine—including wires, motors, and electronic parts—then it would constitute a new physical medium in which robots could evolve. Of course, this would amount to a top-down approach. At the opposite extreme, the most fruitful bottom-up approach is represented by self-assembly, usually the stochastic self-assembly of large numbers of cells or other small modular components that have some kind of affinity for one another. Here the objective is to find a way to make the process programmable. Since most self-assembly is stochastic, the increasing exploration of amorphous material is likely. Lipson thinks that what Rolf Pfeifer calls "morphological computation"—getting the morphology itself to do at least some of the computational work—may offer a promising approach.[43]

Swarm Machines

The various research activities currently devoted to swarm systems and swarm intelligence draw together many of the thematic strands discussed thus far. Ethology, ALife and evolutionary robotics, collective intelligence and distributed processing, autonomous agents and complex adaptive systems—all figure into recent efforts to mimic one of nature's most singular phenomena. Provisionally, a swarm can be defined as a collection of locally interacting organisms or agents whose collective behavior is globally adaptive. By definition, then, swarm systems exhibit both self-organization and emergence. But this raises an obvious problem: What distinguishes a swarm system from any other self-organizing and emergent system? What is it that justifies the term *swarm*? To answer I shall consider three examples, each of which gives to the idea of a swarm system a different twist and thereby constitutes a different kind of computational assemblage. The first proposes a model of distributed intelligence and information processing directly based on social insect behavior; the second, a computational model that simulates the transforming effects of social interaction; and the third, a new kind of autonomous mobile robot with self-assembling capabilities. It is worth noting that as these new swarm systems enter the public realm—mostly in terms of newly developing technologies—they raise disturbing questions about control and the boundaries of the human self-image. Indeed, it is not for nothing that Michael Crichton made swarm technology the centerpiece of *Prey*, his recent techno-thriller novel. Drawing extensively on cutting-edge research, Crichton paints a picture of a new form of autonomous and distributed artificial intelligence gone frightfully out of control.[44] More recently, the evidence Howard Rheingold marshals in *Smart Mobs: The Next Social Revolution* suggests that mobile Internet technology is producing a new kind of human "swarming," with far-reaching social consequences.[45] While the image of the swarm conveys something of the archetypal and atavistic, in these examples it is always harnessed to the development of an almost futuristic technology.

We owe the first scientific formulation of the idea of a swarm system to the entomologist William Morton Wheeler, the founder of the study of social insects. In "The Ant Colony as an Organism," published in the

Journal of Morphology in 1911, Wheeler writes, "Like a cell or the person, [the colony] behaves as a unitary whole, maintaining its identity in space, resisting dissolution ... neither a thing nor a concept, but a continual flux or process."[46] Wheeler was particularly intrigued by the phenomenon of swarming, as when all the bees in a hive collectively depart and go in search of another hive. Who or what decides when to initiate this process? Without understanding the exact mechanism, Wheeler was convinced that the hive itself collectively chooses. This would mean that there is no leader or centralized, organizing agency. Presumably each individual bee simply follows a small set of elementary rules that tells it how to behave in relation to the way its neighboring bees are behaving, very much like the rules Craig Reynolds deduced for flocking behavior in birds. As a result of these countless local interactions, the swarm is able to move out across the landscape as if it were a single body or organism.

That this kind of collective dynamic mobility exhibits a unity of purpose leads entomologists to refer to many insect collectivities as "superorganisms." Ant colonies exhibit even more complex behavior; their collective activities include reproduction, nest or colony building, removal of the dead, and foraging for food. The primary question is, how are these diverse activities organized, given the absence of any centralized command and control structure? In 1959 the French zoologist Pierre-Paul Grassé introduced the concept of *stigmergy*, which provided the basis for an answer. The key idea is that social insects do not communicate among themselves but are incited to do things by signs in or transformations of their environment. Grassé observed that, at first, termites move about completely at random, some of them dropping small, masticated pellets of earth here and there. However, when other termites encounter these pellets, they add their own to the pile, which then quickly grows into a column. If there is only one column in the area the termites will ignore it, but if there are several, the termites then climb to the top and add pellets that form connecting arches. Soon a nest is under full construction, without any explicit plan or supervision. Grassé believed that it was the configuration of pellets in the environment that first stimulated and then became the agency responsible for regulating and coordinating the various activities of social insects. Following Grassé's original work, the concept of stigmergy has been further elaborated, and now includes both direct and indirect forms of interaction. Probably the most familiar example is ants leaving pheromone traces when they forage for food. Since pheromone evaporates, it is not simply the trace but its intensity that operates as a stigmergic marker, temporally modifying the environment and triggering new or different behavior. Because hundreds of ants forage at the same time, the intensity of the double trace (out and back to the nest) not only directs other ants to the location of the food source but to the trail that is shortest in length. Duplicating this natural computational system has led to new strategies in problem solving.

The recent work of Eric Bonabeau, Marco Dorigo, and Guy Théraulaz on swarm intelligence draws deeply on Grassé and the research tradition he inaugurated.[47] They bring to it an innovative awareness of the crucial role played by self-organization, self-assembling, and network dynamics in the collective activities of social insects. Hence their model of social insect behavior is more sophisticated than previous models and points toward new ways of connecting individual behavior with collective performance. But Bonabeau, Dorigo, and Théraulaz also take a further step: their model of self-organization in social insect behavior becomes a way "to transfer knowledge about social insects to the field of *intelligent system design*" (6). Although social insects themselves have very limited cognitive abilities, collectively their self-organizing activities constitute truly remarkable instances of distributed problem solving, as the authors demonstrate. Having created a model of how distributed problem solving works in specific insect societies, they next apply it to the design of swarm-intelligent systems. In practice, this means designing simple agents, including robotic agents, that somehow self-organize to solve problems. To be sure, it is difficult to "program" these systems, precisely because "the paths to problem solving are not predefined but emergent in these systems and result from interactions among individuals and between individuals and their environment as much as from the behaviors of the individuals themselves" (7).

Success, nevertheless, has come on a number of fronts, as Bonabeau and Théraulaz report in their article "Swarm Smarts."[48] For example, the foraging of ants has led to a method for rerouting network traffic in congested telecommunications systems. The method involves modeling the network paths as "ant highways," along which artificial ants (i.e., software agents) deposit and register virtual pheromone traces at the network's nodes, or routers. In another application of the same model, their colleague Marco Dorigo has discovered a method of solving the well-known traveling salesman problem, or how to find the shortest route through a network of cities that passes through each city only once. In a third example, swarm software has been developed that enables a group of robots to cooperate in the transferring of objects from one location to another. In a recent interview Bonabeau comments on the general use of swarm systems:

In social insects, errors and randomness are not "bugs"; rather, they contribute very strongly to their success by enabling them to discover and explore in addition to exploiting. Self-organization feeds itself upon errors to provide the colony with flexibility (the colony can adapt to a changing environment) and robustness (even when one or more individuals fail, the group can still perform its task).

With self-organization, the behavior of the group is often unpredictable, emerging from the collective interactions of all the individuals. The simple rules by which individuals can interact can generate complex group behavior. Indeed, the emergence of such collective behavior out of simple rules is one of the great lessons of swarm intelligence.

This is obviously a very different mindset from the prevailing approach to software development and to managing vast amounts of information: no central control, errors are good, flexibility, robustness (or self-repair). The big issue is this: if I am letting a decentralized, self-organizing system take over, say, my computer network, how should I program the individual ants so that the network behaves appropriately at the system-wide level?

I'm not telling the network what to do, I'm telling little tiny agents to apply little tiny modifications throughout the network. Through a process of amplification and decay, these small contributions will either disappear or add up depending on the local state of the network, leading to an emergent solution to the problem of routing messages through the network.[49]

The major problem, Bonabeau concludes, is selling the concept of swarm intelligence to the commercial world. Managers, he says memorably, "would rather live with a problem they can't solve than with a solution they don't fully understand or control" (4).

A comparable distinction between controllable and uncontrollable systems lies at the center of Kevin Kelly's *Out of Control: The Rise of Neo-biological Civilization*. In essence, Kelly finds, there are two kinds of systems that can be constructed. The first is the familiar mechanical system composed of a long string of sequential operations. This system is linear, predictable, and hierarchical, with a clear chain of command emanating from a central controlling authority. It exhibits the "cold, fast optimal efficiency of machines" (24), and Kelly associates it with the clock. At the opposite extreme is the system ordered as "a patchwork of parallel operations" (21) composed of many autonomous units with a high degree of connectivity between the units. There is no centralized control structure, only the "webby nonlinear causality of peers influencing peers" (22). This is the swarm model, as Kelly calls it, and he associates it with networks, parallel processing, and complex adaptive systems.[50] This kind of system often exhibits the "crisscrossing, unpredictable and fuzzy attributes of living systems" (24). Although Kelly presents a convincing argument that today's machines tend more and more to be of the second type, he acknowledges both the benefits and disadvantages of these swarm systems. On the one hand, they are adaptable, evolvable, resilient, boundless, and capable of generating novelty; on the other hand, they can be nonoptimal, noncontrollable, nonunderstandable, and nonimmediate. In sum, where supreme control is called for, "old clockware" is the way to go, but where "extreme adaptability is required, out-of-control swarmware is what you want" (24).

The fear of a loss of control that Bonabeau evokes above and that Kelly argues against might be interpreted as Western culture's bias against collective intelligence. However, if James Kennedy and Russell C. Eberhart are right, swarm intelligence is the best way to understand human intelligence tout court. In *Swarm Intelligence*, they expressly oppose the view that an isolated individual is an information-processing entity and that thinking or cognition occurs inside an individual's head.[51] It is no accident, of course, that the classic image of thought resembles Kelly's first type of system. Whatever else we do as human beings, we are both the most social of animals and the most intelligent. It makes sense, therefore, to understand intelligence as something that arises from interactions among individuals. Expressed in bullet form: mind is social, human intelligence results from social interaction, and culture and cognition are inseparable consequences of human sociality (xx–xxi). However, as Kennedy and Eberhart explain, they are not referring to the kinds of interaction typically found in multiagent systems, where autonomous agents perform specialized subroutines:

> Agent subroutines may pass information back and forth, but subroutines are not changed as a result of the interaction, as people are. In real social interaction, information is exchanged, but also something else, perhaps more important: individuals exchange rules, tips, beliefs about how to process the information. Thus a social interaction typically results in a change in the thinking processes—not just the contents—of the participants. (xv)

The model Kennedy and Eberhart are looking for explicitly incorporates this change in the participants of a social interaction. At the same time, they want their model to be computational, that is, constituted of algorithms that can be implemented in a computer program. Essentially they want a program that can "simulate societies of individuals, each working on a problem and at the same time perceiving the problem-solving endeavors of its neighbors, and being influenced by those neighbors' successes" (xv). As the authors remind us, a swarm usually refers to a disorganized cluster of moving things (often insects) that while moving irregularly and even chaotically in different directions somehow remain together as a whole. To arrive at their model, Kennedy and Eberhart reconfigure this image in highly abstract, computational terms. If we think of the interacting individuals as particles, and of the swarm as a population of interacting particles that is able to optimize some global objective in and through their collaborative search of a space, then we begin to get a sense of the "particle swarm optimization" model at the heart of their book. Thus while Bonabeau and colleagues define swarm intelligence "to include any attempt to design algorithms or distributed problem-solving devices inspired by the collective behavior of insect colonies and other animal societies" (7), Kennedy and Eberhart draw on a completely different set of sources: the theories and simulations associated with ALife research, evolutionary computation and neural net theory.[52]

Let us briefly consider one of these sources, Mark M. Millonas's research on swarm behavior and phase transitions.[53] Like Bonabeau and his research team, Millonas is interested in how swarms of social insects like ants perform emergent computations. As he points out, both the location of a food source and its utilization "are computed by the self-organization of a column of ants between the nest and the food source" (420). But in contrast to Bonabeau, Millonas seeks to correlate this computational behavior with the behavior of the swarm as a dynamical system. Thus, using data gathered from both laboratory observations of real ant behavior and their computer simulation, Millonas developed a method to apply statistical physics to the flow of organisms, in particular to those changes in its equilibrium that constitute a phase transition. The essential point of his study is that during a phase transition—the same kind of transition that Langton defined as the edge of chaos—the system (i.e., the collective intelligence of the swarm) becomes acutely sensitive to external influences and thus operates as an "information amplifier" (431). In their summary of Millonas's research, Kennedy and Eberhart explain in simple terms the correlation between ant behavior and the formation of a pattern that encapsulates the computation:

> The movements of ants are essentially random as long as there is no systematic pheromone pattern; activity is a function of two parameters, which are the strength of pheromones and the attractiveness of the pheromone to the ants. If the pheromone distribution is random, or if the attraction of ants to the pheromone is weak, then no pattern will form. On the other hand, if a too-strong pheromone concentration is established, or if the attraction of ants to the pheromone is very intense, then a suboptimal pattern may emerge, as the ants crowd together

in a sort of pointless conformity. At the edge, though, at the very edge of chaos where the parameters are tuned correctly ... the ants will explore and follow the pheromone signals, and wander from the swarm, and come back to it, and eventually coalesce into a pattern that is, most of the time, the shortest, most efficient path from here to there. (108)

It was precisely this type of distributed swarm intelligence that Gerardo Beni studied in cellular robotic systems and cellular automata in the late 1980s and early '90s and that leads to my third example: the construction of swarms of robots, or swarm-bots (s-bots). Beni originally defined swarm intelligence as "a property of non-intelligent robots exhibiting collectively intelligent behavior."[54] Although the robots would have no explicit model of the environment, they would be equipped with sensors and could communicate with one another. Since they could only interact with and adapt to the environment collectively, their behavior—and the problem solving it would entail—would be emergent.

These ideas are being realized through the construction of Swarm-bots, a project funded by the Future and Emerging Technologies program of the European Commission.[55] A swarm-bot is a collection of autonomous mobile robots able to self-assemble and self-organize in order to solve problems that cannot be solved by a single robot. Typically, it is composed of twenty to thirty s-bots that are physically interconnected. The connection can be rigid, as implemented by a gripper, or semiflexible, as implemented by a flexible arm (figs. 7.11 and 7.12).[56] Rigid connections are used to form chains that allow a swarm-bot to pass over obstacles or crevices in the terrain, while the flexible connections allow maximum flexibility to each s-bot while still maintaining the swarm-bot configuration. Clearly the main problem is one of coordination: getting the s-bots to act together once they are connected. One team of researchers has made impressive progress using a combination of traction sensors that detect the direction and intensity of the force exerted by the turret on the chassis, and evolutionary programming techniques for evolving the neural network controllers that generate motor outputs in response to sensory inputs for each s-bot.[57] With this combination they were able to get four connected s-bots to move in a single direction together, avoid obstacles, and "spontaneously produce object pushing/pulling behavior when connected to or around a given object."

Although only in the initial stage of development, swarm-bots and swarm systems instantiated in autonomous mobile robots have generated widespread excitement and many new research programs, especially in defense and military-related fields. In one instance, researchers at the

Figure 7.11
Swarm-bot. Image provided by Marco Dorigo, Director of the IRIDIA lab at the Université Libre de Bruxelles.

University of Wyoming have developed small, swarming robots carrying "multi-modal sensory arrays" programmed to detect and disable chemical targets in the so-called war against terrorism.[58] In another, the Defense Department has funded a project to equip a battalion of 120 military robots with swarm intelligence software;[59] a related project is to develop swarming robotic systems that will behave like "killer insects" that can hunt down their prey in bunkers and caves.[60] But no doubt the most futuristic prospects for swarm systems involve a marriage with nanotechnology, which has undergone an astonishingly rapid development in the past few years. As mentioned earlier, Michael Crichton dramatizes this marriage in his novel *Prey*, in which a private research firm

Figure 7.12
Swarm-bots climbing stair. Image provided by Marco Dorigo, Director of the IRIDIA lab at the Université Libre de Bruxelles.

uses predator-prey algorithms to program swarms of nanomachines to evolve more intelligent forms of collective behavior. And reality may not be lagging that far behind. At Rutgers University, computer scientist J. Storrs Halls is developing miniature swarm-bots he calls Foglets.[61] About the size and shape of a human cell, these micromachines have twelve radiating arms that allow them to grip onto other Foglets and form larger structures. Like the swarms in Crichton's *Prey*, Foglets can also merge their computational capacities to create networks of distributed intelligence. While their purpose is the more benign one of simulating different environments by creating wavefronts of light and sound, the difference—in principle, at least—is only a matter of a different set of algorithms.

In conclusion, it is worth recalling once again that the double emphasis on artificial evolution and machinic self-organization so important to contemporary robotics harks back to ideas first proposed in the cybernetics research of the early 1950s. In particular, in his designs for self-reproducing automata, John von Neumann theorized that it would be possible to evolve machines more complex than their makers; and in his "homeostat" experiments W. Ross Ashby showed how machines possessing a "requisite variety" of internal states would inevitably optimize themselves through self-organization.[62] As robotics moves toward the realization of the cybernetic movement's ambition to build machines of a complexity comparable to that of humans, it has increasingly come to rely on what we can now distinguish as a third method for producing complexity. The first, of course, is the completely bottom-up, self-bootstrapping method of nature itself, prodigal and fecund but infinitely slow and wasteful. The second is the top-down method of human technical invention, fully evident in classic AI research. Although faster and more efficient than nature, this method nevertheless seems unable to produce machines of more than limited complexity. The third method combines the top-down with the bottom-up approaches. Widely deployed in ALife simulations, the third method has also become the primary means by which evolutionary robotics has developed out of behavior-based robotics, as top-down strategies like programming and design have been used to create systems that will organize and evolve on their own by taking advantage of the new combinations and creative mixings that emerge from the bottom up. Although still in its infancy, this third method holds the greatest promise for producing machines of higher orders of complexity and consequently more advanced forms of machinic life.

8 Learning from Neuroscience: New Prospects for Building Intelligent Machines

The brain was not a sequential, state-function processor, as the AI people had it. At the same time, it emerged to exceed the chemical sum passing through its neuronal vesicles. The brain was a model-maker, continuously rewritten by the thing it tried to model. Why not model this, *and see what insights one might hook in to.*
—Richard Powers, *Galatea 2.2*

In previous chapters some of the most innovative achievements of the new science of ALife and the new AI have been considered. Yet even with these obvious successes questions have been raised about whether the models of life and intelligence underlying them are sufficiently complex. Perhaps the most notable doubt has come from Rodney Brooks, current director of MIT's Artificial Intelligence Laboratory. Brooks acknowledges that ALife and behavior-based robotic systems built according to bottom-up computational principles are not as robust or lifelike as researchers had hoped.[1] Both robots and artificial life simulations "have come a long way," he states, "but they have not taken off by themselves in the ways we have come to expect of biological systems" (*Flesh and Machines*, 184). Advancing what he calls the "new stuff" hypothesis, Brooks speculates that there may be some "extra sort of 'stuff' in living systems outside our current scientific understanding" ("Matter and Life," 410). If this is true, then the ALife systems and autonomous robots discussed in previous chapters may be limited precisely because the models or assumptions that underlie them are missing some crucial element or understanding of how life and intelligence actually work. In this sense Brooks's hypothesis might better be called the "missing stuff hypothesis." What, specifically, may be lacking? Brooks lists four possibilities:

1. We might just be getting a few parameters wrong in all our systems.

2. We might be building all our systems in too simple environments, and once we cross a certain complexity threshold, everything will work out as we expect.

3. We might simply be lacking enough computer power.

4. We might actually be missing something in our models of biology; there might indeed be some "new stuff" that we need. (*Flesh and Machines*, 184)

After discussing each possibility in some detail, Brooks concludes that the last hypothesis is by far the most likely. He admits, however, that he has no idea what the new stuff might be.[2]

While not explicitly framed as attempts to come up with new stuff, several recent initiatives in AI research have taken a direction that is at least indirectly responsive to Brooks's doubt. These initiatives foreground the biological evolution of the human brain and the specific structure of the neocortex as the most salient facts to consider in understanding intelligence and building intelligent systems. While biological perspectives on human intelligence are hardly new, to appreciate their current significance one only has to peruse the most up-to-date textbooks on AI, which foreground agent theory and give scant if any attention to the biological evolutionary perspective and the latest findings of neuroscience.[3] At the conclusion of chapter 5, after noting some of the apparent methodological limits of what has been the dominant approach in ALife research, I briefly discussed living computation and the attempt to synthesize an artificial protocell as two promising new research paths. In this chapter I attempt something similar for artificial intelligence and the ongoing effort to construct intelligent machines.

While this is obviously not the place for a detailed survey or overview of contemporary AI, a few general observations may be helpful. First, there appears to be wide agreement that, notwithstanding the success of behavior-based robotics and the new AI, the 1990s was a period of weak AI. Or rather, to describe the period in more positive terms, there was a shift in priorities. Rather than attempt to construct human-level machine intelligence, most research was directed toward developing practical AI applications in industry. So much so, in fact, that the acronym AI jokingly came to signify "almost implemented," meaning that once an AI technique had been implemented in a practical application it was no longer AI. An obvious example was the widespread development of smart systems like credit card fraud detection systems, voice and face recognition systems, automated scheduling systems and data mining systems. A notably more ambitious achievement was Remote Agent, an AI system developed by NASA to assume autonomous control of the Deep Space 1 probe. In fact, agent technology became the most important focus of development in the 1990s. Thus far intelligent agents do not exhibit general intelligence but tend to be intelligent only in very specific domains. The Internet, for example, has become a veritable hive of smart agents performing a range of tasks, data mining being the most familiar. But AI algorithms are also deployed in search engines, and AI will assume a large role in the forthcoming Semantic Web. Increasingly, moreover, many commercial software applications contain small AI programs generally known as cognitive assistants.

But however we view the supposed weak AI of the '90s, the defeat of world chess champion Garry Kasparov by the IBM chess computer Deep Blue in 1997 was clearly a signal achievement. Because chess is a rule-bound, highly intellectual activity, producing a machine or program that can defeat a human chess master has been a holy grail or at least marker of progress comparable to the Turing test since AI's inception.[4] Contrary to expectation, however, Deep Blue's victory did not definitively resolve any issues concerning machine versus human intelligence. This was mainly because Deep Blue did not deploy the techniques and computational processes that human players use. Instead of intuitions based on extensive knowledge and experience of the game, Deep Blue relied upon lightning-fast board evaluations at a rate of millions of positions per second and searches fourteen or so moves deep of all possible lines of play. Obviously the searches conducted by expert human players are not nearly so fast or deep and are much more selective. Human players, furthermore, often make winning moves that go against conventional chess wisdom, and their decisions are not bound by probabilities. Kasparov himself raised questions about whether Deep Blue was intelligent in the same way as a human player and whether it represented a new form of intelligence.[5]

In his account of the Deep Blue–Kasparov match, Deep Blue's chip designer Feng-Hsiung Hsu notes that some of Deep Blue's moves resulted from software bugs or the software's inadequacy in dealing with certain board situations like king safety or endgame without queens, inadequacies that had to be fixed on the fly.[6] Indeed, his book makes the strong impression that Deep Blue should be regarded less as a stand-alone entity than a complex computational assemblage within which humans assumed a variety of roles. Feng-Hsiung himself reveals that he and his friends "understood that the [Deep Blue–Kasparov] match was never really 'man versus machine,' but rather 'man as a performer versus man as a tool maker'" (264). Curiously, while a graduate student at Carnegie Mellon, Feng-Hsiung and those around him building computer chess machines felt mainly disdain for AI as a discipline. In contrast, computer game designers and programmers have increasingly found AI to be a very valuable resource. The best computer games now tend to rely upon AI programming techniques like genetic algorithms, neural networks, and fuzzy logic, and their application has begun to displace the development of arresting graphics as the new wave in game development.[7]

No doubt the widespread application of AI in the production of "smart" technology will continue to bring many practical benefits. On the other hand, inasmuch as useful intelligent machines do not have to be built following principles derived from an understanding of how human or animal intelligence actually works, success in this endeavor may also tend to reinforce the long ingrained tendency in AI research to ignore or play down biological models of intelligence. The brain, after all, is assumed to be a complex but messy "kludge," constrained by evolution to repeat variations on earlier solutions or reshaped and "retrofitted" to solve problems for which it wasn't "designed." With the newly invented computer, AI had a powerful and more elegant model at hand. It is hardly surprising, then, that the computer's rapid development became the basis for AI's projected accomplishments and that Moore's law is still touted as the harbinger of AI's bright future. While innovations in hardware will inevitably occur, these optimistic forecasts tend to ignore or play down the difficulties faced in writing—or more likely, evolving—the software necessary for building truly intelligent systems.[8]

However, it now seems increasingly likely that a biological reorientation in the construction of intelligent machines and systems may be the only way to overcome the deficiency that has plagued artificial intelligence for the past fifty years. Biological perspectives on human intelligence are not at all new of course, and (as we saw in chapter 6) mostly derive from the neural net theory inspired by McCulloch and Pitts's classic essay of 1943, which considered how the passage of electrochemical impulses through arrays of connected neurons constituted circuits, or "nets," that could be understood to perform computations. Though neural net theory promised to provide a biological-based alternative to the symbol- and logic-based systems of classic AI, artificial neural nets were found wanting (they could not perform the XOR logical function, for example) and rejected in the late 1960s only to reemerge in a much stronger version in the mid-1980s under the banner of parallel distributed processing, or connectionism, as it came to be called. Over time, unfortunately, neural net theory has been shorn of its biological roots and absorbed as just another computational method in AI's expanding toolbox.

To be fair, researchers often refer to neural net theory as an unfinished revolution. For at least two reasons, however, it was unlikely that neural net theory alone would become a foundation stone in AI's quest to construct intelligent machines. First is the problem of scale. The human brain contains about one hundred billion neurons, with each one connected to thousands of other neurons. Compared with this massive circuit complexity, the neural networks built thus far are extremely simple, though they are capable of recognizing different kinds of patterns, which is a necessary if not sufficient aspect of intelligence. Second, while neuroscience now understands a great deal about neuron behavior and the functioning of many parts of the brain, it doesn't yet have a theory of the whole, and there is still a wide gap between understanding neural mechanisms and understanding the intelligent behavior of whole creatures. In short, the bottom-up view of neuroscience and the top-down view of classic AI do not meet in the middle, where all the interesting behavior—perception, complex movement, and a basic ability to cope with the environment—seems to lie. As Steve Grand puts it, "A human being is not an ant with a natural language interface."[9]

Below I describe three current research initiatives that operate in or are moving toward this middle realm. The first (Eric Baum) serves as a transition from classic AI and cognitive science. The second (Jeff Hawkins) plunges us into a theory of the brain as a memory prediction machine. And the third (Steve Grand) involves the actual building of a working baby android. In conclusion, I consider some recent research in self-modeling (Lipson) and communication (Steels).

The Modular Mind/Brain

In *What Is Thought?* Eric Baum directly confronts AI's failures to deliver on its overhyped promise to produce machines that emulate human thinking and behavior.[10] Unlike most critics of AI, Baum doesn't take these failures as justification for abandoning the premise that human thinking is computational.[11] Rather, he explains what is wrong with the AI programs themselves and what they lack in relation to the programs our brains presumably use. Most typically, AI programs fail to exploit the compact structure inherent in the world, which is precisely what evolution has developed our brains to do. Instead, they resort to sophisticated techniques that yield solutions but don't provide a real understanding of the problem. In a range of examples from Blocks World, games like Go and chess, and the traveling salesman problem, Baum shows how the typical AI approach applies computational tricks like brute force search, branch-and-bound, and evaluation functions to a "problem space" that, because it has been abstracted from the world, no longer exhibits any of the latter's inherent structure. As a result, the formal procedures deployed have little to do with the problem's inherent semantics.

We can easily apply this critique to a range of familiar AI programs. Consider Joseph Weizenbaum's Eliza, which simulates the responses of a psychotherapist to questions posed by the user. (Versions of the program are easily available on the Internet.) It was designed, using pattern-matching algorithms, to return answers that seemed credible, appropriate, and even witty. However, as any persistent questioner soon realizes, the answers, though varied and to some extent unexpected (i.e., randomized), are "canned"—drawn from lists and modified—and never reflect any real understanding of the situation expressed by the user's questions. Here we see a clear example of how the structure of the program has little relationship with the structure of the problem, beyond the obvious fact that it exploits the nature of the encounter (question-answer). As a necessary first step toward achieving real understanding, a genuinely intelligent therapist would have to somehow recognize the nature of the questioner's psychological problem, presumably by extrapolating a pattern from recurrent references to a particular figure—a mother or father, for example, or to a theme, like persecution. In principle, a neural net could be trained to recognize such patterns, though the number and complexity of the training sets would have to be quite large. Nevertheless, if after processing a wide variety of responses from the user the programmed therapist could reliably deliver judgments like "You are suffering from paranoia," or "You are suffering from a guilt complex," then we could reasonably say that the program genuinely describes some aspect of structure in the world outside the program.

Baum calls the code that does inscribe semantic structure "compact," since it embodies, or reflects, the constraints of the process it describes. However, to write code that possesses this semantic dimension is computationally hard, as Baum often reminds us, and AI researchers have not yet learned how to make their programs compact in this sense. In contrast, the programs that underlie the human mind are compact because they are the result of millions of years of evolutionary programming tested and honed by adaptation. Happily for us, the instructions for producing this code are inscribed in our DNA. More specifically, nature's evolutionary programming produces DNA (the source code), which in turn produces the brain's neural circuitry (the executable code), which in turn produces behavior that propagates the species (the output).[12] But Baum is less interested in the details of these processes than in the structural modularity of what he calls the "program of mind."

To conceptualize the program of mind, Baum draws extensively on the field of evolutionary psychology, which understands the human brain as an organ that has evolved in the way it has because the results have made humans more adaptable and therefore better able to survive long enough to reproduce and propagate their genes.[13] Contrary to the view that the mind is a general-purpose computer, evolutionary psychology views it as a collection of many special-purpose computational modules, each having evolved to accomplish a specific behavior or solve a particular problem. Accordingly, evolutionary psychologists liken the mind to a Swiss Army Knife with many small tools—blade, corkscrew, can opener, scissors, and so on—each of which is designed to perform a specific task. Each mental module would thus correspond to one such tool, with modules for vision, touch, hearing, language, mate selection, and many more—as many distinct modules as there are types of behavior required for human survival and propagation of the species. In Baum's more overtly computational language, *mind* is a program comprised of modules that "call" other modules and often reuse the same or similar blocks of code. As we might expect, the granularity of the model—how many modules are there?—remains an open question. The key idea, nevertheless, is that the mental module is the primary unit of evolution. While there is some empirical evidence (which Baum cites), from a computer programmer's point of view this makes perfect sense, since any change to a tightly coupled ensemble of parts or blocks of code—as opposed to quasi-independent modules—would break the system as a whole. Modular evolution allows both repeatability (nature tends to repeat solutions to earlier problems) and flexibility (if newly evolving solutions don't work, they won't break either other subsystems or the system as a whole). Thus from both computational and evolutionary perspectives modularity offers unique advantages.

In Baum's view the evolved modules that constitute the program of mind call computational routines and subroutines from other modules. They are also organized hierarchically. The code at the top of the hierarchy controls speech and action, makes decisions, and is perhaps responsible for "feedback credit assignment" and learning. These upper-level modules, Baum states unexpectedly, "may be deliberately fed misinformation by other modules, specifically to control what we say and do in a manner advantageous to our genes. What we are verbally aware of, then, is the disinformation, not the true information only known to the subconscious processes that direct the flow of information" (423). That lying and self-deception are an intrinsic aspect of human behavior is a familiar idea; that falsification takes place at the information-processing level is something new. To make this idea plausible Baum would have to both posit a computational mechanism that functions somewhat like the Freudian unconscious and establish that these self-deceptions serve biological imperatives—a daunting task, to be sure. Instead, he pursues a different problem, or perhaps it is the same problem considered from a different angle. It follows from the fact that much of human behavior is not preprogrammed (and therefore not predetermined) but learned and adaptive. Thus he focuses on the learning mechanisms and inductive biases with which evolution has equipped us in order to make us more flexible and adaptive in achieving *its* ultimate end, which is to survive and reproduce.

This requires that the modularity thesis also be applied to cultural notions like the self, the will, and our belief in the bedrock reality of experiential qualia (like the redness of red things). Since the purpose of mind is to make decisions about what actions to take, it is more flexible and efficient, Baum argues, to have a "sovereign agent" in charge: "We think of creatures as conscious and endowed with free will because this is a computational module we have: this is much the most compact, effective way to predict actions of sovereign agents such as others and ourselves" (438). Our inner representation of this agent is the self, which (again) is a module that interacts with other modules. For Baum, however, mind is *not* a highly distributed complex of interacting modules or agents without a central authority, as in Marvin Minsky's *The Society of Mind* and Daniel Dennett's *Consciousness Explained*.[14] Essentially the evolutionary origin of the mind means that "it is coordinated to represent a single interest, that of the genes." It is this unity of interest that makes cogent the concept of self: "There are many modules, but they are all working toward the same end, just as the many cells and many organs in the body are" (408). Thus the issue is never really one of truth—either of the self's coherence or the information processing in and through which it is constituted—but whether and how the organism or creature survives as an integral entity.[15]

The metaphoric structure of language provides further evidence for the modular structure of the mind's computational program. Baum points out that basic metaphors pervade all natural languages. For example, in our culture "time is money"; like money, we spend time, save time, waste time, borrow time, and so forth. Such metaphors, he argues, are instances of "code reuse." If "the essence of metaphor is understanding and experiencing one thing in terms of another," then underlying every metaphor is a computational module that is being reused in different contexts. This view of metaphor, Baum believes, can account for how we learn throughout our lifetimes and continually build up the modular structure of our thoughts. We do this "largely by packaging existing instructions into small programs" (229). Metaphorical language, presumably, is at once the means and the result.

Baum, incidentally, does not believe that language and our ability to invent and use symbol systems separate us from all other species. Language does not provide a "new symbolic ability or new reasoning ability" (378); rather, it is the communicative and storage capacity of language that explains our mental superiority. Language constitutes "a huge database of useful code on top of the programs that evolution wired into the genome. The computation that has been done by humankind as a whole in discovering this database of useful code is massive, by some measures arguably comparable to the computation done by evolution itself in creating us" (375). As both a database of useful code and a means of accessing and communicating it (but Baum is not concerned with this distinction), language provides a whole new arsenal of ways by which the human species can extend and perpetuate itself.

Given the extraordinary number of computations required by natural evolution to produce intelligence (or consistently intelligent behavior), Baum is not optimistic about the prospects for AI. Since humans are not particularly good at writing the necessarily compact code, probably the best approach will be to write programs that can recursively write and evolve their own code. Nevertheless, his accomplishments with the Hayek Machine, an evolutionary program that he and Ivan Durdanovic designed to solve different types of computational problems, give some hope that there may be ways to jump-start artificial evolution and leapfrog over the computationally demanding early stages.

Basically, Hayek consists of an evolving population of computational agents (i.e., small computer programs), whose fitness is determined by their contribution to the solving of a larger computational task. Thus, after being randomly generated, the agents undergo a process of selection that increases the performance, or computational "fitness," of the population as a whole. Agents are "motivated" to solve problems through their participation in a world defined in strictly economic terms; that is, there is a reward, or credit assignment, system. Although the agents act separately, their collective interactions eventually solve computational problems that, as a direct consequence, increase the efficiency and wealth of the system as a whole. If, Baum argues, individual agents are rewarded if and only if the performance of the system as a whole improves, then two rules must be imposed: the conservation of money and the absoluteness of property rights. In short, everything must be owned by some agent, and this right must always be respected. Constrained by these two rules, an agent can make money only by increasing the flow of money into the system from the world, which, in Baum's experiments, is usually a problem domain like Blocks World.

To illustrate, suppose that solving a problem entails a particular sequence of actions. In Blocks World the sequence would entail picking up and stacking blocks in the target stack until it exactly matches the object stack. At any moment Blocks World exists in a specific state defined by a particular disposition of the blocks. For each agent, this state is mapped to a set of conditions. Since each agent is capable of computing a simple action, if the condition matches an action that the agent can execute, it will issue a bid in an auction. In effect, the agent "buys" the right to perform the action. More specifically, in order to carry out the action an agent must first bid on the action, win the bid, and then pay the previous owner or agent the amount bid. In this manner agents bid, a sequence of actions ensues, the state of the world improves (e.g., in Blocks World there is progress toward replicating the order of blocks in the object stack), and money flows to the agents responsible. Successful agents thereby accrue money, while unsuccessful ones do not and are eventually removed from the system and replaced by new, randomly generated agents. Although the actions the agents perform accomplish little at first, with increasing feedback from their world the agents begin to make money, and the system as a whole evolves. As Baum reports, after a million or so computational cycles populations of agents evolved that could solve very large instances of Blocks World problems.

While Baum does not claim that the human mind works exactly like this, he thinks the program is "pedagogically useful in illustrating the point that the central auctioneer is a location where summaries come together and yet where little computation is done" (414). In these terms, Hayek illustrates a theory of the brain's functional structure if not its material implementation. If the brain works like a Hayek machine, almost all of the computation would be distributed among a multiplicity of agents; yet there would also be a central location where all the work comes together. This would be the location of the sovereign agent, or self.

Like most cognitive scientists, Baum assumes that the brain is the physical substrate of information-processing mechanisms on which mental software programs run. These programs comprise the *mind*. The information-processing mechanisms themselves are more elemental: they can "add, match a pattern, turn on some other circuit, or do other logical and mathematical operations," as Steven Pinker puts it.[16] At this substrate level, the same elements—neurons in networks—simply combine in different ways to produce different cognitive "programs." Most cognitive scientists (and AI researchers as well) therefore believe that they can ignore this lower level and carry out their analysis of the "programs" at the higher level of organism behavior. However, this assumption (which perpetuates the mind-body split) may not be warranted.

The Neocortex: A Memory Prediction System

In his book *On Intelligence* Jeff Hawkins proposes what he believes to be a more accurate account of how information-processing mechanisms in the brain work, and to what end.[17] At the outset he roundly rejects AI and cognitive science for having failed to adequately explain intelligence and what it means to *understand* something. There are two reasons for this failure: an adherence to the input-output model, which has misled us into thinking that "the brain has nothing to teach us about the mind" (37–38); and the belief that "intelligent behavior should be the metric of an intelligent system" (32). Hawkins therefore agrees with John Searle's famous "Chinese room" critique: it is false to assume that if a machine or program acts intelligently, or does things that require intelligence, then it *is* intelligent. Real intelligence, Hawkins argues, cannot be measured by external behavior; a better metric is "how the brain remembers things and uses its memories to make predictions" (20). And for this the input-output computational model is completely inadequate on three obvious counts. First, time is an essential factor, since information processed by the brain is not only time sensitive but varies dynamically in its very form over time. Second, the brain is saturated with feedback circuits, and connections going back toward sensory input often vastly outnumber forward connections, sometimes by a factor of ten to one. And third, information processing in the brain is hierarchical: it occurs at many interactive levels—six in the neocortex, which is the seat of the cognitive activities that are deemed essential for human intelligence, like perception, memory, language, and reasoning. Since Hawkins's ultimate objective is to understand *real*—not artificial—intelligence in order to build intelligent machines, there is no recourse but to study the human neocortex.

As the outer layer of the mammalian brain, the neocortex comprises about 90 percent of the cerebral cortex. In humans it is about the size of a dinner napkin, 2–4 mm thick, with extensive folds. Appearing roughly ten million years ago, it provided mammals with a memory system, which was tacked on to the sensory paths of the primitive brain. Over time, Hawkins believes, the neocortex began to interact with the motor system of the old brain, until it eventually "usurped most of the motor control from the rest of the brain" (103). This greatly increased the areas of association between sensory and muscle systems, enabling humans to make much more complex movements. Eventually, he claims, "the human cortex [began to direct] behavior to satisfy its predictions" (104) and to assume its primary function as a large, hierarchical memory-prediction machine.

Hawkins's point of departure is the work of neuroscientist Vernon Mountcastle, whose research revealed that all cortical regions perform the same operation. What makes "the vision area visual and the motor area motoric is how the regions of cortex are connected to each other and to other parts of the central nervous system" (51) and not any difference in function or algorithm. If the same algorithm operates throughout the cortex, regardless of particular function or modality—whether to see or hear or control muscle movements—then the cortex must do something "universal that can be applied to any type of sensory or motor system" (52). Before this function can be defined, however, two essential features of the neocortex have to be noted: the extreme flexibility and plasticity of the wiring and the fact that the input signals to the cortex are always and only patterns of electrochemical firings called action potentials, or spikes. These firings occur across the synapses of neurons, and the patterns of spikes are similar in nature whether they constitute signals from the eyes, ears, skin, or muscles. The plasticity of the wiring means that the neocortex can "change and rewire itself depending on the type of inputs flowing into it" (54). In fact, there are many examples of the brain's capacity to rewire itself after damage or the loss of a sensory organ.[18] Equally wondrous and strange, these patterns of electrochemical spikes, stored as complex temporal and spatial sequences, are all that the brain "knows." Our perception of the world is created from these patterns—and from nothing else.

As myriads of patterns continuously stream into the neocortex, it stores them as autoassociative memories, meaning that a whole pattern or memory can be recalled from only a part or fragment. Specifically, these memories are stored, recalled, and recognized as "invariant representations," in a form that captures the essence but not all of the details of repeated experiences. On one level an invariant form is simply the stability and repetition of a specific pattern of neurons firing; but on another it is the face of a parent, a song we know, or a simple physical act—in each case something we will always recognize despite variations over an incredibly wide range of contexts and conditions. Not only are invariant representations constantly forming, but they are also being modified as they are compared with patterns (and sequences of patterns) streaming in from the senses and muscles. The result is a series of predictions about what is about to happen. As we walk through our house or apartment, for example, the neocortex is constantly checking what we actually experience against what it predicts we will experience—simply by comparing sensory input to stored invariant representations. Anything new or missing or different will immediately catch our attention. Indeed, this updating of our experiential reality is so sensitive to sights, sounds, textures, and smells that even slight differences—like the feel of the front doorknob or the door's weight or resistance—will alert us to some possibly significant change. In this way the neocortex operates as a memory prediction machine.

It is often asserted that the brain is a parallel computer, processing multiple streams of information simultaneously. Because neurons pass current much more slowly than do silicon computer chips, they also process information more slowly. But the fact that there are several trillion connections and circuits in the brain enables it to do parallel processing on such a vast scale that it easily overcomes this deficiency in speed. Hawkins is familiar with this argument but completely rejects its relevance because he believes that the brain works not by computing a solution to a problem but by using memory to solve it. Again, the brain (or neocortex) is not a computer but a dynamic memory system. As a simple example, consider the ease with which we can learn to catch a ball, compared to the difficulty of programming a robotic arm to do it. For the robotic arm, catching a ball entails a series of mathematical calculations that must be constantly updated as the ball approaches, and of course the problem is enormously compounded if the robot must also move to intersect with the ball's calculated trajectory. While computers *can* perform these calculations, the human brain uses an altogether different method:

> Your brain has a stored memory of the muscle commands required to catch a ball (along with many other learned behaviors). When a ball is thrown, three things happen. First, the appropriate memory is automatically recalled by the sight of the ball. Second, the memory actually recalls a sequence of muscle commands. And third, the retrieved memory is adjusted as it is recalled to accommodate particulars of the moment, such as the ball's actual path and the position of your body. The memory of how to catch a ball was not programmed into your brain; it was learned over years of repetitive practice, and it is stored, not calculated, in your neurons. (69)

Indeed, this is an essential point: the brain's comparing of any specific catch to an invariant representation of catching a ball and then making the slight adjustments necessary is not accomplished computationally but physically, in the flow and modification of information up and down the six layers of the neocortex.[19] This flow moves through specific columns of cells within the six layers (the column is the basic unit of prediction, Hawkins believes) as well as laterally to adjacent areas and other organs, such as the hippocampus.

It is in and through this process of building up invariant representations, constantly comparing them to incoming sensory patterns and making adjustments, that the brain creates and stores a model of the world as it is, independent from how we see it under changing conditions. It does this by forming and storing invariant representations in all regions of the neocortex and at all levels, in a hierarchy that mirrors the nested structure of the world as we experience it (i.e., there are objects within objects within objects). But instead of images of objects, the neocortex stores sequences of patterns. Say a stranger enters the room. My eyes will immediately scan his or her face. One saccade might yield the pattern eye/nose/eye/mouth/eye/ear, while another might yield the same elements in a different order. For the neocortex, however, they constitute the "same" sequence, which is matched with an invariant representation of "face." Note also the nested hierarchy of objects: the face is both part of the larger head and formed of smaller parts—two eyes, a mouth, nose, and ears. But all are invariant representations composed of sequences of patterns. Moreover, if one kind of pattern always seems to accompany another in time, the cortex interprets them as related and gives them what Hawkins refers to as "a group name." It is this group name—not the individual patterns—that is passed up the cortical hierarchy to higher regions. In this way cells in associated groups cause other groups of cells to fire.

Information is thus registered differently at different cortical levels. In the primary sensory regions many cells are active, as the incoming patterns ceaselessly change. At higher levels fewer cells are active and the cell-firing patterns are more stable. This is because as information is passed upward predictable sequences are collapsed into "named groups," whereas when a pattern moves down the hierarchy it is "unfolded" into sequences. Let's say a student is called upon to recite a poem. Assuming she has memorized it, this pattern unfolds into a sequence of phrases, each of which unfolds into a sequence of words. At this point the unfolding pattern splits and travels down both the auditory and the motor sections of the cortex. Following the motor path, each word is unfolded into a memorized sequence of phonemes and then, at the lowest level, each phoneme is unfolded into a sequence of muscle commands to make sounds. The lower you go in the hierarchy, the faster the patterns are changing. If the student has to type the poem instead, it is the same sequence of patterns down the hierarchy but at the appropriate level the motor cortex would take a different path: words would be unfolded into letters (not phonemes), and the letters would unfold into muscle commands to her fingers.

In Hawkins's account of the memory-prediction system, perception and behavior, intention and expectation, are almost the same, for there is little difference between intending to move your arm and predicting it will move. But what happens if the cortex can't find any memory or invariant form that matches an incoming pattern of sequences? Then confusion sets in, and

> the unexpected pattern will keeping propagating up the cortical hierarchy until some higher region can interpret it as part of its normal sequence of events. The higher the unexpected pattern needs to go, the more regions of the cortex get involved in resolving the unexpected input. Finally, when a region somewhere up the hierarchy thinks it can understand the unexpected event, it generates a new prediction. This new prediction propagates down the hierarchy as far as it can go. If the new prediction is not right, an error will be detected, and again it will climb up the hierarchy until some region can interpret it as part of its currently active sequence. (159)

For Hawkins this process concludes happily with a "eureka!"—the high-level prediction is found. However, those who are nagged by a vague feeling that some areas of their brains are searching for patterns that may never be found can find respite in the assurance that the process is eventually dampened by the onward press of life.

Hawkins's whole book vectors toward the question, Can an intelligent machine be built from the principles that underlie the workings of the neocortex, as Hawkins himself has described them? The first step, he says, is to start with a set of senses that extract patterns from the world, and they don't have to be equivalent to human senses. The second is to attach to the senses a hierarchical memory system that works like the neocortex. And the third is to train the memory system so that the machine can begin to build a model of the world as seen through its senses. The machine will then "see analogies to past experiences, make predictions of future events, propose solutions to new problems, and make this knowledge available to us" (209). The only serious obstacles, he believes, are technological: creating a very large memory capacity and making massively parallel connections among millions of artificial neurons. But for Hawkins, who was trained as a silicon chip designer and has already invented the Palm Pilot as well as a transcription system (Graffiti) for entering data into it, these obstacles are not insurmountable, since the basic principles are already well understood. However, in order to grasp more exactly how we will get from the first to the third step—that is, from building a memory system to seeing an intelligence emerge that can model the world of its senses—we have to turn to the research of Steve Grand, who has already built something that (almost) does this.

Emergent Intelligence in a Baby Android

The machine is an android robot that Grand has named Lucy, after one of our hominid ancestors well known to anthropologists. The first version, Lucy I, consists of a head with mouth, eyes and ears, upper arms, and a torso; Lucy II, unfinished but well along in construction, will be a fully mobile autonomous robot.[20] What Grand hopes to achieve with Lucy is not human-level intelligence but emergent "mammal-style intelligence," specifically a rudimentary intelligence that emerges from the self-organization and learning that occurs in an array of neural modules. Presumably to emphasize Lucy's mammal-like intelligence, Grand has made her loosely resemble a short orangutan by giving her a face and soft cover of hairy skin. But since what makes intelligence possible in mammals is the special architecture of the neocortex, he has attempted to recreate this architecture in a simplified but nonetheless functional form. To make Lucy intelligent, in short, he has given her a simulated biological brain.

There is, consequently, some overlap and plenty of agreement between Hawkins and Grand about the key role of the neocortex. This is important, since neither seems to know anything about the other's work. In contrast to Hawkins's book, however, Grand's *Growing Up with Lucy* makes us acutely aware of how difficult the construction of a genuine intelligence actually is or will be, despite Grand's obvious technical ingenuity and inventiveness. Grand's book is also often amusing. Subtitled "How to Build an Android in Twenty Easy Steps," on the first page he admits, "I lied about the number of steps." Even so, the book covers a wide range of material, deftly anchoring the Lucy project not only in neuroscience and neural net theory, but AI, robotics, evolutionary theory, and machine vision.

Grand is the inventor (software designer and programmer) of the best-selling computer game Creatures, which is not really a conventional game but an artificial environment in which little cartoonlike "Norns" are born and, if given proper care, flourish.[21] These creatures exhibit an astonishing range of behavior. They are capable of learning about their environment, either by being shown things by their owners or by learning from their own mistakes. They can interact with their owners, using simple language, as well as with one another. They can form relationships and produce offspring, which inherit their neural and biochemical structure from their parents and are capable of open-ended evolution over time. These capabilities makes Creatures a landmark in the deployment of ALife theory and methodology, for what underlies the Norns' complex behavior is Grand's understanding of ALife simulation and his very effective instantiation of the theory of emergence.

In *Creation: Life and How to Make It*, Grand draws out some of the implications of ALife theory for the construction of virtual worlds. Recall that Langton argued that because ALife machines and programs would instantiate and realize the essential processes of living systems, they would constitute instances of *real life*, not just a *modeling* of these processes. This amounts to a strong theory of ALife, in contrast to weak theories, according to which ALife simulations are only lifelike replicas. As Grand points out, such a strong, or "real," instance of artificial life would emerge as a second-order simulation from the interactions of first-order simulations. According to this logic, a computer simulation of a bird is not a bird. Perhaps this is why graphics designer Craig Reynolds calls his bird simulations "boids." But a group of boids interacting according to the abstracted and programmable rules of flocking behavior would be an instance of real flocking. Similarly, a computer simulation of an atom or biological cell would not be a real atom or cell; however, a simulation that accurately captured the actual dynamics of atomic interaction in molecules or cellular interaction in living systems would be a real instance of these processes. In short, a simulation of a living thing is not alive; a simulation of intelligence is not intelligent; but—and this is a central premise of strong ALife research—intelligent, living things can be made out of simulations. Grand proceeded from exactly this assumption when he created Creatures: "Each norn is composed of thousands of tiny simulated biological components, such as neurons, biochemicals, chemoreceptors, chemoemitters and genes. The norns' genes dictate how these components are assembled to make complete organisms, and the creatures' behaviour then emerges from the interactions of those parts, rather than being explicitly 'programmed in'" (Web site, "Norns: The First Generation," under "Creatures").

In this sense, the Norns' behavior is comparable to behavior in an emergent system. This is not surprising, given that the theory of ALife simulation (discussed in previous chapters) is based on a notion of emergence derived from nonlinear dynamical systems theory. Recall that in emergent systems a multitude of simple interacting parts or agents can spontaneously self-organize, producing orderly and complex behavior at the global level. The interactions are deterministic but nonlinear, usually characterized by large positive feedback, the amplification of fluctuations, and the massive acquisition and transfer of information at local levels. In emergent systems—insect colonies, swarm systems, the immune system, and the market economy are typical examples—the behavior of the whole is more complex than that of the constituent parts. Consciousness is also said to be emergent, inasmuch as it arises from the interactions of neurons in multiple networks in the brain and cannot be reduced to the specific terms of these interactions. In contrast to these systems, however, the complex behavior of the Norns is intentionally emergent: the first-order simulations of neurons, biochemicals, genes, and so forth interact at local levels, producing rich, unpredictable lifelike behavior at a global level. This higher-level behavior constitutes a second-order simulation of life.

The Lucy project is an attempt to build an intelligent creature that "lives" in the actual physical world, rather than in a virtual world on a personal computer. What Lucy and Grand's Creatures share is a foundation rooted in the theory of emergence. In his public presentations of Lucy, in fact, nothing has so irritated Grand as the persistent question, But what is she programmed to do? Consequently, much of his book attempts to explain an only apparent anomaly: even though a set of computer programs make Lucy's behavior possible, the intelligence she displays emerges from the interactions among thousands of simulated neurons that self-organize into neural modules capable of learning. Grand agrees therefore with Hawkins that AI has completely missed the boat, for it has always been looking in the wrong direction: "Strangely, most neuroscientists who study the cerebral cortex already know this: they know that the cortex is a highly structured, arraylike, self-configuring machine, and they know that understanding the cortical architecture is the key to understanding the mind" (51–52). Grand also agrees that "the brain [is] an all-purpose prediction machine" (128), and, like Hawkins, obligingly walks us through several thickets of neuroscience. But Grand's task is more difficult, since at the same time he must explain how certain of the functions described can be achieved analogously with mechanical and electronic hardware. There are differences, of course: where Hawkins emphasizes the role of invariant representations, Grand focuses on cortical maps (areas of the cortex associated with a specific function) and invariant transforms. Grand is also much more concerned with the details of how the cortex processes sensory information (both visual and auditory) and how these functions can be emulated. However, both give primary emphasis to the massive and profuse feedback circuitry that makes the standard computational input-output model so useless: yes, information does arrive from the senses, but that information is accompanied by the information expected as well as information going out to the muscles to modify or alter the reception of further incoming information. In fact, information is simultaneously cascading up and down the cortical hierarchy and laterally across adjacent and related areas in such a profusion of patterns that Grand's proposal to model and replicate these processes with digital circuits and servo-motors is staggering—and yet it works.

Consider what Lucy can actually do: she can recognize and point to simple objects like an apple or banana. Admittedly, this does not seem like complex behavior. But no human child can recognize a banana until it has been taught to. And once it knows what a banana is, it will recognize it from whatever angle it appears, whatever the lighting, whether the banana is green, ripe, yellow, or rotten. It is essential to note that Lucy has *not been programmed* to recognize a banana; rather, like a child she has *learned* to recognize simple objects like bananas and therefore can point to one and identify it if it appears in her visual field. And, like a child, Lucy will learn to perform more complex actions over the coming years. What matters at this early stage is that she is physically functional: she has eyes that can see (though dimly), ears that hear (though badly); she can make sounds and move her eyes, mouth, head, and arms. Some of these movements can be seen on the brief video available on Grand's Web site.

Grand has devoted a great deal of attention and labor to Lucy's physical abilities because movement and "multimodal interactions" are essential for learning. Not only do we see depth because we can reach out with our arms and vice versa, but establishing correlations among different physical activities is a necessary part of the brain's development. So much so that Grand is committed to making Lucy's body emulate a biological body as closely as possible; in fact, he spent a year trying to make her muscles animal-like in tensile strength and dexterity. This devotion, however, has nothing to do with a slavish attempt to imitate nature; rather, it reflects something fundamental about the brain itself, as Grand explains:

> Brains don't talk in ASCII code—they talk in muscle tensions and retinal signals. Our senses have co-evolved with our brains, so that each depends upon the other. Trying to understand or even replicate the functions of the brain using video cameras and electric motors is futile. It is important to modify these technologies to make them appear as similar as possible to their biological equivalents. (Web site, "Rationale")

In short, building Lucy with technologies that work as similarly as possible to their biological equivalents seems to be the best way to construct her from the bottom up.

This principle is applied to building Lucy's every function—to her muscles as much as her brain, which cannot be conceived and built as something extra or apart from her capacity to move and sense things in the world. In fundamental respects this approach draws on the lessons of cybernetics: the "brain" is first and foremost what coordinates physical activities in the world; it is very much situated and embodied. But where Grand goes beyond both cybernetics and subsumption robotics is in his emphasis on learning: a mammalian brain must be capable of making certain basic connections itself, in order to be able to alter its responses and invent new behavior when confronted with novel situations in the environment. Grand makes a key point in this regard in contrasting the two evolutionary paths taken by insects and mammals. As he puts it, "An insect can learn by degree but not by kind: an individual insect can learn its way back to its own nest using landmarks that are unique to its particular circumstances, but it can't choose to do anything other than navigate with this circuitry, since most of the configuration details were set in place genetically, as a result of evolution" (48).[22] Grand thus notes that the great success of insects is directly due to the intelligence of evolution itself. Their nervous system is built upon and has evolved from a highly generalized, repetitive, and easily modified block structure. When confronted with ecological change, insects have rapidly evolved into new niches, with only slight alternation of their nervous systems. For mammals, on the other hand, adaptation and survival depends to a much greater extent on individual learning and invention—think of the many different ways that mammals modify their environment. The basic building block of a mammal's nervous system is therefore structurally different. In contrast to the fixed nature of the insect's nervous system, the mammal's is essentially a highly plastic structure.

Grand guides us into thinking about this difference in the following way. A single neuron can't do much by itself, but blocks of neurons can be programmed by evolution to do a variety of tasks. This is basically evolution's insect solution. But suppose that "an even more powerful building block structure happened to evolve, which could rewire *itself* to perform a range of tasks. Instead of being wired up by genes, just suppose that this array of building blocks could become wired up *by the very information that is flowing into it*" (48, author's emphasis). As we already know (recall Hawkins's point of departure in Vernon Mountcastle's research), this capacity "to wire itself up" in order to accomplish a range of different processing tasks turns out to be a primary feature of the neocortex. The same neural tissue can process either visual or audio stimuli, or change from one to the other if the input signals are changed. As Grand points out, there is a fairly common electronic device, a programmable logic array (PLA), which operates in a roughly analogous manner. A PLA is basically a large, generalized block of logic circuits that can be configured and programmed to do specific tasks by making and breaking connections among the circuits within the block. Nature does something similar with basic neural building blocks. In some cases (like insects), the configurations in these natural PLAs are set by evolution; in others (like mammals), the configurations are made and remade over time according to the dictates of the creature's own individual learning experience.

Nature, then, seems to work with three fundamental neural architectures to solve the problems of behavior and control. The first and simplest is to use the neuron itself as a basic building block in relatively simple networks, a solution that suffices for very primitive creatures. The second is to use repeated circuits of neurons, roughly similar to the PLAs, as the basic building blocks. This solution—the insect solution mentioned above—offers the advantage that these blocks can be easily configured by evolution to perform somewhat different tasks. The third solution is a more complex version of the second. Here a similar but higher-level kind of building block is used but its functionality is not set by evolution;

rather, the block has the capacity to reconfigure itself during the creature's lifetime.

Yet Grand thinks that in humans and primates there may be a fourth level built upon this underlying architecture but greatly augmenting its power. This fourth level has to do with our ability to take control of and manipulate certain functional capacities that operate unconsciously and serve other ends.[23] This additional capacity depends on and makes use of another fundamental feature of the cerebral cortex (or neocortex): it is made up of neural building blocks that wire themselves up in many different cortical maps, or regions, devoted to the accomplishment of specific tasks. The process of vision, for example, occurs in different stages. Incoming signals from the optic nerve are first processed in a map called V1 located at the back of the brain; then they are passed to other maps for further processing. Muscle movements are driven by a pattern of neuron activity in another map, the primary motor cortex, or M1. And so on. These specialized maps, which are largely self-organizing, carry out much of our learned but unconscious activity. At the same time, Grand notes, we are able "to make voluntary decisions, and to initiate, alter or suppress our behavior at will . . . [and to] perform mental tasks for which no specialized and rigidly structured circuitry is likely to exist" (50). Grand believes that when we do these things, we may be "switching information around among existing maps, altering the flow and hence making use of their specializations in new ways" (50). Thus he postulates a fourth, volitional level of brain structure that may control the "flow of signals around groups of preexisting self-configured maps in a highly flexible and general-purpose way" (50). One thing is certain in any case: the fundamental building blocks of the mind are neither the symbol manipulations of classic AI nor the simple pattern recognition mechanisms of its old rival, neural net research. The clue to the mystery of what generates intelligence, rather, is to be found in the basic circuits (each composed of only a few thousand neurons) that make up the cerebral cortex. "Understand the structure of these circuits," Grand proclaims, "what they do to the signals that enter them (from all directions) and what rules control their self-organization, and we will be well on our way towards understanding the mind" (51).

Lucy is primarily a test bed for working out this fundamental idea. As Grand states, "I'm essentially on the lookout for a proto-machine: a generalized neural architecture that can spontaneously self-organise into a variety of specialized machines, driven only by the nature of the signals supplied to it" (Web site). To put it the other way around, there must be specialized machines—for vision, hearing and making sounds, muscle movement and proprioception—but they must be variations on a core machine or set of basic principles. On the software level, Lucy's brain consists of a large neural network of 50,000 (in the case of Lucy II, 100,000) virtual neurons, or rather, as Grand clarifies, "neural columns, each of which is made from a small circuit of neurons." This neural network runs on a computer, a parallel processing multiple instructions–multiple data (MIMD) machine that Grand built expressly for the task. It is constructed of five boards, one each devoted to vision, hearing, voice, muscle/proprioception, and communication with a PC. (Lucy II's brain, made of three interconnected boards refashioned from desktop PCs, will be more powerful.) What really matters, however, is not so much the details of the computational machine as how it is configured. About the particulars, Grand remains reticent (understandably, given that the project is still being developed) and only summarizes the success he has found thus far. With Lucy I the object was

> to find a single neural architecture that is capable of self-organizing into a wide range of virtual machines, capable of the manifold tasks involved in seeing, hearing, thinking and moving. By the end of Phase I of the project [Lucy's] brain consisted of a number of "modules" (cortical maps), each of which performed a different function. Each map was unique, but nevertheless they all had enough in common to offer exciting hints about a universal architecture. (Web site, "Lucy Mk1")

An essential feature of this universal architecture is that it maintains a state of dynamic tension and is completely bidirectional. It is most definitely not an input-output device. In fact, if we want to think about this architecture in terms of feedback, we should think of the incoming sensory signals themselves as the feedback rather than the input (an idea that of course recalls Maturana and Varela's concept of autopoiesis). According to Grand, "The brain acts in order to keep the bottom-up and the top-down ... signal streams in balance. It does this in the manner of a nonlinear servomotor, comparing the top-down intention/expectation with the bottom-up sensory information and trying to reconcile the two by causing either a change in the internal model or in the outside world" (Web site, "The Science Bit"). Cyberneticists of the 1950s like Ross Ashby and Grey Walter would have understood this perfectly. But to understand how the brain is at once an information processor and dynamical system—the two faces of the same process, somehow inextricable—requires a contemporary perspective.

Thinking about this universal architecture has led Grand to a bold but not unreasonable speculation. Unlike most roboticists, he considers the imagination to be essential to intelligence:

> The key to Lucy's brain is imagination. Many people still think of the brain as a passive receptor of information—as if raw data comes in from our eyes and other senses and gradually gets refined into more symbolic form (e.g. recognition of a face or a word) until finally all the streams come together somewhere and somehow give rise to an active response. I think of perception as a very much more active process. As conscious beings we don't live in the real world—we live in a virtual world inside our heads. Most of the time this world is closely synchronized to the external world—our model matches reality, tracks it and predicts it. When we dream or when we imagine things (including making plans and rehearsing scenarios) we disconnect from the real world and let the model run on its own. The same mechanisms are at work in both cases, but there's no synchronization with reality going on when we dream or think. The model is the crucial thing, and perception is an active process of using this model to predict, hypothesize about and correct for the data coming in from our senses—"filling in" when the data is incomplete and "being surprised" when reality fails to live up to the model.... Lucy's brain is designed around a key set of hunches about how such a mechanism can be made using (simulated) neurons and biochemicals, and how something similar might have evolved in nature. (Web site, "FAQ")

So, in summary, what do we find in the middle ground, in this space where top-down and bottom-up processes somehow come together? In contrast to the highly abstract models of classic AI, which were conceived in terms of symbol systems and the frames and scripts of knowledge representation, the middle ground is completely physical, but it is complexly physical, since it is where changes and movements on different levels are coordinated. This coordination requires a memory-prediction system and a model of the world. But as Grand insists, it also requires a way of disconnecting from the world—of letting the model run on its own, as in acts of imagination and dream—and then reengaging with it. Hence his Lucy project also entails looking for a mechanism that generates mental imagery and "imagination." In this respect his project is unique and of special interest not only to scientists and engineers but also to those who study philosophy, literature, and the arts.

Conclusion: Self-Modeling and Communication in Robots

Grand's argument for the necessity of an internal model for Lucy—and by implication any robot, if it is to be capable of truly complex behavior—cannot help but recall Rodney Brooks's decisive rejection of modeling in behavior-based robotics. In developing a new approach to robotics based on a subsumption architecture (see chapter 7), Brooks argued explicitly against the *sense-model-plan-act* approach that characterized traditional robotics and AI more generally. Indeed, for Brooks and his followers the rallying cry was, "The world is its own best model." For robotics, this meant that cognition was already implicit in sensor-actuator articulations, a view that was closely aligned with Varela and Maturana's theory that cognition is a structural coupling with an aspect of the environment. Yet as Grand compellingly suggests, evolutionarily speaking, this approach is equivalent to the insect solution. In order for more complex behavior to occur, learning and modeling are necessary, and these require a much more complex neural substrate than insects have or could even develop.

Grand has not been alone in initiating a return to the concept of a model in robotics. Hod Lipson's successful efforts to coevolve simulations and robot controllers by continually feeding back the sensory data recorded by physical robots to populations of simulators that were then further evolved is directly relevant, and in chapter 7 I discussed how this work could be applied to the development of self-modeling mechanisms for robots. The basic idea is to evolve models that match the robot's own physical observations (i.e., sensory data). This would entail seeking actions that would make the models disagree as well as actions whose only goal is to extract data from the environment. Instead of using this information to evolve a controller, it would be used by the robot to evolve a model of itself, that is, of its own morphology and dynamic interactions with the environment. Needless to add, this model would be completely unlike the abstract maps of the "world" as a geometric 3-D space proposed by early roboticists following classic AI's sense-model-plan-act approach.

Recently Lipson and his colleagues Josh Bongard and Victor Zykov have taken a remarkable step in the development of this new kind of model.[24] Constructed for a four-legged robot that could push itself across the floor, the model is actually a series of models that are synthesized, updated, and replaced as the robot generates and evaluates new actions. When half of one its legs was removed, thanks to its continuous self-modeling, the robot was able to infer this change and adapt by generating an alternative gait by which it could continue to propel itself forward. In a large number of experiments, this new adaptive behavior was tested against two different baselines of randomly generated behavior, with very positive results.

The self-modeling process is based on having multiple competing internal models that generate actions to maximize disagreement between predictions of new sensory data. The process is broken down into three algorithmic components for modeling, testing, and prediction, which are continuously executed whether the robot is moving or at rest. Initially the robot performs an arbitrary action and records the resulting sensory data. The model synthesis component (using a stochastic optimization method) then synthesizes a set of fifteen candidate models that explain the action-sensory data in a causal relationship. Next, the action synthesis component generates a new action that is most likely to garner the most information from the robot. This is accomplished by searching for an actuation pattern that will produce the maximum disagreement among the predicted sensor signals. The robot then executes this action, and the resulting sensory data is used as additional information by the model synthesis component to assess the model once again. After sixteen iterations of this process, the most accurate model is used by the action synthesis component to generate a desired behavior, which is then performed. If, however, the robot senses an unexpected sensor-motor pattern or an external signal resulting from a change of its morphology, the cycle of modeling and exploratory actions is again initiated, and eventually a new, compensatory action is taken to recover functionality. This is exactly what happened when half of one of the robot's legs was removed. Throughout the tests, functional behavior generated by the model (as measured by distance traveled from a point of origin) remained significantly higher than randomly generated behavior.

While Bongard, Zykov, and Lipson think it is unlikely that biological organisms maintain explicit models like the one they present here, their method may well shed light "on the processes by which organisms actively create and update self-models in the brain, how and which sensor-motor signals are used to do this, what form these models take, and the utility of multiple competing models" (1121). Moreover, by presenting an operational model of how a robot can use directed exploration to acquire and evolve predictive self-models that will enable it to continue functioning after damage or a change in its morphology, this work clearly opens a new path toward achieving a higher level of machine intelligence.

What is particularly impressive about this accomplishment in robotic self-modeling is how it begins to reclaim and develop a feature or functionality always deemed essential for artificial intelligence, but without returning to the disembodied abstractions of classic AI. Instead, this development builds on and realizes a new synthesis of ideas we have seen throughout this book at work in cybernetics (most notably in Ross Ashby), ALife simulations, behavior-based robotics, and artificial evolution. At the same time, there is something genuinely new. In an appreciative commentary on Bongard, Zykov and Lipson's achievement published in the same issue of *Science*, ALife scientist Christoph Adami emphasizes that the robot's updating of its own model (by continually reviewing its sensory data) is like dreaming—or maybe *is* dreaming.[25] Like us, Adami writes, robots "need to constantly ascertain where they are in the world, and like us, they work better if they have an accurate sense of self" (1093). The means that Bongard and colleagues have devised to achieve this may one day lead to a discipline of experimental robot psychology that could, for example, "record the changes in the robot's artificial brain as it establishes its beliefs and models about the world and itself, and from those infer not only its cognitive algorithms, but also witness the emergence of a personality" (1094). Significantly, both Adami and Bongard and colleagues refer to the research initiated by neuroscientists Francis Crick and Christof Koch on the neural correlates of consciousness.

In the context of AI, references to modeling and consciousness invariably raise the question of language and whether or not some kind of turn—or indeed return—to symbolic thinking might be necessary in order to develop higher cognitive functions in robots. Luc Steels's recent work in robot communication and the origins and evolution of language takes an innovative approach to this old problem, one that portends exciting new possibilities.[26] Whereas classic AI treated language primarily as a static representational system, Steels understands language as a living system—a complex adaptive system, to be exact. From every aspect and perspective, language is constantly changing and adapting as the needs and social interactions of its users change. Thus language is really a rich and dynamically self-organizing ecology of competing forms that continually demand accommodation, negotiation, and the creation of new forms. Evolutionary linguistics and the ALife approach to language are both predicated on this recognition. The former has "a specific focus (understand the origins of language and meaning), a specific hypothesis (language is a complex adaptive system), and a specific methodology (construct artificial systems as a way to develop and test theories)" (308). The latter attempts to evolve artificial languages with natural language-like properties, and thereby explore the space of possible languages in the same way ALife explores the space of possible life forms. As Steels has understood more clearly than anyone, this new approach leads to a bottom-up investigation into how communication protocols, as well as a lexicon and syntax, can emerge and under what specific conditions this emergence becomes possible.[27] In a series of experiments performed with robotic agents grounded in real-world situations through a sensorimotor apparatus, he shows how these components self-organize, evolve, and are complexified in and through the agents' interactions. Thus, in Steels's experiments, the communication systems that arise are developed by the agents themselves; they are not designed and programmed in by an external observer.

Much of this research is based on simple language games played between robots or among robots and humans.[28] At least two agents, a speaker and a hearer, as well as a concrete context consisting of agents, objects, and a situation, are required. Steels's Talking Head experiment, for example, is a guessing game in which the two agents are pan-tilt cameras facing a whiteboard on which various colored geometric shapes are displayed. Software in the computers to which the agents are connected enables them to segment visual images into a few basic categories like relative location, shape, color, and size. The game is initiated when one agent (the speaker) points to a particular "topic" (a red square, let's say), and gives the hearer a verbal hint. The verbal hint would be a random sound that the agent intends to be associated with the selected topic. Steels's gives as an example the utterance "malena," which might mean [UPPER EXTREME-LEFT LOW-REDNESS]. Looking in the direction pointed to, the hearing agent then tries to guess which topic the speaker means. Depending on whether the hearer guesses correctly or not, a learning mechanism forms an association between the topic and the utterance or a new category (like redness or squareness) is created. As the game proceeds, a shared lexicon and a shared ontology (the category distinctions) are gradually built up and stored, thanks to a positive feedback loop between use and success. More precisely, two dynamic systems—the evolving lexicons and the evolving ontologies—are coupled to each other. From this simple setup the experiment was expanded to include robots located in cities all over the world connected through the Internet. This also allowed human users to interact with the agents. In a four-month period in 1999, some 400,000 games were played among a population of just under 2,000 agents, and a total of 8,000 words and 500 concepts were created. The experiment thus demonstrates how a population of agents is able to generate and self-organize a shared lexicon based on perceptually grounded categorizations of aspects of the world.

The Talking Head experiments have been followed by further experiments with AIBO "pet" robots that describe to each other events involving a ball and with the humanoid Sony Dream Robot (SDR). The latter is an operationally autonomous robot that can walk on two legs, possesses stereo vision, surround audition, real-time adapted dynamical trajectory planning, vision-based navigation, and many other features.[29] As such, it reflects the enormous advances made since the early days of behavior-based robotics in the late 1980s. It also points to the next challenge, which is to find operational models of how populations of robotic agents equipped with a cognitive apparatus and a sensorimotor system (a body) can cocreate shared communication systems that are grounded in their own activities. In this endeavor, Steels argues, research in the new field of evolutionary linguistics "provides the first timid steps towards a radically new approach, in which language-like communication systems autonomously evolve in embodied agents through grounded language games" (311). As in the case of human natural languages, when this perspective is applied to robotic communication it shifts the focus to how the agents themselves participate actively in the invention and propagation of a dynamically evolving language. (Propagation here is viral and memetic and not simply genetic.) Success in robotics research oriented by an evolutionary linguistics approach can thus not only bring about further advances in the complexity of human-robot and robot-robot interactions but also yield new insights into the origin and evolution of language and communication systems more generally.

Like Bongard, Zykov, and Lipson's work on robotic self-modeling, Steels's work on the development of robotic communication systems draws on research that has been an essential part of this book's trajectory, particularly in ALife, complex adaptive systems, evolutionary theory, and behavior-based robotics. At the same time, by making the assumptions and methods from evolutionary linguistics and the ALife approach to language the working tools of current robotics, Steels has been able to forge a fruitful new research path into territory until now forbidden to behavior-based robotics: the realm of symbolic, representational thinking instantiated in natural human language. To be sure, Steels is fully aware of the extraordinary complexity of human natural language. He notes, for example, that there are no "convincing simulations of grammaticalization," which in human natural language appears to be too deeply grounded in aspects of human culture and embodiment to be incorporated into artificial systems ("Evolving Grounded Communication for

Robots," 311). He also realizes that the nascent or protolanguage of robotic communication does not necessarily have to be anything like human language; indeed, it cannot be. But while the singularity of human language may impede translation or a full transcoding into a computational medium, Steels shows how that bugaboo of classic AI, the symbol-grounding problem, can easily be handled by robots—*pace* the objections of philosophers—when the symbols emerge as both objects and the means of social interaction.

As new capacities of robotic machines, self-modeling and communication mark high levels of complexity and intelligence that inevitably lead further. These capacities, that is, are not simply large notches on a ruler of measurable achievement—they are markers of a new threshold. A machine that possesses an internal model of its own morphology and dynamic functioning is a very complex machine. A machine that can communicate information with another machine about a world of shared experience is obviously a great deal more intelligent than one that cannot. But with these particular new capacities the terms complexity and intelligence begin to signify something new—not only known and accountable actions but future possible actions for which we can only wait in eager anticipation. With these newly complex and intelligent machines, the world becomes a different place, inhabited by new forms of complexity that are also new forms of life.

Notes

Introduction

1. I take the term *machinic phylum* from French philosophers Gilles Deleuze and Félix Guattari but use it in a somewhat different sense, as I explain in chapter 3. Less discordant to French ears, the term *machinic* itself unsettles the opposition between the natural and the artificial.

2. As noted in the preface, I use the term *machinic life* to designate forms of nascent life that emerge in and through technical interactions in a human-constructed environment. ALife refers specifically to the productions (machines and programs) of the scientific research discipline established in 1987 following the first conference devoted to "the synthesis and simulation of living systems" organized by Christopher Langton; more generally, artificial life refers to the whole spectrum of humanly constructed lifelike objects and processes—automata, computer viruses, robots, and so forth, which of course include those of ALife.

3. Compare Langton: "The key concept in [Artificial Life] is emergent behavior. Natural life emerges out of the organized interactions of a great number of nonliving molecules, with no global controller responsible for the behavior of every part. Rather, every part is a behavior itself, and life is the behavior that emerges from out of all the local interactions among individual behaviors. It is this bottom-up, distributed, local determination of behavior that AL employs in its primary methodological approach to the generation of lifelike behaviors." From "Artificial Life," in *Artificial Life*, ed. Christopher G. Langton (Reading, Mass.: Addison-Wesley, 1989), 2–3.

4. For discussion, see John Maynard Smith, *The Problems of Biology* (Oxford: Oxford University Press, 1986), chap. 1, and Lynn Margulis and Dorian Sagan, *What Is Life?* (New York: Simon and Shuster, 1995), chap. 9. The whole issue is of fundamental concern to ALife. See, for example, Mark A. Bedau, "The Nature of Life," in *The Philosophy of Artificial Life*, ed. Margaret Boden (Oxford: Oxford University Press, 1996), 332–357.

5. Here I allude to Herbert Simon's *The Sciences of the Artificial* (Cambridge, Mass.: MIT Press, 1969), even though it is concerned only with the methodology and conceptual framework of artificial intelligence. Focusing on issues of rationality and design, Simon touches on biology and evolutionary process only near the book's conclusion. More significant for the development of ALife as a second new "science of the artificial" is John Holland's *Adaptation in Natural and Artificial Systems* (Cambridge, Mass.: MIT Press, 1975), which brings both perspectives into play in his concept of the "genetic algorithm" and approach to machine learning. Simon remains important, nevertheless, for his emphasis on computer simulation and the organization of complexity.

6. In attempting to determine the ultimate meaning or essence of modern technology, the German philosopher Martin Heidegger returned to the Greek concepts of *phusis* and *technē*, which constitute a primordial first instance of the opposition between nature and technology (or craft) and are valorized as such. Yet, inasmuch as both terms designate forms of *poiēsis*,

which Heidegger understood in the Greek sense of a "bringing forth into presence," their difference cannot be one of simple opposition. Heidegger develops this position in "The Question Concerning Technology," in *The Question Concerning Technology and Other Essays* (New York: Harper Torchbook, 1977), 3–35.

7. Michel Foucault, *The Order of Things* (New York: Vintage Books, 1973; orig. pub. 1966), 128.

8. See Michel Foucault, *The Birth of the Clinic* (New York: Vintage Books, 1975; orig. pub. 1963).

9. The oldest narrative of artificial life seems to be the story of the Golem in Jewish folklore. Interestingly, though made of clay, like humans, the Golem was an informational entity, at least in the sense that in many versions what animated it was an inscription on its forehead or under its tongue, often of the name of God.

10. Critiques of molecular biology based on its too exclusive reliance on information theory and overly literalist understanding of coding are now well known. For an exposition of this point of view, see Lily E. Kay, *Who Wrote the Book of Life? A History of the Genetic Code* (Stanford, Calif.: Stanford University Press, 2000).

11. Bernard Stiegler, *La technique et le temps*, vol. 2, *La désorientation* (Paris: Galilée, 1996), 173–187. See, for example, the section "La synthèse biologique: quand faire c'est dire," where the title itself announces this performative dimension.

12. This is not, however, a case of "pure science" being corrupted or perverted by its application—the usual cover story for our refusal to think the "question of technics" bequeathed by Heidegger in *The Question Concerning Technology and Other Essays*. Rather, it is an instance in which a supposedly autonomous entity is first posited as existing outside the technical mediations that allow us to conceive of the entity and gain access to it and then found to be violated when those same technical mediations denied in the original positing are manipulated for other ends. By denying the constitutive role of technical mediation, or technics—in my terms, by denying that molecular biology constitutes a complex material and discursive assemblage whose discourse about life is conjoined with and dependent on several highly advanced computational technologies—one sets up a situation in which the technical can only return (like Freud's "return of the repressed") as a corrupting manipulation. In a sense never explored, molecular biology's denial of its technological dependency is mirrored and inverted by ALife's explicit claim to instantiate a form of "real" life through technological manipulation.

13. See Lynn Helena Caporale, *Darwin in the Genome: Molecular Strategies in Biological Evolution* (New York: McGraw-Hill, 2003). Caporale also includes a chapter on "jumping genes," or transposons (i.e., genes that move from one location to another on a chromosome), which were a major early challenge to the orthodox view of the genome. Discovered by Barbara McClintock, for which she received the Nobel Prize in 1983, transposons are now the object of intense study because of their relation to genome evolution and retroviruses. As discussed in chapter 5, ALife experiments with mutations in the "genome" of digital organisms have shown several interesting parallels.

14. Stuart A. Kauffman, *The Origins of Order: Self-Organization and Selection in Evolution* (New York: Oxford University Press, 1993), 10 (author's emphasis).

15. See note 3, above.

16. See *La technique et le temps*, vol. 1, *La faute d'Épiméthée* (Paris: Galilée, 1994), esp. 43–94. Unfortunately, Stiegler himself does not consider either cybernetics or artificial life in his discussion of the technical system.

17. See "On Computable Numbers, with an Application to the *Entscheidungsproblem*," in *Proceedings of the London Mathematical Society*, ser. 2, 42 (1936): 230–265, reprinted in *The Essential Turing*, ed. B. Jack Copeland (Oxford: Oxford University Press, 2004), 58–90. Of particular importance is the passage that begins: "We may compare a man in the process of computing a real number to a machine which is only capable of a finite number of conditions q1, q2, ... qR, which will be called *internal states*."

18. As those working in neurophysiology and early neuroscience were aware, this architecture bore no discernable resemblance to the architecture of the human or mammalian brain. It seems that Turing himself was aware of this discrepancy. In "Intelligent Machinery" (1948), a remarkable essay never published during his lifetime, Turing postulated what he called "unorganized machines," which, unlike organized machines (i.e., Turing machines), had the capacity to learn. In effect, these unorganized machines amounted to a different type of computational assemblage. See Christof Teuscher, *Turing's Connectionism* (London: Springer, 2002), for detailed discussion.

19. I discuss these differences and the story of their rivalry in chapter 6.

20. W. Ross Ashby, "Principles of the Self-Organizing System," in *Principles of Self-Organization*, ed. Heinz von Foerster and George W. Zopf Jr. (New York: Pergamon Press, 1962), 270 (author's emphasis).

21. The research on neural nets conducted in the early 1980s, particularly that of the physicist J. J. Hopfield, was also crucial in this respect. See chapter 6 for discussion.

22. Unfortunately, a history of cellular automata has yet to be written. See Andrew Ilachinski, *Cellular Automata: A Discrete Universe* (Singapore: World Scientific, 2001) for a broad review of much CA research.

23. Stephen Wolfram, "Statistical Mechanics of Cellular Automata," *Reviews of Modern Physics* 55 (July 1983): 601–644.

24. *Cellular Automata: Proceedings of an Interdisciplinary Workshop, Los Alamos March 7–11, 1983*, ed. Doyne Farmer, Tommaso Toffoli, and Stephen Wolfram (Amsterdam: North-Holland, 1984).

25. Ibid., 197. The importance of this new capacity for visual simulation cannot be emphasized enough. As we'll see in chapter 3, the computer played an analogous role in the development of chaos theory. Toffoli and Margolus would go on to develop a series of CAMs, which they describe in their influential book *Cellular Automata Machines* (Cambridge, Mass.: MIT Press, 1987), and in their article "Programmable Matter: Concepts and Realization," *Physica D* 47 (1991): 263–272.

26. I discuss this theory at length in chapter 4.

27. See Samuel Butler, *Erewhon* (London: Penguin Classics, 1985), "The Book of the Machines," chaps. 23–25.

28. Hans Moravec, "Human Culture: A Genetic Takeover Underway," in *Artificial Life*, 167–199. See also his book *Mind Children* (Cambridge, Mass.: Harvard University Press, 1988).

29. See N. Katherine Hayles, *How We Became Posthuman* (Chicago: University of Chicago Press, 1999), Andy Clark, *Natural-Born Cyborgs* (Oxford: Oxford University Press, 2003), and Francis Fukuyama, *Our Posthuman Future: Consequences of the Biotechnology Revolution* (New York: Farrar, Straus & Giroux, 2002).

30. This is not at all to underestimate the tremendous value of what has been achieved by applying biological methods (mating, reproducing and evolving) to produce more efficient algorithms, as in the case of genetic algorithms. See Melanie Mitchell and Stephanie Forrest, "Genetic Algorithms and Artificial Life," *Artificial Life* 1, no. 3 (1994): 267–289.

31. See chapter 5 for an elaboration of this point.

32. Stephen Jay Gould, *Wonderful Life: The Burgess Shale and the Nature of History* (New York: Norton, 1989).

33. John H. Holland, *Hidden Order: How Adaptation Builds Complexity* (Reading, Mass.: Addison-Wesley, 1995), 41.

34. John H. Holland, *Adaptation in Natural and Artificial Systems* (Cambridge, Mass.: MIT Press, 1998), 186–195.

35. Thomas S. Ray, "Selecting Naturally for Differentiation: Preliminary Evolutionary Results," *Complexity* 3, no. 5 (1998): 25–33.

36. Mark Bedau, "Four Puzzles about Life," *Artificial Life* 4, no. 2 (1998): 125–140.

37. On animats, see *From Animals to Animats: Proceedings of the First International Conference on Simulation of Adaptive Behavior*, ed. Jean-Arcady Meyer and Stewart W. Wilson (Cambridge, Mass.: MIT Press, 1991). On software agents, see Andrew Leonard's lively (but no longer up-to-date) *Bots: The Origin of New Species* (San Francisco: HardWired, 1997).

38. See Karl Sims, "Evolving 3D Morphology and Behavior by Competition," in *Artificial Life IV*, ed. Rodney A. Brooks and Pattie Maes (Cambridge, Mass.: MIT Press, 1994), 28–39.

39. For example, in *Lamarck's Signature: How Retrogenes Are Changing Darwin's Natural Selection Paradigm* (Reading, Mass.: Perseus Books, 1998) Edward J. Steele, Robyn A. Lindley, and Robert V. Blanden argue that in certain cases a specific acquired immune response can be inherited.

40. See *Technological Evolution as an Evolutionary Process*, ed. John Ziman (Cambridge: Cambridge University Press, 2000), particularly the chapters Ziman himself contributes: "Evolutionary models for technological change" and "Selectionism and complexity."

41. *Technological Evolution as an Evolutionary Process*, 6.

42. For Ziman, the most surprising feature of ALife "models" is the degree to which they exhibit "many familiar characteristics of bio-organic evolution" (48). Hence their great significance for evolutionary theory.

43. Published in *Artificial Life II*, ed. Christopher G. Langton (Reading, Mass.: Addison-Wesley, 1992). Also see D. H. Ackley and M. L. Litman, "A Case for Lamarckian Evolution," in *Artificial Life III*, ed. Christopher G. Langton (Reading, Mass.: Addison-Wesley, 1994). Ackley and Litman report that their computer simulations demonstrate a "tremendous increase in speed and solution quality possible with distributed Lamarckian evolution." The latter, in fact, results in a "yeasty dynamical system."

44. Harold J. Morowitz, *The Emergence of Everything* (Oxford: Oxford University Press, 2002). While the agents involved and the principles of selection differ from one stage to the next, the underlying process, Morowitz argues, remains the same. Even so, with the birth of human reflexivity and the control of the physical environment by a single species, something truly novel comes about. This forces Morowitz, a respected biophysicist, to conclude with a hasty attempt to reconcile science and religion.

45. Deleuze and Guattari develop their theory of becoming in *A Thousand Plateaus* (Minneapolis: University of Minnesota Press, 1987; orig. pub. 1980). Of course Darwinian evolution and becoming are not intrinsically opposed. The Belgian philosopher Isabelle Stengers has written: "In contrast to the physical and chemical sciences, which deal with the behavior of beings [or entities] that they define, Darwinian evolution deals with becomings, and puts the human species under the sign of a becoming" (my translation). See "Pour une approche spéculative de l'évolution biologique," in *L'évolution*, with Pierre Sonigo (Les Ulis Cedex A, France: EDP Sciences, 2003), 88.

46. This is especially true of viruses. In this regard, see Manfred Eigen's "Viral Quasispecies," *Scientific American*, July 1993, 42–49.

47. See chapter 5 for discussion of this phenomenon, which has also appeared spontaneously in ALife experiments.

48. I begin to explore this theme in relation to chaotic systems in chapter 3.

49. See Fredkin's essays on digital philosophy, available on his Web site, http://www.digitalphilosophy.org. The more neutral, normative view is that computational theory is a fundamental component of our present understanding of the "fabric of reality," as David Deutsch argues in *The Fabric of Reality* (New York: Penguin Books, 1997).

50. Stephen Wolfram, *A New Kind of Science* (Champaign, Ill.: Wolfram Media, 2002). Wolfram's position on the ultimate status of computation is a little difficult to pin down. On the one hand, he states: "It is possible to view every process that occurs in nature or else-

where as a computation"; on the other (in chapter 9), he takes very seriously the "hypothesis" that underlying our universe is a "simple program" often yielding great complexity.

51. We are already cyborgs, Andy Clark argues compellingly in *Natural-Born Cyborgs* (New York: Oxford University Press, 2003). Surprisingly, Clark shows little interest in the burgeoning new field of computer-brain interfaces (BCI) as such. While most current BCI research is primarily prosthetic in orientation, no doubt it will eventually turn to the construction of assemblages in which neural interfaces serve augmentative ends.

52. Compare Danny Hillis on the construction of his parallel-processing "connection machine": "Instead of optimizing the algorithm to match the operation of the von Neumann machine, we could make a machine to match the algorithm." W. Daniel Hillis, *The Connection Machine* (Cambridge, Mass.: MIT Press, 1985), 14.

Chapter 1

1. Norbert Wiener, *Cybernetics; or, Control and Communication in the Animal and the Machine* (Cambridge, Mass.: MIT Press, 1948, 1961), 11.

2. The best known instance is Wiener himself, who worked on the control system for an antiaircraft gun called the AA Predictor. For a detailed discussion, see Peter Galison, "The Ontology of the Enemy: Norbert Wiener and the Cybernetic Vision," *Critical Inquiry* 21 (Autumn 1994): 228–266.

3. Originally published in the *Bell System Technical Journal* 27, no. 3 (1948), Shannon's theory subsequently appeared in book form under the same title, *The Mathematical Theory of Communication* (Urbana: University of Illinois Press, 1949). Warren Weaver, who contributed a long introduction, is listed as coauthor.

4. Apparently Shannon believed that this difference was complementary rather than contradictory and merely reflected a difference in their points of view. At least this is what he states in a letter to Wiener dated October 13, 1948 (box 2.85, Wiener Papers, MIT Institute Archives and Special Collections, Cambridge, Mass.). I consider Shannon's theory in more detail in chapter 3.

5. Norbert Wiener, *The Human Use of Human Beings: Cybernetics and Society* (New York: Houghton Mifflin, 1950).

6. Claude E. Shannon, "Presentation of a Maze-Solving Machine," in *Cybernetics: Circular Causal and Feedback Mechanisms in Biological and Social Systems, Transactions of the Eighth Conference, March 15–16, 1951* (New York: Josiah Macy Jr. Foundation, 1952), 173–180. Although Shannon never refers to it as anything but "a maze-solving machine," the other Macy conference participants called it a mouse and the term caught on.

7. With the development and widespread use of computers in science and engineering, of course, simulation would acquire a very different meaning.

8. The paper was published in *Philosophy of Science* 10 (1943): 18–24.

9. See J.-C. Beaune, *L'automate et ses mobiles* (Paris: Flammarion, 1980), chap. 6.

10. The view I present differs from or at least qualifies N. Katherine Hayles's argument in *How We Became Posthuman* (Chicago: University of Chicago Press, 1999) that cybernetics—specifically Wiener and Shannon—effected a "disembodiment" of information, by defining it independently of its material substrate. But although the cyberneticists defined information as a mathematical function, "disembodied" was not the way they understood it. However, as I later say below and discuss in chapters 2 and 5, what Hayles means by disembodiment does occur in the early formulations of Artificial Intelligence, most notably in those of Allen Newell and Herbert Simon.

11. W. Ross Ashby, *Introduction to Cybernetics* (London: Chapman and Hall, 1956), 1–2.

12. W. Ross Ashby, *Design for a Brain* (London: Chapman and Hall, 1952). Chapter 3 is entitled "The Organism as Machine."

13. Even further along, Ashby will discuss "Markovian regulators," giving a range of examples including "the rat that knows its way about the warehouse" (236) and can always make its way back to the nest. I return to Markovian models of behavior in chapter 2 in relation to psychoanalysis.

14. Although I draw loosely on Ashby here, the example that follows is my own. It is intended only as a general illustration, not as a model of a specific machine.

15. Subsequently, automata theory would rapidly develop into the study of abstract machines considered in relation to formal language theory. Whereas Marvin Minsky's *Computation: Finite and Infinite Machines* (Englewood Cliffs, N.J.: Prentice-Hall, 1967) is primarily concerned with machines, Richard Y. Kain's *Automata Theory: Machines and Languages* (New York: McGraw-Hill, 1972) introduces the linguistic model. Contemporary automata theory is based almost exclusively on the mathematical formalisms of set theory. See chapter 2 for further discussion.

16. John von Neumann, "The General and Logical Theory of Automata," in *Cerebral Mechanisms of Behavior: The Hixon Symposium*, ed. Lloyd A. Jeffress (New York: John Wiley & Sons, 1951), 2. The papers collected here were presented at the Hixon Symposium at Caltech in September 1948.

17. John von Neumann, *The Computer and the Brain* (New Haven: Yale University Press, 1958).

18. McCulloch and Pitts's essay, which was seminal for the cybernetics movement, first appeared in the *Bulletin of Mathematical Physics* 5 (1943): 115–133; it has been republished in McCulloch's *Embodiments of Mind* (Cambridge, Mass.: MIT Press, 1965), 19–39. All quotations are taken from the latter.

19. See chapter 2 for discussion of Turing machines, and chapter 6 for more on neural nets.

20. It is interesting that this inversion—that when confronting true complexity it is easier to describe the underlying structure than the resultant behavior—is mobilized in the contemporary theory of "bottom-up" emergent complexity, though the connection with von Neumann is seldom made. See chapter 5.

21. John von Neumann, *Theory of Self-Reproducing Automata*, ed. and completed by Arthur W. Burks (Urbana: University of Illinois Press, 1966), 26.

22. John von Neumann, "Probabilistic Logics and the Synthesis of Reliable Organisms from Unreliable Components," in *Automata Studies*, ed. W. R. Ashby, C. E. Shannon, and J. McCarthy (Princeton: Princeton University Press, 1956).

23. For example, in his book *The Cybernetics Group* (Cambridge, Mass.: MIT Press, 1991), Steve Heims only mentions Ashby once (on page 73).

24. See, for example, Jean-Pierre Dupuy, *Aux origines des sciences cognitives* (Paris: Editions La Découverte, 1994), 38, 162, 166.

25. W. Ross Ashby, "Homeostasis," in *Cybernetics: Circular Causal and Feedback Mechanisms in Biological and Social Systems*, Transactions of the Ninth Conference, March 20–21, 1952, New York, ed. Heinz von Foerster (New York: Josiah Macy Jr. Foundation, 1953), 73–108, here 73.

26. In a footnote Ashby refers to his book *Design for a Brain* and his article, "The Nervous System as Physical Machine, with Special Reference to the Origin of Adaptive Behavior," *Mind* 56, no. 221 (1947): 44–59.

27. Andre Pickering, "Cybernetics and the Mangle: Ashby, Beer and Pask," available at http://www.soc.uiuc.edu/faculty/pickerin/cybernetics.pdf.

28. Pickering also discusses these tortoises in "The Tortoise against Modernity: Cybernetics as Science and Technology, Art and Entertainment," available at http://www.soc.uiuc.edu/faculty/pickerin/tortoises.pdf.

29. Walter J. Freeman, untitled biographical essay on W. Grey Walter, available at http://sulcus.Berkeley.edu/wjf/CI.GreyWalter.pdf, p. 2.

30. W. Grey Walter, *The Living Brain* (New York: W. W. Norton, 1963), 125. As Freeman notes in his biographical essay, the two activities also correspond to "two cognitive operations . . . characteristic of animal (and human) behavior" (2).

31. W. Grey Walter, "An Imitation of Life," *Scientific American*, May 1950; "A Machine That Learns," *Scientific American*, August 1951. In subsequent quotations from these articles the title followed by the page number will be inserted directly in the text.

32. Owen E. Holland, "Grey Walter: The Pioneer of Real Artificial Life," in *Artificial Life V*, ed. Christopher G. Langton and Katsunori Shimohara (Cambridge, Mass.: MIT Press, 1997), 34–41.

33. Rodney Brooks, *Flesh and Machines: How Robots Will Change Us* (New York: Pantheon, 2002), 21.

34. W. R. Ashby, "Principles of the Self-Organizing Dynamic System," *Journal of General Psychology* 37 (1947): 125–128.

35. At the first of these conferences von Foerster presented a paper "On Self-Organizing Systems and Their Environments," and at the second Ashby presented "Principles of the Self-Organizing System," a lengthier reconsideration of the topic of self-organization. The papers presented at these three conferences are collected in, respectively, *Self-Organizing Systems*, ed. Marshall C. Yovits and Scott Cameron (New York: Pergamon Press, 1960); *Principles of Self-Organization*, ed. Heinz von Foerster and George W. Zopf Jr. (New York: Pergamon Press, 1962); and *Self-Organizing Systems 1962*, ed. Marshall C. Yovits, George T. Jacobi, and Gordon D. Goldstein (Washington, D.C.: Spartan Books, 1962).

36. In an interview published in the French journal *CREA* 8 (November 1985) devoted to "Généalogies de l'auto-organisation," von Foerster states that initially the Biological Computer Laboratory had little to do with cybernetics, and was mostly concerned with "the computational principles of living organisms" (257). It was only some years later, with the participation of Humberto Maturana and Francisco Varela and the exposure of their theory of "autopoiesis," that a second order cybernetics focused on cognition "was seen as a possibility" (258). See the section on autopoiesis in chapter 4 for further discussion.

37. "Principles of the Self-Organizing System" was published in von Foerster and Zopf's *Principles of Self-Organization* (1962).

38. Shalizi, *Notebooks*, "W. Ross Ashby," available at http://www.cscs.umich.edu/~crshalizi/notebooks/ashby.html.

39. The paper was published in Yovits and Cameron's *Self-Organizing Systems* (1960).

40. See, however, Henri Atlan's two books, *L'organisation biologique et la théorie de l'information* (Paris: Hermann, 1972) and *Entre le crystal et la fumée: Essai sur l'organisation du vivant* (Paris: Editions du Seuil, 1979), which make a strong case for the richness of von Foerster's essay in information-theoretic terms.

41. Kevin Kelly, *Out of Control: The Rise of Neo-biological Civilization* (Reading, Mass.: Addison-Wesley, 1994), 453–454.

42. See Foerster's *Observing Systems* (Seaside, Calif.: Intersystems Publications, 1982) and *Cybernetics of Cybernetics* (Future Systems, 1995). The latter was also the name of a talk given at the University of Illinois, Urbana.

43. Kelly's blithe dismissal of the self-referential moment in cybernetics is highly problematic on a number of counts. For a corrective view, see N. Katherine Hayles *How We Became Posthuman* (Chicago: University of Chicago Press, 1999), chap. 6, "The Second Wave of Cybernetics: From Reflexivity to Self-Organization."

44. Von Neumann's paper, "First Draft of a Report on the EDVAC" (first circulated in 1945), lays out the first detailed diagram of the architecture of the modern digital computer. Although von Neumann does not claim credit for the idea of the "stored program" (usually given to Presper Eckert and John Mauchly, whose team was working on ENIAC when von Neumann joined them), he did develop a formal design that would enable it to work. The design called for an arithmetic unit for performing calculations, a control unit, a random access memory where both data and instructions were stored, and input and output devices.

For further details, see Herman H. Goldstine, *The Computer from Pascal to von Neumann* (Princeton: Princeton University Press, 1972), 191ff.

45. A. Newell and H. A. Simon, "Computers in Psychology," in *Handbook of Mathematical Psychology*, ed. R. D. Luce, R. R. Bush, and E. Galanter, vol. 1 (New York: Wiley, 1963), 385–386.

46. See Claude E. Shannon, "Programming a Computer for Playing Chess," *Philosophy Magazine* 41 (March 1950), 356–375. For an account of Samuel's checkers program see Daniel Crevier, *AI: The Tumultuous History for the Search for Artificial Intelligence* (New York: Basic Books, 1993), 220–222.

47. In fact, in "The Bandwagon," *IEEE Transactions on Information Theory* 2, no. 3 (1956), Shannon expressed serious reservations about the widespread extension of information theory to other disciplines.

48. In these terms one could argue that early AI was at once a progression and a regression. (In my discussion in chapter 6, I prefer Deleuze and Guattari's terms "decoding" and "recoding.") In any event, the Cartesian duality can be avoided by understanding the matter-symbol distinction in the terms provided by Howard Pattee, i.e., that physical processes are determined by the laws of physics and structures of control are determined by a semiotic or symbolic "syntax," where neither is reducible to the other and both are necessary for a complete description of a living organism. But whereas Pattee develops his theory in terms of the relationship of physics and biology, I understand these complementary aspects as defining the nature of an assemblage, a term I also adopt from Deleuze and Guattari. See Pattee, "Evolving self-reference: matter, symbols and semantic closure" in *Communication and Cognition—Artificial Intelligence* 12, nos. 1–2 (1995): 9–27, as well as chapter 3.

49. Valentino Braitenberg, *Vehicles: Experiments in Synthetic Psychology* (Cambridge, Mass.: MIT Press, 1984), 2.

50. Rendered by the artist Maciek Albrecht, the drawings "illustrate only a few of the many marvelous 'creatures' inspired by Valentino Braitenberg's text" (84).

Chapter 2

1. Quoted by Steve Heims, *The Cybernetics Group* (Cambridge, Mass.: MIT Press, 1991), 122.

2. The essay first appeared in the *Bulletin of Mathematical Physics* 5 (1943): 115–133, and was republished in McCulloch's *Embodiments of Mind* (Cambridge, Mass.: MIT Press, 1965), 19–39.

3. Lawrence Kubie, "Repetitive Core of Neuroses," *Psychoanalytic Quarterly* 10, no. 23 (1941): 23–43.

4. Warren McCulloch, "The Past of a Delusion," in *Embodiments of Mind*, 276–306.

5. All references are to *The Seminar of Jacques Lacan: Book II, The Ego in Freud's Theory and in the Technique of Psychoanalysis 1954–1955*, trans. Sylvana Tomaselli (New York: W. W. Norton, 1988). Occasionally phrases are taken from the original French edition, *Le séminaire livre II: Le moi dans la théorie de Freud et dans la technique de la psychanalyse* (Paris: Éditions du Seuil, 1978).

6. Lacan alludes to this work, which was discussed at the 9th Macy Conference, in the second seminar, p. 89.

7. Elisabeth Roudinesco, *Jacques Lacan*, trans. Barbara Bray (New York: Columbia University Press, 1997), 305. Strangely, Roudinesco makes no mention of cybernetics or information theory in her otherwise extensive biography.

8. See the interview with Heidegger in *Der Spiegel* (September 1966) as well as Heidegger's *The End of Philosophy* (Chicago: University of Chicago Press, 2003). Heidegger's anathema toward modern science and technology is particularly evident in "The Question Concerning Technology," first published in *Vorträge und Aufsätze* in 1954. It was while writing the first

version of this essay for the Club of Bremen lectures in 1949 that Heidegger became aware of cybernetics and information theory. With the advent of cybernetics, he came to think, Western philosophy comes to an end (in calculation) though its end also coincides or could coincide with the beginning of a new era of thinking.

9. The lecture, discussed below, is published with the seminar (294–308).

10. The term first appears on page 31 in the English translation.

11. A. M. Turing, "On Computable Numbers, with an Application to the *Entscheidungsproblem*," in *Proceedings of the London Mathematical Society*, 2nd ser., 42 (1936): 230–265; reprinted in *The Essential Turing*, ed. B. Jack Copeland (Oxford: Oxford University Press, 2004), 58–90.

12. See Marvin L. Minsky, *Computation: Finite and Infinite Machines* (Englewood Cliffs, N.J.: Prentice-Hall, 1967), for a discussion of the difference between finite-state machines and Turing machines.

13. I refer here to George Boole's *The Laws of Thought*, an influential book on logic and probability first published in England in 1854.

14. There can be no doubt that Lacan was thinking specifically of Grey Walter's cybernetic tortoises (discussed in chapter 1). Walter described them in his two popular articles published in *Scientific American* (in May 1950 and August 1951), and they were discussed extensively by Pierre de Latil in his *La pensée artificielle*, which Lacan knew. The behavior of these machines suggests very clearly that they were capable of "the mirror stage."

15. Jacques Lacan, *Écrits* (Paris: Editions du Seuil, 1966), 11–61.

16. It is also important to note Lacan's familiarity with game theory, which he alludes to in the seminar. In later writings he also considers several instances of Prisoner's Dilemma. Though it is difficult to gauge its specific importance, in all likelihood Lacan was familiar with John von Neumann and Oscar Morgenstern's *Theory of Games and Economic Behavior* (Princeton: Princeton University Press, 1944), which discusses the game of odd and even as a two-person zero-sum game. Prisoner's Dilemma, invented in 1950 by RAND scientists Albert W. Tucker and Melvin Dresher, was widely discussed at the time because it thwarts commonsense reasoning. See William Poundstone's fascinating account in *Prisoner's Dilemma* (New York: Anchor Books, 1992). See Steve Heims, *The Cybernetics Group* (Cambridge, Mass.: MIT Press, 1991), 109–110, and Jean-Pierre Dupuy, *Aux origines des sciences cognitives* (Paris: Editions La Découverte, 1994), 132–133 and passim, for brief accounts of von Neumann's theory and the attacks on it at the Macy Conferences.

17. In a well-known critique of Lacan's reading of Poe's story Jacques Derrida argues that this tripartite structuration of the subject is everywhere exceeded by the proliferation of doubles and that these subject positions are given in advance by the privileging in psychoanalysis of the phallic signifier and the Oedipal triangulation. While Derrida's critique seems cogent on this point, it fails to address the more fundamental issue, which is precisely the emergence of the symbolic through binary coding. See Jacques Derrida, *La carte postale* (Paris: Flammarion, 1980), 441–524.

18. The workings of this unconscious are not random or unstructured but follow a peculiar kind of order or logic, first ferreted out by Freud in *The Interpretation of Dreams* (1900). Basically, Freud showed that the cathexes of unconscious desire are determined by mechanisms of displacement, condensation, isolation, and denial. For Lacan, however, it is commonly assumed that "the unconscious is structured like a language." We shall see below what this actually means and what particular force is leant to this "like." Meanwhile, it should be emphasized that over the course of his writings Lacan offered several different understandings of the relationship of language to the unconscious and that the example of cybernetics enabled him to formulate his first full understanding.

19. In *La technique et le temps*, vol. 2 (Paris: Galilée, 1996), Bernard Stiegler criticizes the ideality of the Turing machine (its infinite tape) and its implicit denial of the subject's temporality (its retentional finitude). A reasonable question (which I cannot address here) is whether Lacan, by distinguishing between the remembering subject and the integrations of a symbolic order that establish a machinic memory, resolves this problem or merely elides it.

20. Friedrich A. Kittler, *Gramophone, Film, Typewriter* (Stanford, Calif.: Stanford University Press, 1999, orig. pub. 1986), 15. Not incidentally, the phrase "nature ... stopped not writing itself" is taken from Lacan's twentieth seminar, published as *Encore* (Paris, Editions du Seuil, 1975), 55.

21. Friedrich Kittler, "The World of the Symbolic—The World of the Machine," in *Literature, Media, Information Systems*, ed. John Johnston (Amsterdam: G + B Arts International, 1997), 134.

22. Friedrich A. Kittler, *Discourse Networks 1800/1900* (Stanford, Calif.: Stanford University Press, 1990), 284.

23. "Discourse network" is the English translation of Kittler's German term *Aufschreibesystem*, a neologism meaning "a notation- or writing down system" that he borrows from Daniel Paul Schreber's *Memoirs of My Nervous Illness* (1903). For discussion of the term as a critical concept, see Kittler's *Discourse Networks 1800/1900* as well as my introduction to *Literature, Media, Information Systems.*

24. Friedrich Kittler, "The World of the Symbolic," 141.

25. Claude E. Shannon, *The Mathematical Theory of Communication* (Urbana: University of Illinois Press, 1963; orig. pub. 1949), 40.

26. G. A. Miller, "Finite Markov Processes in Psychology," *Psychometrika* 17 (1952): 149–167. Significantly, in a few years Miller would become one of the founders of cognitive science.

27. As noted in chapter 1, Ashby shows how the apparently unpredictable behavior of an insect that lives in and about a shallow pond, hopping to and fro among water, bank, and pebble, can be described as a "Markovian machine" in which the state transitions correspond to those of a stochastic transformation (*An Introduction to Cybernetics*, 165–166). While at any particular moment the insect's behavior is determined by seemingly random events in the environment, the probabilities of its movements clearly reveal an overall pattern.

28. Lawrence M. Ward, *Dynamical Cognitive Science* (Cambridge, Mass.: MIT Press, 2002), 14. In the section "Markovian Analysis of Behavior" (13–16) Ward provides a clear example of how "states as categories of behavior, say reading the TV Guide, watching a TV show and getting a snack" can be analyzed in Markovian terms using a simple transition matrix. What is most striking is the clear pattern of return or repetition of specific states.

29. Michael A. Harrison's *Introduction to Formal Language Theory* (Reading, Mass.: Addison-Wesley, 1978) draws attention to the historical context of the late 1940s and 1950s. Formal language theory, Harrison notes, is important not only for the analysis of both natural and programming languages but for models of biological systems.

30. The first two are taken from Lacan's *Écrits* (48 and 57); the third, from John E. Hopcroft and Jeffrey D. Ullman's *Introduction to Automata Theory, Languages, and Computation* (Reading, Mass.: Addison-Wesley, 1979), 16.

31. The question of whether Lacan's later formalisms, specifically what he calls the "algorithm of desire" elaborated in relation to the operations of metaphor and metonymy and the signification of the phallus, can also be analyzed in these terms goes beyond the framework of the present discussion.

32. I refer to Chomsky's "Three Models for the Description of Language," *IRE Transactions of Information Theory*, 1956, IT-2, 113–124. Republished in *Readings in Mathematical Psychology*, vol. 2, ed. R. Duncan Luce et al. (New York: John Wiley and Sons, 1965), 105–124.

33. Noam Chomsky, *Syntactic Structures* (The Hague: Mouton, 1957), 18–19. It should be clear that the machine to which Chomsky refers is a finite-state automaton.

34. Chomsky, "On Certain Formal Properties of Grammars," *Information & Control* 1 (1959), also republished in *Readings in Mathematical Psychology*, 2:125–155. The distinctions separating these levels of languages and machines are quite technical, mostly having to do with how symbol strings are parsed and accessed in memory.

35. These examples are taken from Richard Y. Kain's *Automata Theory: Machines and Languages* (New York: McGraw-Hill, 1972), 38–39.

36. It was Claude Shannon, in his master's degree at MIT (1937), who first demonstrated that arrays of on-off switches could instantiate and evaluate the logical propositions formulated by George Boole in *Laws of Thought* (1854), as well as perform basic mathematical operations. For further details, see Daniel Crevier, *AI: The Tumultuous Search for Artificial Intelligence* (New York: Basic Books, 1993), 11–18.

37. The essay first appeared in *Mind* 59 (1950): 433–460. Underlying the test is an operational notion of intelligence. Thus, an artificial intelligence would be judged according to its capacity to imitate a putatively human response to questions presented to it by teletype. In the first version of what Turing called the "imitation game," a man would attempt to pass himself off as a woman by answering appropriately; in the second version a machine would take his place. The idea of intelligence as successful simulation clearly stems from Turing's earlier thesis on computation, according to which a universal Turing machine can simulate the operations of any particular Turing machine. Perhaps what most strikes contemporary readers is the double displacement of woman (first by a man, then by a machine). For an analysis of the "imitation game" from a feminist perspective, see Warren Sack, "Replaying Turing's Imitation Game," at http://web.media.mit.edu/~wsack/console-ing.html.

38. Allen Newell and Herbert A. Simon, "Computer Science as Empirical Inquiry: Symbols and Search," in *Mind Design*, ed. John Haugeland (Cambridge, Mass.: MIT Press, 1981), 35–66. The physical symbol system hypothesis is discussed at greater length in chapter 6.

39. As shown in chapter 6, another way of understanding natural human language arises in neural net and dynamical systems theory. They are alternative in that they provide a different account of the rule-bound nature of language.

40. Although Lacan does not pursue the question of the origins of the symbolic order, there seem to be two possibilities consistent with what he says here and in the seminar: either the symbolic, like language, arises because of a linguistic or syntactic capacity hardwired into the brain (Chomsky's position), or it arises from the evolution of a primitive but self-organizing counting or notation system. In the 1980s, with the revival of neural net theory in connectionism, alternatives to the Chomsky-influenced "language of thought" hypothesis (as formulated by Jerry Fodor) were proposed. More recently, rejecting the connectionist alternative, Steven Pinker has sought to rehabilitate this hypothesis by supplying the missing and crucial evolutionary perspective that would answer the question, How could the human capacity for syntax (or computation) have come to be an innate or hard-wired feature of the brain? See Pinker, *How the Mind Works* (New York: W. W. Norton, 1997), chaps. 2 and 3.

41. Whereas Friedrich Kittler consolidates these designations into a single order—i.e., "[the real] forms that residue or waste which can be caught neither in the mirror of the imaginary nor in the grids of the symbolic: physiological accident, stochastic disorder of bodies" (*Literature, Media, Information Systems*, 45), the Lacanian theorist Slavoj Žižek privileges the first, making the encounter with the real the originating kernel of individual and collective phantasy. See *Looking Awry: An Introduction to Jacques Lacan through Popular Culture* (Cambridge, Mass.: MIT Press, 1991).

42. Quoted by Bruce Weber, "Computer Defeats Kasparov, Stunning the Chess Experts," *New York Times*, May 5, 1997.

43. Quoted by Steven Levy, "Big Blue's Hand of God," *Newsweek*, May 19, 1997.

44. This quotation and the following are taken from Bruce Weber, "Swift and Slashing, Computer Topples Kasparov," *New York Times*, May 12, 1997.

45. This may be the best place to mention Gilbert Simondon, another French thinker influenced by cybernetics, who also demands a different understanding of the human-machine relationship. In *Du mode d'existence des objets techniques* (Paris: Aubier, 1958), published only three years after Lacan's seminar, Simondon calls for "a new consciousness of the sense of technical objects.... Culture has constituted itself as a system of defense against technics; yet, this defense is presented as a defense of humanity, supposing that technical objects do not contain human reality" (9).

46. On Wiener's limits, see N. Katherine Hayles, "Liberal Subjectivity Imperiled: Norbert Wiener and Cybernetic Anxiety," in *How We Became Posthuman* (Chicago: University of Chicago Press, 1999), 84–112.

47. This schema, developed by von Neumann in his *Theory of Self-Reproducing Automata* (Urbana: University of Illinois Press, 1966), is discussed in chapter 4.

48. Republished in *Multimedia: From Wagner to Virtual Reality*, ed. Randall Packer and Ken Jordan (New York: W. W. Norton, 2001), 57–63.

49. Claude Shannon, "A Chess-Playing Machine," *Scientific American*, February 1950, 48.

50. Rodney A. Brooks, "Elephants Don't Play Chess," *Robotics and Autonomous Systems* 6 (1990): 3–15.

Chapter 3

1. Originally published as *Anti-Oedipe* (Paris: Minuit, 1972) and *Mille plateaux* (Paris: Minuit, 1980). Except where noted, references are to *Anti-Oedipus* and *A Thousand Plateaus*, the English translations published by the University of Minnesota Press in 1983 and 1987, respectively.

2. Manual DeLanda, *War in the Age of Intelligent Machines* (New York: Zone Books, 1991) and *A Thousand Years of Nonlinear History* (New York: Zone Books, 1997).

3. David Campbell, Jim Crutchfield, Doyne Farmer, and Erica Jen, "Experimental Mathematics: The Role of Computation in Nonlinear Science," *Communications of the ACM* 28, no. 4 (April 1985): 374–384. For mathematically detailed surveys, see Steven Strogatz, *Nonlinear Dynamics and Chaos* (Reading, Mass.: Addison-Wesley, 1994), and Daniel Kaplan and Leon Glass, *Understanding Nonlinear Dynamics* (New York: Springer, 1995).

4. See N. Katherine Hayles, *Chaos Bound* (Ithaca, N.Y.: Cornell University Press, 1990), for a discussion of the putative differences between the two strands.

5. Ilya Prigogine and Isabel Stengers, *Order out of Chaos* (New York: Bantam, 1984), xvi. The book was originally published in French in 1979 under the title *La nouvelle alliance.*

6. James Gleick, *Chaos: Making a New Science* (New York: Viking, 1987).

7. The central texts were published in various science journals; many are available in Predrag Cvitanovic's *Universality in Chaos: A Reprint Selection* (New York: Adam Hilger, 1989).

8. DeLanda, *War in the Age of Intelligent Machines*, 234–237. Deleuze discusses singularities in chapter 15 of *The Logic of Sense* (New York: Columbia University Press, 1990; orig. pub. 1969), but I think his use of this mathematical concept is more clearly revealed in his earlier book, *Difference and Repetition* (New York: Columbia University Press, 1995; orig. pub. 1968). Deleuze takes the term from Albert Lautman's *Le problème du temps* (Paris: Hermann, 1946), where it designates a point that remains constant throughout a series of topological transformations. Even though its value remains unknown or indeterminate, it thus defines a particular "problem-field." For Deleuze, following Gilbert Simondon's usage in *L'individu et sa genèse physico-biologique* (Paris: Presses Universitaires de France, 1964), the term designates a turning point or sudden inflection, and thus a radical change of state. In this perspective Deleuze conceives of the "event" as a novel (re)distribution of singularities and of "individuation" as the actualization of singularities that are only latent or virtual.

9. This feature has made the book important for theories of hypertext in studies of electronic literature. See Stuart Moulthrop, "Rhizome and Resistance: Hypertext and the Dreams of a New Culture," in *Hyper/Text/Theory*, ed. George P. Landow (Baltimore: Johns Hopkins University Press, 1994), 299–319, as well as my essay "Are Rhizomes Scale-Free? Network Theory and Contemporary American Fiction," in *The Holodeck in the Garden: Science and Technology in Contemporary American Fiction*, ed. Peter Freese and Charles B. Harris (Normal, Ill.: Dalkey Archive Press, 2004), 53–71.

10. Gilles Deleuze and Claire Parnet, *Dialogues* (New York: Columbia University Press, 1987; orig. pub. 1977), 70.

11. Louis Hjelmslev, *Prolegomena to a Theory of Language* (Madison: University of Wisconsin Press, 1969). Hjelmslev offers an alternative to the structuralist linguistic theory based on Ferdinand de Saussure's signifier-signified relationship, which Deleuze and Guattari attack in *Anti-Oedipus.*

12. *Kafka: Toward a Minor Literature* (Minneapolis: University of Minnesota Press, 1986; orig. pub. 1975), 19.

13. Henri Bergson, *Creative Evolution* (New York: Random House, 1944; orig. pub. 1907 as *L'évolution créatrice*).

14. For a similar conception, see Jesper Hoffmeyer and Claus Emmeche, "Code-Duality and the Semiotics of Nature," in *On Semiotic Modeling* (Berlin and New York: Mouton de Gruyter, 1991), 117–166.

15. H. H. Pattee, "How Does a Molecule Become a Message?" *Developmental Biology*, suppl. 3 (1969): 1.

16. H. H. Pattee, "Evolving Self-Reference: Matter, Symbols and Semantic Closure," *Communication and Cognition—Artificial Intelligence* 12, nos. 1–2 (1995): 9–27. As discussed in chapters 5, this principle becomes crucial in one critique of ALife.

17. H. H. Pattee, "Cell Psychology: An Evolutionary Approach to the Symbol-Matter Problem," *Cognition and Brain Theory* 5, no. 4 (1982): 325–341.

18. Richard Dawkins, *The Selfish Gene* (Oxford: Oxford University Press, 1976, 1989).

19. See R. Abraham and C. Shaw, *Dynamics, the Geometry of Behavior*, 2nd ed. (Reading, Mass.: Addison-Wesley, 1992), for a nonmathematical pictorial exposition of the dynamical systems model.

20. See Strogatz, *Nonlinear Dynamics and Chaos*, and Kaplan and Glass, *Understanding Nonlinear Dynamics*, for specific examples.

21. Their discussion of the transformations that lead from one "regime of signs" to another (see *A Thousand Plateaus*, 111–148) may be the sole exception.

22. The "Royal McBee" was a modestly priced ($40,000) stored program, vacuum-tube computer first produced in 1954 for use in laboratories and engineering facilities.

23. Edward N. Lorenz, *The Essence of Chaos* (Seattle: University of Washington Press, 1993), 14. As Lorenz himself recounts, the term *butterfly effect* has a "cloudy history," possibly originating in the title of one of his lectures, "Does the Flap of a Butterfly's Wings in Brazil Set Off a Tornado in Texas?" or possibly deriving from the shape of the attractor that now bears his name and that explains in mathematical terms the behavior of a weather system.

24. It was published in the *Journal of Atmospheric Sciences* 20 (1963): 130–141.

25. *Mathematical Intelligencer* 11, no. 1 (1989): 66.

26. Stephen H. Kellert, *In the Wake of Chaos* (Chicago: University of Chicago Press, 1993), 128–134.

27. See James P. Crutchfield, J. Doyne Farmer, Norman H. Packard, and Robert Shaw, "Chaos," *Scientific American*, December 1986, 46–57.

28. In particular, see N. H. Packard, J. P. Crutchfield, J. D. Farmer, and R. S. Shaw, "Geometry from a Time Series," *Physical Review Letters* 45, no. 9 (September 1, 1980): 712–716.

29. See, for example, James Crutchfield, "Space-Time Dynamics in Video Feedback," *Physica D* 10 (1984): 229–245.

30. Ross Ashby was certainly among them. As he acknowledges, Shannon's theory was essential to his discussion of the stochastic, or nondeterminate, machines he calls Markovian machines in *An Introduction to Cybernetics* (1956). However, according to Garnett P. Williams, *Chaos Theory Tamed* (Washington, D.C.: Joseph Henry Press, 1997), 390, the

Russian mathematician A. N. Kolmogorov was the first to apply Shannon's theory as a whole to dynamical systems, in a paper published in 1959. As Williams points out, Shannon's "information entropy ... cannot by itself identify chaos" and therefore requires further elaboration.

31. The diagram and subsequent quotations are taken from Claude Shannon and Warren Weaver, *The Mathematical Theory of Communication* (Urbana: University of Illinois Press, 1963).

32. On these and other questions, see *Maxwell's Demon: Entropy, Information, Computing*, ed. Harvey S. Leff and Andrew F. Rex (Princeton: Princeton University Press, 1990), especially the essays by Charles Bennett and Rolf Landauer.

33. For details, see Thomas M. Cover and Joy A. Thomas, *Elements of Information Theory* (New York: John Wiley & Sons, 1991).

34. Laplace's argument for complete predictability in his *Treatise on Celestial Mechanics* (1799–1825) is of course based on the application of Newton's laws. He also discusses determinism in the second chapter of his *A Philosophical Essay on Probabilities*. It was originally intended to serve as an introduction to his *Analytical Theory of Probability*, which became a landmark in the field. Laplace devoted so much attention to probability precisely because—as an iron-clad determinist—he recognized that human beings lacked the "vast intelligence" necessary for grasping the complete predictability of everything. This was, in fact, the classical position: indeterminacy appeared only *because of* a lack in human knowledge.

35. Florin Diacu and Philip Holmes, *Celestial Encounters* (Princeton: Princeton University Press, 1996), 29–43.

36. Henri Poincaré, *The Foundation of Science: Science and Method* (Lancaster, Pa.: Science Press, 1946; orig. pub. 1903), 397.

37. Robert Shaw, "Strange Attractors, Chaotic Behavior, and Information Flow," *Zeitschrift für Naturforschung*, Teil A, Band 36a (1981): 80–112.

38. Gleick: "The structures that provided the key to nonlinear dynamics proved to be fractal" (114); also see 139–141. The term "fractal" was coined by the French mathematician Benoit Mandelbrot, who was the first to recognize and study this strange new "geometry of nature," as he called it. In 1977 he presented his theory of fractals in *Les objets fractals: Forme, hasard et dimension* (translated into English as *Fractals: Form, Chance and Dimension*). It was expanded and revised as *The Fractal Geometry of Nature* (New York: W. H. Freeman, 1982).

39. That highly dissipative, chaotic physical systems provide direct evidence of time irreversibility is of course Ilya Prigogine's major thesis, argued in great detail in *Order out of Chaos*. It should be apparent that Shaw's work thus bridges the putative gap between "the two strands" of chaos theory.

40. Robert Shaw, *The Dripping Faucet as a Model Chaotic System* (Santa Cruz: Aerial Press, 1984).

41. More precisely, physical behaviors describable by linear equations are the exception rather the rule. As a consequence, nonlinear science and the computers that make nonlinear phenomena accessible to study assume a much greater importance. On this issue, see Campbell, Crutchfield, Farmer, and Jen, "Experimental Mathematics: The Role of Computation in Nonlinear Science."

42. A. N. Kolmogorov extended Shannon's theory by proposing a method to calculate information entropy's rate of change. Y. Sinai refined the definition and proof. For details, see Garnett P. Williams, *Chaos Theory Tamed*, chap. 26.

43. Again, it is the same as mutual information. In terms of entropies, mutual information equals receiver entropy minus the receiver's conditional entropy, or $I(x'/x) = H(x') - H(x'/x)$.

44. It is worth recalling here Walter J. Ong's remark in *Orality and Literacy* (London: Methuen, 1982): "What is distinctive of modern science is the conjuncture of exact observation and exact verbalization: exactly worded descriptions of carefully observed complex

objects and practices" (127). This suggests that modern science itself is a specific kind of assemblage: a conjunction of the observation of bodies and their mathematically quantified relations.

45. James P. Crutchfield, "Reconstructing Language Hierarchies," in *Information Dynamics*, ed. Harald Atmanspacher and Herbert Scheingraber (New York: Plenum Press, 1991), 45.

46. Crutchfield and Karl Young, "Computation at the Onset of Chaos," in *Complexity, Entropy and the Physics of Information*, ed. Wojciech H. Zurek (Redwood City, Calif.: Addison-Wesley, 1990), 230.

47. Crutchfield discusses these necessary extensions in "The Calculi of Emergence," *Physica D* (1994): 18–19.

48. In "The Calculi of Emergence," Crutchfield notes that a duality "between languages as sets and automata as functions which recognize sets, runs throughout computation theory" (18).

49. Compare Crutchfield: "When investigating nonlinear processes, one concludes that the existence of chaotic, deterministic behavior precludes the detailed comparison of theoretical models to experimental data. The conventional picture of inexorable improvement of models only applies to non-chaotic behavior." J. P. Crutchfield and B. McNamara, "Equations of Motion from a Data Series," *Complex Systems* 1 (1987): 417.

50. See "The Period-Doubling Route to Chaos" in Kaplan and Glass, *Understanding Nonlinear Dynamics*, 29–33, and Strogatz, *Nonlinear Dynamics and Chaos*, 353–357.

51. Crutchfield provides an illustration of a Bernoulli-Turing machine in "Calculi of Emergence," 24.

52. A. N. Kolmogorov, "Three Approaches to the Concept of the Amount of Information," *Problems of Information Transmission* 1, no. 1 (1965): 4–7. Algorithmic complexity is measured by the length of the algorithm to which a symbol string or sequence can be compressed. For example, the sequence 0011001100110011 can be compressed to the formula: write 0011 four times. On the other hand, a genuinely random string cannot be compressed at all, only rewritten as given.

53. Crutchfield and Young, "Inferring Statistical Complexity," 105.

54. I consider Crutchfield's later work in evolving cellular automata and evolutionary theory in chapter 5.

Chapter 4

1. The first paper was published in *Cerebral Mechanisms in Behavior: The Hixon Symposium*, ed. Lloyd A. Jeffress (New York: John Wiley and Sons, 1951); the second and third, in John von Neumann, *Theory of Self-Reproducing Automata*, edited and completed by Arthur W. Burks (Urbana: University of Illinois Press, 1966).

2. See the conference proceedings, *Artificial Life*, ed. Christopher G. Langton (Reading, Mass.: Addison-Wesley, 1989).

3. Practitioners and theorists often refer to Artificial Life as ALife, A-life or simply AL.

4. Daniel Dennett, "Artificial Life as Philosophy," in *Artificial Life: An Overview*, ed. Christopher G. Langton (Cambridge, Mass.: MIT Press, 1995), 291.

5. Daniel Dennett, "Cognitive Science as Reverse Engineering," *Brainchildren* (Cambridge, Mass.: MIT Press, 1998), 256.

6. As previously noted, although ALife is one of the new "sciences of the artificial," its relationship to the natural or biological realm is more dynamic and transformative than anything Herbert Simon envisions in *The Sciences of the Artificial* (Cambridge, Mass.: MIT Press, 1969). Primarily concerned with rationality and design—but also with the organization of complexity—he remains conceptually close to symbolic AI, and relegates biology

and evolutionary process to the margins. In contrast, John Holland's *Adaptation in Natural and Artificial Systems* (Cambridge, Mass.: MIT Press, 1998; orig. pub. 1975) provides not only groundbreaking new tools for ALife—most notably the genetic algorithm—but an approach more conducive to ALife's founding assumptions.

7. Immanuel Kant, *The Critique of Judgment*, trans. Werner S. Pluhar (Indianapolis: Hackett, 1987), 253.

8. Kevin Kelly, *Out of Control: The Rise of Neo-biological Civilization* (Reading, Mass.: Addison-Wesley, 1994). Kelly describes his book as "an update on the current state of cybernetic research" (453) but doesn't offer much historical perspective on the new blurring of boundaries he details.

9. I draw here on Arthur W. Burks's account in *Essays on Cellular Automata*, ed. Arthur C. Burks (Urbana: University of Illinois Press, 1970), 3–83, as well as von Neumann's *Theory of Self-Reproducing Automata*.

10. Following von Neumann's work, in the late 1960s the Princeton mathematician John Conway invented a particular cellular automaton he called the Game of Life, which made this complexity directly visible. The game was popularized by Martin Gardner's "Mathematical Games: The Fantastical Combinations of John Conway's New Solitaire Game 'Life,'" *Scientific American*, October 1970, 112–117. For a fascinating account of the game and its scientific ramifications, see William Poundstone's *The Recursive Universe* (Chicago: Contemporary Books, 1985).

11. See Christopher G. Langton, "Self-Reproduction in Cellular Automata," *Physica D* 10, nos. 1–2 (1984): 135–144.

12. *Cellular Automata: Proceedings of an Interdisciplinary Workshop, Los Alamos March 7–11, 1983*, ed. Doyne Farmer et al. (Amsterdam: North Holland, 1984), vii–viii. See also *Theory and Applications of Cellular Automata*, ed. Stephen Wolfram (Singapore: World Scientific, 1986), as well as Tommaso Toffoli and Norman Margolus's influential *Cellular Automata Machines: A New Environment for Modeling* (Cambridge, Mass.: MIT Press, 1987). For a more up-to-date survey, see Andrew Ilachinski, *Cellular Automata: A Discrete Universe* (Singapore: World Scientific, 2001).

13. Christopher G. Langton, "Artificial Life," in *Artificial Life*, 41.

14. Langton's idea that life comprehends a *virtual* dimension because it arises from nonlinear interactions among many physical parts has never been adequately emphasized. The best discussion of the implications, at least to my knowledge, is Steve Grand's *Creation: Life and How to Make It* (Cambridge, Mass.: Harvard University Press, 2001). I consider Grand's own important work in ALife and robotics in chapter 8.

15. Langton, preface to *Artificial Life*, xv.

16. In *Artificial Life: A Report from the Frontier Where Computers Meet Biology* (New York: Vintage Books, 1992), Steven Levy calls von Neumann the "father" and Langton the "midwife" (93). The most detailed historical account of the development of Artificial Life up to 1992, Levy's book is the source of the biographical details that follow.

17. "Artificial Life," in *Artificial Life*, 1–47. The citations that follow are taken from the extended version reprinted in *The Philosophy of Artificial Life*, ed. Margaret A. Boden (Oxford: Oxford University Press, 1996). In a few instances, however, I refer to the original version published in 1989. In those cases the date is followed by the page number.

18. Early neural net research was largely abandoned when Marvin Minsky and Seymour Papert showed that these networks could not perform the XOR computational function. In the mid-1980s it reappeared under the name of connectionism and again became a viable alternative to symbolic AI when the addition of another layer of neurons and other discoveries nullified Minsky and Papert's demonstration. I discuss these developments in chapter 6.

19. Of course, Langton was only one of many whose work with CA gave him a new sense of the limits of traditional computers and programs. In their *Cellular Automata Machines*, Toffoli and Margolus make a similar point in regard to simulation: "In this context ordinary

computers are of no use.... On the other hand, the structure of the cellular automaton is ideally suited for realization on a machine having a high degree of parallelism and local and uniform connections" (8). Furthermore, Daniel Hillis, who designed one of the world's largest parallel-processing computers, was profoundly influenced by CA research, as he reveals in "The Connection Machine: A Computer Architecture Based on Cellular Automata," a paper he presented at the 1983 conference on CA.

20. See the description of the Swarm Simulation System at http://www.swarm.org.

21. For example, Steven Johnson's *Emergence: The Connected Lives of Ants, Brains, Cities and Software* (New York: Scribner, 2001) fails to acknowledge Langton's essential contribution.

22. Here Langton anticipates the development of "artificial chemistries," which became an important thread in ALife research. For an early and important example, see Walter Fontana, "Algorithmic Chemistry," in *Artificial Life II*, ed. Christopher G. Langton (Reading, Mass.: Addison-Wesley, 1992), 159–200; for a theoretical overview, see Peter Dittrich et al., "Artificial Chemistries—A Review," *Artificial Life* 7, no. 3 (2001): 225–275.

23. The three conferences held at Los Alamos—on cellular automata (1983), "Evolution, Games and Learning: Models for Adaptation in Machines and Nature" (1986), and "Emergent Computation" (1990)—discussed in the introduction, further in this chapter, and in chapter 5 were instrumental in the development of a number of different computational assemblages as well as for the theory that made them possible.

24. See, in particular, Langton, "Self-Reproduction in Cellular Automata," *Physica D* 10 (1984): 135–144; and "Studying Artificial Life with Cellular Automata," *Physica D* 22 (1986): 120–149.

25. In these terms Dennett's notion of Artificial Life as "bottom-up reverse engineering" cited at the chapter's outset is somewhat misleading.

26. In computer science a recursive function is one that calls itself either directly or through another function. Recursion is often used to solve complex problems by breaking the problem into a set of simpler problems, which are then solved by applying the same method over and over again.

27. See Langton, "Self-Reproduction in Cellular Automata," *Physica D* 10, nos. 1–2 (1984): 135–144, for more details.

28. In *Creation: Life and How to Make It*, Steve Grand usefully distinguishes between first- and second-order simulations (79). A boid is a first-order simulation; flocking, a second-order simulation. See chapter 8 for further discussion.

29. See Holland, *Adaptation in Natural and Artificial Systems*, as well as Melanie Mitchell, *An Introduction to Genetic Algorithms* (Cambridge, Mass.: MIT Press, 1996).

30. Using this method, he would later produce computer programs that spontaneously emerge, self-replicate, and evolve self-improving versions. See John R. Koza, "Artificial Life: Spontaneous Emergence of Self-Replicating and Evolutionary Self-Improving Computer Programs," in *Artificial Life III*, ed. Christopher G. Langton (Reading, Mass.: Addison-Wesley, 1994), 225–262.

31. The essay was published in *Artificial Life II* (41–91) and covers some of the material Langton presented in his doctoral dissertation, "Computation at the Edge of Chaos: Phase-Transition and Emergent Computation" (University of Michigan, 1991).

32. As an example from popular culture, consider this description of the stock market by the main character (a mathematician) in Darren Aronofsky's film π (pi): "The universe of numbers that represents the global economy, millions of human hands at work, billions of minds, a vast network, screaming with life, an organism, a natural organism." Needless to point out, this vast network of hands, minds, and numbers also constitutes a striking computational assemblage.

33. The phrase "programmable matter," as Langton acknowledges, is taken from T. Toffoli and N. Margolus, *Cellular Automata Machines* (Cambridge, Mass.: MIT Press, 1987).

34. Tommaso Toffoli and Norman Margolus, "Programmable Matter: Concepts and Realization," *Physica D* 47 (1991): 263–272.

35. See Langton, "Studying Artificial Life with Cellular Automata," for a more detailed description of these changes.

36. A first-order phase transition usually refers to an abrupt jump from one state or kind of behavior to another, whereas a second-order, or critical, phase transition denotes a smoother transition with much longer transient times.

37. Langton cites Jim Crutchfield's work on "Computation at the Onset of Chaos" (discussed in chapter 3) as providing further evidence for this claim.

38. For example, challenging the information theoretical model in this and other respects is a central aspect of Maturana and Varela's theory of autopoiesis, discussed further below.

39. Langton's "Computation at the Edge of Chaos: Phase Transitions and Emergent Computation," *Physica D* 42 (1990): 12–37, treats the informational process more explicitly.

40. Langton explains the elimination of gases from consideration as follows: "As it is possible to continuously transform liquids into gases and vice-versa without passing through a phase transition, they are taken to constitute a single, more general phase of matter: fluids" (83).

41. As discussed in chapter 5, the biologist Stuart Kauffman essentially agrees, having found supporting evidence in his own work with gene networks and autocatalytic sets. Kauffman believes that life flourishes at the edge of chaos because evolution takes it there, thereby providing the conditions for further evolution.

42. The book was published in English in 1980, together with Maturana's earlier book, *The Biology of Cognition*, under the collective title *Autopoiesis and Cognition: The Realization of the Living* (Dordrecht, Holland: D. Reidel, 1980).

43. I refer to the revised and expanded French version, *Autonomie et connaissance: Essai sur le vivant* (Paris: Éditions du Seuil, 1989). The prior English version was first published by North Holland Press in 1979.

44. Having picked up the idea from von Neumann, Varela worked with cellular automata in the early 1970s, long before "the Artificial Life wave hit the beach," as he puts it in the "The Emergent Self," in *The Third Culture*, ed. John Brockman (New York: Touchstone Books, 1995), 211.

45. The conference papers were published in an influential volume entitled *Toward a Practice of Autonomous Systems*, ed. Francisco J. Varela and Paul Bourgine (Cambridge, Mass.: MIT Press, 1992).

46. Of course, this is true for many biologists; Richard Dawkins, for example, has been a vocal proponent of defining life explicitly in terms of information machines. See *The Blind Watchmaker* (New York: Norton, 1996), 112.

47. Francisco Varela, "Heinz von Foerster, the Scientist, the Man," *Stanford Humanities Review* 4, no. 2 (1995).

48. See *Cahiers du CREA*, no. 8 (November 1985), which is devoted to "Genealogies of Self-Organization," as well as von Foerster's "On Self-Organizing Systems and Their Environment," in *Observing Systems* (Salinas, Calif.: Intersystems Publications, 1981). The essay was first published in 1960.

49. This research began in collaboration with J. Y. Lettvin, W. S. McCulloch, and W. H. Pitts, and is summarized in "What the Frog's Eye Tells the Frog's Brain," in McCulloch's *Embodiments of Mind* (Cambridge, Mass.: MIT Press, 1989; orig. pub. 1965).

50. See the section on "The Nervous System as a Closed System" in *Autopoiesis*, 127–128.

51. As discussed further in chapter 6, this expansion of cognition to include most aspects of the living system is essential to Varela's notion of enaction and his revisionary view of cognitive science.

52. Maturana explains in the introduction that the book grew out of his former student Varela's demand for a more formal definition of circular organization (xvii); they also agreed that *autonomy* should replace *circular organization.*

53. The essay appears in his book *Chaosmose* (Paris: Galilée, 1992). Quotations are taken from *Chaosmosis*, trans. Paul Bains and Julian Pefanis (Bloomington: Indiana University Press, 1995), 39–40.

54. For further examples, see Maturana and Varela's *The Tree of Knowledge* (Boston: Shambhala Publications, 1987).

55. This diagram of the CA model is found in Varela, Maturana, and R. Uribe, "Autopoiesis: The Organization of Living Systems, Its Characterization and a Model," *BioSystems* 5, no. 4 (May 1974): 187–196.

56. Barry McMullin and Francisco J. Varela, "Rediscovering Computational Autopoiesis," in *Fourth European Conference on Artificial Life*, ed. Phil Husbands and Inman Harvey (Cambridge, Mass.: MIT Press, 1997), 38–47.

57. See Langton, "Self-Reproduction in Cellular Automata," for a full account.

58. *Toward a Practice of Autonomous Systems*, xi. See also chapters 6 and 7, where the influence of this orientation on contemporary robotics is discussed.

59. See chapter 6 for a detailed discussion of the shift from symbolic computation to neural net modeling.

60. This is more apparent in his book, *Invitation aux sciences cognitives* (Paris: Editions du Seuil, 1996; orig. pub. 1988), where Varela suggests that enaction supersedes emergence and the computational bias of connectionism.

61. For details, see *Invitation aux sciences cognitives* and chapter 6. In brief, Varela argues that with his own theory of enaction cognitive science arrives at a fully noncomputational and therefore fully acceptable model.

62. Quoted by Stefan Helmreich in *Silicon Second Nature* (Berkeley: University of California Press, 1998), 225.

63. See, for example, Claus Emmeche, "Life as an Abstract Phenomenon: Is Artificial Life Possible?" in *Toward a Practice of Autonomous Systems*, 466–474, and Alvaro Moreno et al., "Universality without Matter?," in *Artificial Life IV*, ed. Rodney Brooks and Pattie Maes (Cambridge, Mass.: MIT Press, 1994), 406–410.

64. See Walter Fontana et al., "Beyond Digital Naturalism," in *Artificial Life: An Overview*, ed. Christopher G. Langton (Cambridge, Mass.: MIT Press, 1995), 211–227, and "Algorithmic Chemistry," in *Artificial Life II*, 159–200.

65. Christopher Langton, introduction to *Artificial Life II*, 5–6.

66. In the context of ALife the best known list of life's essential features is to be found in J. Doyne Farmer and Alletta d'A. Belin, "Artificial Life: The Coming Evolution," in *Artificial Life II*, 815–837. See also Mark Bedau's review of the attempts to define life in "The Nature of Life," in *The Philosophy of Artificial Life*, 332–357.

67. While it is true that infants share the mother's immune system through her milk in the early stages of human and mammalian life, the idea of "completeness" in relation to the immune system is misleading.

68. See Francisco J. Varela and Mark R. Anspach, "The Body Thinks: The Immune System in the Process of Somatic Individuation," in *Materialities of Communication*, ed. Hans Ulrich Gumbrecht and K. Ludwig Pfeiffer (Stanford, Calif.: Stanford University Press, 1994), 273–285.

69. Francisco J. Varela, Antonio Coutinho, Bruno Dupire, and Nelson N. Vaz, "Cognitive Networks: Immune, Neural, and Otherwise," in *Theoretical Immunology, Part Two* (Redwood City, Calif.: Addison-Wesley, 1988), 363.

70. Alan Perelson makes this comparison in "Toward a Realistic Model of the Immune System," in *Theoretical Immunology, Part Two*, 396.

71. The conference was held in 1984 at the Los Alamos National Laboratory. Proceedings were later published in *Physica D* 22 (1986).

72. *Evolution, Games and Learning: Models for Adaptation in Machines and Nature*, ed. Doyne Farmer et al. (Amsterdam: North Holland, 1986), 190. Basically, an antibody molecule consists of two polypeptide chains, each of which is coded for by a slightly different gene library. Thus there is a "huge combinatorial amplification in the number of different antibody types that can be formed from a small number of gene libraries, each containing a limited number of genes."

73. David E. Goldberg, *Genetic Algorithms in Search, Optimization, and Machine Learning* (Reading, Mass.: Addison-Wesley, 1989), 223.

74. See, for example, Eric Baum's "Hayek Machine," which is described in chapter 8.

75. See M. Mitchell Waldrop, *Complexity: The Emerging Science at the Edge of Order and Chaos* (New York: Simon and Schuster, 1992), 238. Waldrop reports that Langton was fearful of the negative associations and did not want to attract hackers to Los Alamos.

76. It was published in *Scientific American*, March 1985, 14–23. In *Artificial Life II*, however, Langton published Eugene Spafford's article "Computer Viruses—A Form of Artificial Life?" While Spafford acknowledges that science has much to learn from studying computer viruses, he is disturbed that their "origin is one of unethical practice" (744).

77. Even the boundaries between science and games are not always clear. Within a few years of its appearance, Dewdney's computer game Core Wars was rewritten and "repurposed" by a small group of scientists led by Steen Rasmussen in order to investigate "the emergence and evolution of cooperative structures in computational chemistry." (See S. Rasmussen, C. Knudsen, R. Feldberg, and M. Hindsholm, "The Coreworld: Emergence and Evolution of Cooperative Structures in a Computational Chemistry," in *Emergent Computation*, ed. Stephanie Forrest [Cambridge, Mass.: MIT Press, 1991], 111–134.) And the influence has worked the other way as well, with ALife research leading to the commercial (and presumably controlled and legal) production of digital "life" in many games and simulations. Notable early instances include SimLife (produced by Maxis in 1993), which was widely hailed as an important commercial prototype for simulating an ecology of dynamically interactive artificial organisms, and Steve Grand's Creatures (1993), which presented a much more sophisticated use of ALife concepts and methodology. While on the surface these various activities are "authorized" according to social and institutional criteria, on a deeper level they could be collectively described as the performative activity of machinic, anorganic life, as it is explored, "worked," and further extended along lines of continuous variation by "silicon probeheads" who resemble the nomadic metallurgists of an earlier era.

78. Mark A. Ludwig, *Computer Viruses, Artificial Life and Evolution* (Tucson, Ariz.: American Eagle Publications, 1993).

79. Jeffrey O. Kephart, "A Biologically Inspired Immune System for Computers," in *Artificial Life IV*, ed. Rodney A. Brooks and Pattie Maes (Cambridge, Mass.: MIT Press, 1994), 130–139.

80. Quoted by Lesley S. King, "Stephanie Forrest: Bushwacking through the Computer Ecosystem," *SFI Bulletin* 15, no. 1 (Spring 2000): 26–29.

81. A. Somayaji, S. Hofmeyr, and S. Forrest, "Principles of a Computer Immune System," in *1997 New Security Paradigms Workshop* (New York: Association for Computing Machinery, 1998), 75–82.

82. In this regard, see the discussion of David Ackley's ideas for a computational system based on principles of "living computation" in chapter 5.

Chapter 5

1. In generating its own strong and weak theories, ALife repeats a pattern evident in AI. In strong AI, a machine or program is understood to exhibit genuine, autonomous intelligence,

whereas weak AI only models, or simulates, intelligence. However, Alan Turing argued that a successful simulation is effectively equivalent to real intelligence. Hence his operational test for AI hinges on our inability to discern the difference. In practice, weak AI usually means the building of software that can perform some special or limited cognitive function, as in expert systems, whereas strong AI seeks to achieve human-level general intelligence.

2. Quoted by Charles E. Taylor, in "'Fleshing Out' Artificial Life II," in *Artificial Life II*, ed. Christopher G. Langton (Reading, Mass.: Addison-Wesley, 1992), 31.

3. Introduction to *Toward a Practice of Autonomous Systems*, ed. Francisco J. Varela and Paul Bourgine (Cambridge, Mass.: MIT Press, 1992), xi.

4. The most notable precursor was Steen Rasmussen and his group's computer simulator VENUS, which was inspired by the computer game Core Wars. In VENUS, stable cooperative structures of code emerged from an artificial prebiotic soup of instructions when processed by a small rule set. However, none of these structures could replicate and evolve. See Rasmussen et al., "The Core World: Emergence and Evolution of Cooperative Structures in a Computational Chemistry," *Physica D* 42 (1990): 111–134.

5. Dennett discusses natural selection as nature's algorithm in *Darwin's Dangerous Idea* (New York: Touchstone, 1995), 48–60.

6. For these biographical details I draw upon Steven Levy's *Artificial Life: The Quest for a New Creation* (New York: Pantheon Books, 1992). Ray himself describes Tierra in detail in "An Approach to the Synthesis of Life," in *Artificial Life II*, 371–401.

7. Thomas S. Ray, "Evolution and Complexity," in *Complexity: Metaphors, Models, and Reality*, ed. George A. Cowan et al. (Reading, Mass.: Addison-Wesley, 1994), 166.

8. Christopher Langton, "Artificial Life," in *Artificial Life*, 91.

9. Langton refers here to Niles Eldredge and Stephen J. Gould's well-known thesis published in their essay "Punctuated Equilibria: An Alternative to Phyletic Gradualism," in *Models of Paleobiology*, ed. T. J. M. Schopf (San Francisco: Freeman Cooper, 1972), 82–115.

10. Thomas S. Ray, "An Approach to the Synthesis of Life," 395.

11. To cite two examples, Stefan Helmreich, in *Silicon Second Nature* (Berkeley: University of California Press, 1998), claims that the rhetoric of ALife (and Ray's in particular) is determined by unconscious projections of Western white hetero-sexist males, and N. Katherine Hayles, in *How We Became Posthuman* (Chicago: University of Chicago Press, 1999), chap. 9, questions the anthropomorphizing of many of Ray's assertions.

12. Fritjof Capra, *The Web of Life* (New York: Anchor Books, 1996), 225.

13. See her book *Darwin in the Genome* (New York: McGraw-Hill, 2003).

14. See Lynn Margulis and Dorion Sagan, *Microcosmos: Four Billion Years of Evolution from Our Microbial Ancestors* (New York: Summit Books, 1986).

15. Two well-known examples are Margulis's theory of symbiogenesis, which understands the creation of new life forms through the merging of the genomes of different species, and James Lovelock's Gaia hypothesis, according to which the earth itself, or at least its atmosphere and the circulation of gases and minerals through it, is pictured as a self-organizing, self-regulating supersystem or superorganism, alive in the sense that it appears to exhibit will, or purpose, in the cooperation of subsystems directed toward the higher end of sustaining the system as a whole.

16. While Daniel Dennett rightly argues that Kauffman deepens Darwin (in *Darwin's Dangerous Idea*, 220–227), Dennett seriously underestimates the importance of self-organization and the dynamic emergence of order by attempting to understand them as principles of design and metaengineering.

17. Kauffman discusses "order for free" in "Antichaos and Adaptation," *Scientific American*, August 1991, 78–84. The NK model was initially developed in relation to fitness landscapes, with adaptation seen as a random walk through a parameter space defined by NK variables. For details, see Stuart A. Kauffman, *The Origins of Order: Self-Organization and*

Selection in Evolution (New York: Oxford University Press, 1993). For a general summary, see David J. Depew and Bruce H. Weber, *Darwinism Evolving* (Cambridge, Mass.: MIT Press, 1995), chap. 16, which is devoted to Kauffman's work.

18. Stuart A. Kauffman, "The Sciences of Complexity and 'Origins of Order,'" in *Principles of Organization in Organisms*, ed. J. Mittenthal and A. Baskin (Reading, Mass.: Addison-Wesley, 1992), 306. We now know that the human genome contains less than 30,000 genes, which doesn't necessarily invalidate Kauffman's general idea.

19. In the chaotic regime behavior is so drastically altered by minor perturbations that variations would propagate immediately throughout the system but could not accumulate and modify the system's structure. Conversely, deep in the ordered regime small perturbations only propagate locally, at best altering only a few neighboring elements.

20. As recounted by M. Mitchell Waldrop, in *Complexity: The Emerging Science at the Edge of Order and Chaos* (New York: Simon and Shuster, 1992), 302.

21. Jacques Monod, *Chance and Necessity* (New York: Vintage Books, 1972; orig. pub. 1970).

22. Stuart Kauffman, *At Home in the Universe: The Search for the Laws of Self-Organization and Complexity* (New York: Oxford University Press, 1995), 25.

23. Waldrop, *Complexity*, 304–306. Bak explains his theory in *How Nature Works: The Science of Self-Organized Criticality* (New York: Springer-Verlag, 1996).

24. A power law expresses some quantity **a** as a power of another quantity **b**, usually in inverse proportionality, as in $\mathbf{a} = \mathbf{1/b^n}$. Power laws appear commonly in both nature and culture; for example, in the relation between size and frequency of earthquakes and large cities, and in word distributions in natural languages. See Bak (1996) as well as Manfred Schroeder, *Fractals, Chaos, Power Laws* (New York: Freeman, 1991) for further discussion.

25. See Stuart Kauffman and Sonke Johnson, "Co-evolution to the Edge of Chaos: Coupled Fitness Landscapes, Poised States, and Co-evolutionary Avalanches," in *Artificial Life II*, 325–370.

26. Roger Lewin, *Complexity: Life at the Edge of Chaos* (New York: Macmillan, 1992), 104.

27. Thomas S. Ray, "Selecting Naturally for Differentiation: Preliminary Evolutionary Results," *Complexity* 3, no. 5 (1998): 26.

28. Thomas Ray, Abstract, p. 1, http://www.hip.atr.co.jp/~ray/pubs/reserves/node1.html. The Web site also contains links to publications about Internet Tierra.

29. Thomas S. Ray and Joseph Hart, "Evolution of Differentiated Multi-threaded Digital Organisms," in *Artificial Life VI*, ed. Christoph Adami et al. (Cambridge, Mass.: MIT Press, 1998), 3.

30. John H. Holland, *Adaptation in Natural and Artificial Systems* (Cambridge, Mass.: MIT Press, 1992), 184–185. This material, as well as material on classifier systems (discussed in the previous chapter), was added to the earlier 1975 edition.

31. Holland doesn't discuss the details of the implementation. In the early 1990s Terry Jones implemented a UNIX version of Echo at the Sante Fe Institute, which still maintains an Echo Web site with links to relevant articles and a newer, downloadable version. See http://www.santefe.edu/projects/echo/#research.

32. See Holland, *Adaptation*, 195, for comments on this particular feature.

33. In this sense they also illustrate life as described by Richard Dawkins in *The Blind Watchmaker* (New York: W. W. Norton, 1987): "What lies at the heart of every living thing is not a fire, not warm breath, not a 'spark of life.' It is information, words, instructions. If you want a metaphor, don't think of fires and sparks and breath. Think, instead, of a billion discrete, digital characters carved in tablets of crystal. If you want to understand life, don't think about vibrant, throbbing gels and oozes, think about information technology" (112).

34. Stefan Helmreich makes this critique in *Silicon Second Nature*, 165.

35. "Echoing Emergence: Objectives, Rough Definitions, and Speculations for ECHO-Class Models," in *Complexity: Metaphors, Models, and Reality*, ed. George A. Gowan et al. (Reading, Mass.: Addison-Wesley, 1994). This volume gathers papers and discussions from a conference held at the Santa Fe Institute devoted to the science of complexity and specifically to features of complex adaptive systems.

36. After a series of empirical tests on the most current computer implementation of Echo, Richard M. Smith and Mark A. Bedau concluded that Echo "lacks the diversity of hierarchically organized aggregates that typify complex adaptive systems." See "Is *Echo* a Complex Adaptive System?" *Evolutionary Computation* 8, no. 4 (2000): 419–442.

37. Holland actually refers to three mechanisms—parallelism, competition, and recombination—that he thinks will be the basic building blocks for a general theory of complex adaptive systems (*Adaptation*, 197). He also emphasizes the importance of an "internal model" that can anticipate the future, through "look-ahead" or expectation mechanisms. While he gives several examples from nature, it is not evident how Echo or Tierra model their environments in these terms. Below we shall see how Christoph Adami addresses this issue in terms of information about the environment inscribed in an organism's genome.

38. See Holland, *Hidden Order: How Adaptation Builds Complexity* (Reading, Mass.: Addison-Wesley, 1995), 9–10.

39. David Ackley and Michael Litman, "Interactions between Learning and Evolution," in *Artificial Life II*, 487–509. The article is grouped together with seven other articles in a large section entitled "Learning and Evolution."

40. For the most part, learning now falls within the domain of AI, both generally and in contemporary robotics (see chapter 7).

41. The importance of dynamic hierarchies was recognized early in ALife research. See Nils Baas, "Emergence, Hierarchies, and Hyperstructures," in *Artificial Life III*, ed. Christopher G. Langton (Reading, Mass.: Addison-Wesley, 1994), 515–537. More recently, Steen Rasmussen et al., "*Ansatz* for Dynamical Hierarchies," *Artificial Life* 7, no. 4 (2001), stimulated a call for papers on dynamical hierarchies. These were published in *Artificial Life* 11, no. 4 (Fall 2005).

42. James P. Crutchfield, "Is Anything Ever New? Considering Emergence," in *Complexity: Metaphors, Models, and Reality*, 519. This essay is a shorter version of "Calculi of Emergence," discussed briefly in chapter 3 in relation to Crutchfield's project of ϵ-machine reconstruction.

43. Luc Steels makes a similar distinction between what he calls first-order and second-order emergence. Whereas Crutchfield (a physicist) is interested in the complexity of dynamical systems that support information processing, Steels works in the field of behavior-oriented (as opposed to knowledge-oriented) AI and robotics. For Steels, "an emergent behavior leads to emergent functionality if the behavior contributes to the system's self-preservation and if the system can build further upon it." "The Artificial Life Roots of Artificial Intelligence," in *Artificial Life: An Overview*, ed. Christopher G. Langton (Cambridge, Mass.: MIT Press, 1995), 78. Steels's concept of emergent functionality is discussed in chapter 7.

44. James P. Crutchfield, "When Evolution Is Revolution—Origins of Innovation," in *Evolutionary Dynamics*, ed. James P. Crutchfield and Peter Shuster (Oxford: Oxford University Press, 2003), 101–133.

45. John Horgan, "From Complexity to Perplexity," *Scientific American*, June 1995, 105–109.

46. As discussed in chapter 3, H. H. Pattee addresses this question in more rudimentary terms in his seminal essay, "How Does a Molecule Become a Message?"

47. First published in *Physica D* 42 (1990), the proceedings are now available in *Emergent Computation*, ed. Stephanie Forrest (Cambridge, Mass.: MIT Press, 1991).

48. Proceedings were published in *Physica D* 10 (1984) and *Physica D* 22 (1986), respectively.

49. "Evolving Cellular Automata with Genetic Algorithms: A Review of Recent Work," in *Proceedings of the First International Conference on Evolutionary Computation and Its Applications* (Moscow: Russian Academy of Sciences, 1996). Rajarshi Das is also listed as a coauthor.

50. See Crutchfield, "Statistical Complexity" and "Computation at the Onset of Chaos," which are discussed in chapter 3.

51. Melanie Mitchell, "Computation in Cellular Automata: A Selected Review" in *Nonstandard Computation*, ed. H. G. Shuster and T. Gramss (Weinheim: VCH Verlagsgesellschaft, 1997).

52. Melanie Mitchell, James P. Crutchfield, and Peter T. Hraber, "Dynamics, Computation, and the 'Edge of Chaos': A Re-examination," in *Complexity: Metaphors, Models, and Reality*, 503.

53. Crutchfield and Mitchell, "The Evolution of Emergent Computation," SFI Technical Report 94-03-012, p. 1. This article also appears in *Proceedings of the National Academy of Sciences.* USA 92 (1995), 10742–10746.

54. See their replies to Tim van Gelder's "The Dynamical Hypothesis in Cognitive Science," in *Behavior and Brain Sciences* 21 (1998): 635 and 645–646.

55. See Chris Adami and C. Titus Brown, "Evolutionary Learning in the 2D Artificial Life System 'Avida,'" in *Artificial Life IV*, ed. Rodney A. Brooks and Pattie Maes (Cambridge, Mass.: MIT Press, 1994), 377–381, as well as Adami's *Introduction to Artificial Life* (New York: Springer-Verlag, 1998), esp. 50–53, chap. 9, and "The Avida User's Manual" in the appendix. Contrary to its title, Adami's book is not a general introduction to the field of ALife but to his own physics-oriented, information theoretical approach.

56. The second version of Avida, which is included in Adami's book, was written by Charles Ofria. A third version is now available from Adami's Digital Life Laboratory Web site at http://dll.Caltech.edu/avida/.

57. See Adami, *Introduction*, 302–304, and "The Avida User's Manual," 297ff, for a description of these options.

58. "Evolution of Biological Complexity" (coauthored with Charles Ofria and Travis C. Collier), *Proceedings of the National Academy of Science* 97 (2000), 4463–4468.

59. See *Introduction*, 320–322, for details about the kinds of mutations Avida supports.

60. Adami reviews these basic differences in "What Is Complexity?" *BioEssays* 24, no. 12 (2002): 1085–1094.

61. For a discussion of biological complexity in these terms, see the three articles by D. W. McShea that Adami cites: "Metazoan Complexity and Evolution: Is There a Trend?" *Evolution* 50 (1996): 477–492; "Functional Complexity in Organisms: Parts as Proxies," *Biology and Philosophy* 15 (2000): 641–668; and "The Hierarchical Structure of Organisms: A Scale and Documentation of a Trend in the Maximum," *Paleobiology* 27 (2001): 405–423.

62. The best-known opponent no doubt is Stephen Jay Gould, who argues in *Full House* (New York: Three Rivers Press, 1996) that apparent trends toward "progress" or evolutionary complexity can be explained by the "drunkard's walk" model. Gould argues that the appearance of "progress" is due to the arbitrary fixing of boundaries.

63. Christoph Adami and N. J. Cerf, "Physical Complexity of Symbolic Sequences," *Physica D* 137 (2000): 68.

64. "Physical Complexity of Symbolic Sequences," 68–69. The embedded quotation is from Charles A. Bennett's "Universal Computation and Physical Dynamics," *Physica D* 86 (1995): 268.

65. See Richard E. Lenski, Charles Ofria, Travis C. Collier, and Christoph Adami, "Genome Complexity, Robustness and Genetic Interactions in Digital Organisms," *Nature*, August 12, 1999, 661–664; Claus O. Wilke, Jia Lan Wang, Charles Ofria, Richard E. Lenski, and Christoph Adami, "Evolution of Digital Organisms at High Mutation Rates Leads to Survival of the Flattest," *Nature*, July 19, 2001, 331–333; and Richard E. Lenski, Charles

Ofria, Robert T. Pennock, and Christoph Adami, "The Evolutionary Origin of Complex Features," *Nature*, May 8, 2003, 139–144.

66. See A. N. Pargellis, "The Spontaneous Generation of Digital 'Life,'" *Physica D* 91 (1996): 86–96, and "The Evolution of Self-Replicating Computer Programs," *Physica D* 98 (1996): 111–127.

67. Pargellis continues: "In the studies mentioned here, a cell can not write its own code into memory allocated for another cell's child" (90). I take this to mean that a *cell* can't do this but that a *virus* can.

68. As reported by Henry Bortman in "Survival of the Flattest: Digital Organisms Replicate," *Astrobiology Magazine*, October 7, 2002, available at http://www.space.com/scienceastronomy/generalscience/digital_life_021007.html.

69. The term *quasi species* was coined by Manfred Eigen, whose article "Viral Quasispecies," *Scientific American*, July 1993, 42–49, provides the background to this discussion.

70. Mark A. Bedau, Emile Snyder, C. Titus Brown, and Norman H. Packard, "A Comparison of Evolutionary Activity in Artificial Evolving Systems and in the Biosphere," *Fourth European Conference on Artificial Life*, ed. Phil Husbands and Inman Harvey (Cambridge, Mass.: MIT Press, 1997), 125–134.

71. Evita is an ALife simulation platform similar to Tierra and Avida. In Bugs, however, the organisms move about on a spatial grid by means of a simulated sensorimotor mechanism that allows them to "sense" resource sites, move toward them, and replenish themselves. These resources are necessary for the organism to pay existence and movement "taxes" as well as to reproduce. Thus Bugs simulates a form of metabolism in addition to reproduction and evolution. It was first introduced into ALife research by Norman Packard in "Intrinsic Adaptation in a Simple Model for Evolution," in *Artificial Life*, ed. Langton, 141–155.

72. See Bedau et al., "A Comparison of Evolutionary Activity," 129, for a full explanation.

73. Mark A. Bedau, Emile Snyder, and Norman H. Packard, "A Classification of Long-Term Evolutionary Dynamics," in *Artificial Life VI*, ed. Christoph Adami et al. (Cambridge, Mass.: MIT Press, 1998), 228–237. Smith and Bedau's examination of Echo as a complex adaptive system (mentioned above in note 36) follows this work.

74. In regard to this work, see Bedau et al., "Open Problems in Artificial Life," *Artificial Life* 6, no. 4 (2000): 363–376.

75. See Tim Taylor and John Hallam, "Studying Evolution with Self-Replicating Computer Programs," in *Fourth European Conference on Artificial Life*, 550–559.

76. As Taylor reports in "From Artificial Evolution to Artificial Life" (doctoral diss., University of Edinburgh, 1999).

77. H. H. Pattee, "Evolving Self-Reference: Matter, Symbols, and Semantic Closure," *Communication and Cognition—Artificial Intelligence* 12, nos. 1–2 (1995): 9–27.

78. H. H. Pattee, "Artificial Life Needs a Real Epistemology," in *Advances in Artificial Life: Third European Conference on Artificial Life*, ed. A. Moran et al. (Berlin: Springer, 1995), 36.

79. Luis Mateus Rocha, "Evolution with Material Symbol Systems," *Biosystems* 60 (2001): 95–121.

80. Thomas S. Ray, "Selecting Naturally for Differentiation: Preliminary Evolutionary Results," *Complexity* 3, no. 5 (1998): 25–33.

81. See Larry Yaeger, "Computational Genetics, Physiology, Metabolism, Neural Systems, Learning, Vision, and Behavior or PolyWorld: Life in a New Context," in *Artificial Life III*, 263–298.

82. It is important to note that the digital creatures in Steve Grand's Creatures, released commercially in 1993 as an "artificial life game," possess a comparable complexity and even more sophisticated neural systems. The primary difference, of course, is that, unlike

PolyWorld, Creatures is not a platform developed specifically for scientific experiments. I discuss Creatures and Grand's other work in chapter 8.

83. David H. Ackley, "Real Artificial Life: Where We May Be," in *Artificial Life VII*, ed. Mark A. Bedau et al. (Cambridge, Mass.: MIT Press, 2000), 487–496.

84. A more detailed treatment of this parallel would necessarily require that software development be considered in relation to hardware development, or rather, in keeping with the natural analogy, the two would be considered together as forming a coevolutionary relay. To my knowledge, no history of computers or computation has attempted to do this.

85. Steen Rasmussen et al., "Transitions from Nonliving to Living Matter," *Science*, February 13, 2004.

86. Claire M. Fraser et al., "The Minimal Gene Complement of Mycoplasma Genitalium," *Science*, October 20, 1995, 397.

87. See Justin Gillis, "Scientists Planning to Make New Form of Life," *Washington Post*, November 21, 2002, p. A01.

88. Philip Ball, "Artificial Cells Take Shape," http://www.nature.com/news/2004/041206/pf/041206-2_pf.html.

89. The discovery was made by P. L. Luisi, who is also engaged in protocell research. See Rasmussen and Chen's discussion of "the lipid world" in Rasmussen and Chen et al., "Bridging Nonliving and Living Matter," *Artificial Life* 9 (2003): 269–303. They also discuss the protocell research based on the use of lipids by J. Szostak, A. Pohorille, and D. Deamer.

90. In addition to the article cited in the previous note, see Rasmussen and Chen et al., "Proto-Organism Kinetics: Evolutionary Dynamics of Lipid Aggregates with Genes and Metabolism," *Origins of Life and Evolution of the Biosphere* 34 (2004): 171–180; and David Castelvecchi, "A New Game of Life," http://davidecastelvecchi.com/prptocells.html.

Chapter 6

1. Melanie Mitchell, "A Complex-Systems Perspective on the 'Computation vs. Dynamics' Debate in Cognitive Science," *Behavioral and Brain Sciences* 21 (1998): 645–646.

2. Tim van Gelder, "The Dynamical Hypothesis in Cognitive Science," *Behavioral and Brain Sciences* 21 (1998): 615–665.

3. See Esther Thelen and Linda B. Smith, *A Dynamic Systems Approach to the Development of Cognition and Action* (Cambridge, Mass.: MIT Press, 1994), as well as *Mind as Motion: Explorations in the Dynamics of Cognition*, ed. Robert F. Port and Timothy van Gelder (Cambridge, Mass.: MIT Press, 1995). Several examples from the latter are discussed below.

4. For a detailed exposition of how the dynamical systems approach is applied to cognitive science, see Lawrence M. Ward, *Dynamical Cognitive Science* (Cambridge, Mass.: MIT Press, 2002).

5. James P. Crutchfield makes the same argument in his work on "computational mechanics." See his response to van Gelder's article in *Behavioral and Brain Sciences* 21 (1998): 635.

6. Parts of the book were originally written in English and are included in Varela's *The Embodied Mind*, coauthored with Evan Thompson and Eleanor Rosch (Cambridge, Mass.: MIT Press, 1991). More conceptually focused and compact, *Invitation aux sciences cognitives* (Paris: Éditions du Seuil, 1996), is more useful here.

7. For examples of the Heideggerian critique, see Hubert L. Dreyfus, *What Computers Can't Do* (New York: Harper and Row, 1972), and Terry Winograd and F. Flores, *Understanding Computers and Cognition* (Norwood, N.J.: Abex, 1986).

8. *The Embodied Mind*, 208. There Varela devotes about five pages to Brooks's work (208–213). His alliance with Brooks and the new AI is also signaled in his essay, "The Reenchantment of the Concrete," which serves as the prologue to Brooks and Luc Steels's

edited collection, *The Artificial Life Route to Artificial Intelligence: Building Embodied, Situated Agents* (Hillsdale, N.J.: Lawrence Erlbaum Associates, 1995).

9. In *Artificial Minds* (Cambridge, Mass.: MIT Press, 1995), Stan Franklin also groups Varela and Brooks together on the basis of their antirepresentational stance. As Franklin sees it, the issue of representation provides the nexus for what he calls the third AI debate. The first debate was animated by the question, can machines think? the second by the question, which is preferable, the computer or the brain as a model for the mind? The third debate is complicated by the fact that the term *representation* covers a wide range of phenomena, from mental representations to various cultural inscriptions, schemas, artifacts, and so forth.

10. Howard Gardner, *The Mind's New Science: A History of the Cognitive Revolution* (New York: Basic Books, 1985), 6. Along with these two features, Gardner lists three more: cognitive science de-emphasizes emotions as well as historical and cultural context; cognitive science is interdisciplinary; and cognitive science gives special attention to epistemological issues basic to the Western philosophical tradition.

11. Daniel Crevier, *AI: The Tumultuous History of the Search for Artificial Intelligence* (New York: Basic Books, 1993), 248.

12. Patrick Winston, *Artificial Intelligence* (Reading, Mass.: Addison-Wesley, 1984).

13. Walter J. Ong, *Orality and Literacy: The Technologizing of the Word* (London: Methuen, 1982), discusses the role of writing in research as diverse as that of Jack Goody, Adam Parry, and Jacques Derrida.

14. Edwin Hutchins, *Cognition in the Wild* (Cambridge, Mass.: MIT Press, 1995).

15. Bernard Stiegler, *La technique et le temps*, vol. 2, *La désorientation* (Paris: Galilée, 1996), 188 (my translation).

16. On the repression of the origins of cognitive science in cybernetics, see Jean-Pierre Dupuy, *Aux origines des sciences cognitives* (Paris: Éditions La Découverte, 1994).

17. In *Embodying Technesis: Technology beyond Writing* (Ann Arbor: University of Michigan Press, 2000), Mark Hansen argues compellingly that modern machines or technologies produce effects (and affects) that cannot be captured and assimilated by the resources of language—that cannot, in short, be rendered in discourse.

18. Warren McCulloch and Walter Pitts, "A Logical Calculus of the Ideas Immanent in Nervous Activity," reprinted in McCulloch's *Embodiments of Mind* (Cambridge, Mass.: MIT Press, 1965), 35. Turing himself would no doubt have contributed to this alternative development if his essay "Intelligent Machinery" (1948) had been published in his lifetime. There Turing discusses what he calls "unorganized machines," which, unlike organized machines (such as Turing machines), are not designed for a definite purpose and are compared to a random arrangement of neurons. Because of their capacity to acquire a purpose and to learn, Turing suggests that "the cortex of the infant is an unorganized machine" (424). See Alan Turing, "Intelligent Machinery," in *The Essential Turing*, ed. B. Jack Copeland (Oxford: Oxford University Press, 2004).

19. *Emergent Computation*, ed. Stephanie Forrest (Cambridge, Mass.: MIT Press, 1990). The essays by J. Doyne Farmer and S. Harnad explicitly consider connectionism.

20. See Friedrich Nietzsche, *The Birth of Tragedy* (New York: Vintage Books, 1967) and *On the Genealogy of Morals* (New York: Vintage Books, 1989).

21. Seymour Papert, "One AI or Many?" in *The Artificial Intelligence Debate: False Starts, Real Foundations*, ed. Stephen R. Graubard (Cambridge, Mass.: MIT Press, 1988), 3–4.

22. Like Varela's *Invitation aux sciences cognitives*, Stan Franklin's *Artificial Minds* presents a three-stage history; in Franklin's view, however, the "jury's still out" on the essential questions raised at each of these stages.

23. Paul N. Edwards, *The Closed World: Computers and the Politics of Discourse in Cold War America* (Cambridge, Mass.: MIT Press, 1996).

24. As Sadie Plant relevantly notes, "There is always a point at which technologies geared towards regulation, containment, command, and control, can turn out to be feeding into the

collapse of everything they once supported." Plant, *Zeros and Ones: Digital Women and the New Technoculture* (New York: Doubleday, 1997), 143.

25. See, for example, Philip Mirowski's *Machine Dreams: Economics Becomes a Cyborg Science* (Cambridge: Cambridge University Press, 2002). Taking the term "cyborg science" from Donna Haraway (but without her politically vectored ambivalence), Mirowski applies the perspective developed by Edwards and others to post–Second World War economic theory. At various points Mirowski refers confusingly to cybernetics, information theory, molecular biology, artificial intelligence, cognitive science, and ALife as "cyborg sciences," by which he really means the ideological capture and harnessing of the new technologies of computation for purposes like social control and national resource allocation. In short, in Mirowski's "cyborg sciences" there is very little real science, which has been replaced by ideology.

26. The quotation is from the Rockefeller grant proposal for funding the conference, cited by Pamela McCorduck in *Machines Who Think* (Natick, Mass.: A. K. Peters, 2004; orig. pub. 1979), 111. The authors were John McCarthy, Marvin Minsky, Nathaniel Rochester, and Claude Shannon. Among those who also attended the conference were Arthur Samuel, Oliver Selfridge, Ray Solomonoff, Allen Newell, and Herbert Simon.

27. Quoted by Howard Gardner, *The Mind's New Science*, 29. Gardner also cites a number of similar responses.

28. Published in the *Psychological Review* 63 (1956): 81–97.

29. George Miller, Eugene Galanter, and Karl Pribram, *Plans and the Structure of Behavior* (New York: Holt, Rinehart and Winston, 1960), 7.

30. In *A History of Psychology* (Englewood Cliffs, N.J.: Prentice-Hall, 1994), T. H. Leahey summarizes this new idea: "What began to emerge in the 1950s was a new conception of the human being as machine, and a new language in which to formulate theories about cognitive processes. People could be described, it seemed, as general-purpose computing devices, born with a certain hardware, and programmed by experience and socialization to behave in certain ways. The goal of psychology would be the specification of how human beings process information: the concepts of stimulus and response would be replaced by the concepts of information input and output, and theories about mediating [stimulus-response] chains would be replaced by theories about internal computations and computational states" (267).

31. Allen Newell and Herbert A. Simon, "The Logic Theory Machine," *IRE Transactions of Information Theory*, IT-2, 3 (1956): 61–79. Cliff Shaw was equally involved in the construction of the Logic Theorist, as Simon acknowledges in his informative autobiographical account, *Models of My Life* (New York: Basic Books, 1991), 201–211.

32. Compare Simon: "We needed a higher-level language, congenial to the human programmer, which would do automatically much of the 'housekeeping' in the computer and which would be translated automatically by the computer itself into machine language. And memory structures would have to be highly modifiable" (*Models of My Life*, 204). Simon adds that IPL embodied "most of the ideas of what is now called object-oriented programming" (212) and was the ancestor of John McCarthy's LISP, which became the standard AI language for thirty years.

33. The problems were taken from Russell and Whitehead's *Principia Mathematica*. When informed by Simon of their accomplishment Russell had written back that he was "delighted to know that *Principia Mathematica* can now be done by machinery" (*Models of My Life*, 208).

34. Arthur Samuels's checker-playing program (1955–1959), another milestone in early AI, deploys a similar search strategy. However, it not only searches for appropriate moves but also "learns" from its experience, thanks to a reward system that weighted successful choices. In fact, the program learned to beat Samuels himself and several expert players. See John Holland's description in *Emergence: From Chaos to Order* (Reading, Mass.: Addison-Wesley, 1998), 53–80.

35. Claude E. Shannon, "A Chess-Playing Machine," *Scientific American*, February 1950, 48–51.

36. Allen Newell, J. C. Shaw, and Herbert Simon, "Chess-Playing Programs and the Problem of Complexity," *IBM Journal of Research and Development* 2 (October 1958): 320–335. They state, for example, that "if one could devise a successful chess machine, one would seem to have penetrated the core of human intellectual endeavor."

37. Andrew Hodges, *Alan Turing: The Enigma* (New York: Simon and Shuster, 1983), and B. Jack Copeland, *The Essential Turing*, 353.

38. Herbert Simon and Allen Newell, "Heuristic Problem-Solving, the Next Advance in Operations Research," *Journal of the Operations Research Society of America* 6, no. 1 (1958): 6.

39. Allen Newell and Herbert A. Simon, "Computer Science as Empirical Inquiry: Symbols and Search," reprinted in *Mind Design*, ed. John Haugeland (Cambridge, Mass.: MIT Press, 1981), from which I cite here. It was first published in *Communications of the Association of Computing Machinery* (March 1976).

40. Allen Newell, "Physical Symbol Systems," *Cognitive Science* 4 (1980): 135–183.

41. Edwards, in *The Closed Mind* (250–252), makes the same point, emphasizing that for the cyberneticists the subject was always "the embodied mind" (240). Brooks, in *The Artificial Life Route to Artificial Intelligence* (37–39), also calls attention to the emphasis on situated, embodied intelligence among the cybernetic researchers, in contrast to the abstracting tendency of AI.

42. Allen Newell, "Intellectual Issues in the History of AI," in *The Study of Information*, ed. Fritz Machlup and Una Mansfield (New York: John Wiley & Sons, 1983), 188.

43. General Problem Solver, Newell and Simon's next program after Logic Theorist, explicitly focused on "the psychology of human thinking" by simulating the way a college student in engineering would solve a problem in symbolic logic, through means-ends analysis: "GPS deals with a task environment consisting of *objects* which can be transformed by various *operators*; it detects *differences* between objects; and it organizes the information about the task environment into *goals*. Each goal is a collection of information that defines what constitutes goal attainment, makes available the various kinds of information relevant to attaining the goal, and relates the information to other goals" (284). Allen Newell and Herbert Simon, "GPS, A Program that Simulates Human Thought," in *Computers and Thought*, ed. Edward Feigenbaum and Julian Feldman (New York: McGraw-Hill, 1963), 279–293.

44. In a summary passage Edwards suggests that Newell and Simon "began to displace the cybernetic computer-brain analogy with the even more comprehensive and abstract computer-*mind* metaphor of artificial intelligence" (252).

45. John Haugeland, *Artificial Intelligence: The Very Idea* (Cambridge, Mass.: MIT Press, 1985), 2.

46. Hubert L. Dreyfus and Stuart E. Dreyfus, "Making a Mind versus Modeling the Brain," in *The Artificial Intelligence Debate: False Starts, Real Foundations*, 15–43.

47. For details, see Edward A. Feigenbaum and Pamela McCorduck, *The Fifth Generation* (Reading, Mass.: Addison-Wesley, 1983), although the authors do not adequately confront the project's failure.

48. Daniel Crevier, *AI: The Tumultuous History*, 241–242.

49. That natural language is not a simple or single coding of human experience, in other words, that multiple codes are involved, was well understood by linguists. However, as we shall see, the most advanced linguistic theory of the period, Chomsky's generative grammar, did not lead to a rethinking of the nature of coding in language but to a consolidation of the computational approach.

50. See Noam Chomsky, *On Language* (New York: New Press, 1998), esp. 94–95, for a clear discussion of this double objective.

51. Jerry Fodor, *The Language of Thought* (New York: Crowell, 1975).

52. Recently the linguist Steven Pinker has attempted to remedy this deficit in *How the Mind Works* (New York: W. W. Norton, 1997). Working in the Chomsky-Fodor line, Pinker

attempts to integrate the computational theory of mind with the perspective of evolutionary psychology, and specifically the natural selection of replicators.

53. An earlier version, entitled "Artificial Intelligence: Subcognition as Computation," appeared in *The Study of Information* (1983). The version I discuss was published in Hofstadter's book, *Metamagical Themas* (New York: Basic Books, 1985). In a postscript, Hofstadter discusses his subsequent discovery of the connectionist parallel distributed processing models of Rumelhart, Smolensky, et al., which I consider in the next section.

54. As discussed in chapter 2, this is essentially Jacques Lacan's position, although he arrives at it by an altogether different route.

55. Brooks, "Intelligence without Reason," in *The Artificial Life Route to Artificial Intelligence*, 30–34.

56. For example, by Ray Kurzweil, in *The Age of Spiritual Machines* (New York: Viking, 1999) and *The Singularity Is Near* (New York: Viking, 2005). Tellingly, Kurzweil does not say very much about the software that will have to be developed for strong AI.

57. Donald Hebb, *The Organization of Behavior* (New York: John Wiley, 1949). This came to be known as Hebb's rule, and the neurons that fired together were called cell assemblies.

58. In the *Web of Life* (New York: Anchor Books, 1996), Fritjof Capra writes: "In the 1950s scientists began to actually build models of such binary networks [like those of McCulloch and Pitts], including some with little lamps flickering on and off at the nodes. To their great amazement they discovered that after a short time of random flickering, some ordered patterns would emerge in most networks. They would see waves of flickering pass through the network, or they would observe repeated cycles. Even though the initial state of the network was chosen at random, after a while those ordered patterns would emerge spontaneously, and it was that spontaneous emergence of order that became known as 'self-organization'" (83–84). Though not specific here, Capra is presumably referring to Rosenblatt (among others), who had presented a paper on the Perceptron at one of the conferences on "self-organizing systems" organized by Marshall C. Yovits.

59. Frank Rosenblatt, "The Perceptron: A Probabilistic Model for Information Storage and Organization in the Brain," *Psychological Review* 65 (November 1958): 386–408; reprinted in *Minds, Brains, and Computers: The Foundations of Cognitive Science*, ed. Robert Cummins and Denise Cummins (Oxford: Blackwell Publishers, 2000), 179–197. All quotations that follow refer to the reprinted version.

60. Like each cell in a cellular automaton, each neuron is thus a single computational module that takes its own state and the states of all the other modules in its network as its input, applies a "rule" or condition, then delivers an output based on the result. And like the cells in a CA, the behavior of no single neuron matters; it is only the dynamic pattern that emerges from their interaction that determines the behavior of the network as a whole.

61. Selfridge had demonstrated an earlier version of Pandemonium to Allen Newell in the summer of 1954. It made such a strong impression on Newell that it "turned" his life, and he began working on artificial intelligence. See McCorduck's account in *Machines Who Think* (157–159). It is also important to note that both Rosenblatt's Perception and Selfridge's Pandemonium are much closer in spirit (and as computational assemblages) to cybernetic machines than to new AI programs like Logic Theorist.

62. O. G. Selfridge, "Pandemonium: A Paradigm for Learning," in *Symposium on the Mechanization of Thought Processes*, ed. D. V. Blake and A. M. Uttley (London: HMSO, 1959).

63. Pamela McCorduck makes this point in *Machines Who Think*, 106.

64. The essay, which had circulated in various versions since 1956, was published as "Steps toward Artificial Intelligence," *Proceedings of the Institute Radio Engineers* 49, no. 1 (January 1961): 8–30.

65. Marvin Minsky and Seymour Papert, *Perceptrons: An Introduction to Computational Geometry* (Cambridge, Mass.: MIT Press, 1988; orig. pub. 1969).

66. Their work was published in *Parallel Distributed Processing: Explorations in the Microstructure of Cognition*, 2 vols. (Cambridge, Mass.: MIT Press, 1986). Widely hailed as the "Bible of connectionism," it stirred a great deal of excitement and renewed interest in neural net research and the connectionist approach.

67. William Bechtel and Adele Abrahamsen, *Connectionism and the Mind* (Oxford: Basil Blackwell, 1991), 16–17.

68. Mitchell Resnick argues compellingly for this view in *Turtles, Termites, and Traffic Jams: Exploration in Massively Parallel Microworlds* (Cambridge, Mass.: MIT Press, 1994), 6–19.

69. For details, see Rumelhart, Hinton, and Williams, "Learning Internal Representations by Error Propagation," in *Parallel Distributed Processing*, 1:319–322.

70. For an overview of these developments, see Jack D. Cowan and David H. Sharp, "Neural Nets and Artificial Intelligence," in *The Artificial Intelligence Debate*, 85–121.

71. The essay was published in the *Proceedings of the National Academy of Sciences* 79, no. 8 (1982): 2554–2558.

72. G. T. Sejnowski and D. Ackley, *Boltzmann Machines: Constraint Satisfaction Networks that Learn* (Carnegie-Mellon University, Tech. Rep. CMU CS 84-111, 1984).

73. Terrence Sejnowski and Charles Rosenberg, "Parallel Networks that Learn to Pronounce English Text," *Complex Systems* 1, no. 1 (1987): 145–168.

74. D. E. Rumelhart and J. L. McClelland, "On Learning the Past Tenses of English Verbs," in *Parallel Distributed Processing*, 2:216–271.

75. The example and discussion that follow are based on Paul Churchland's clear and accessible account in *The Engine of Reason, the Seat of the Soul* (Cambridge, Mass.: MIT Press, 1995), chap. 2.

76. Churchland gives many examples of how vector coding works, for several senses as well as faces. For example, considering the difference between the acuteness of the canine olfactory sense compared to the relative poverty of the human's, he points out that "should a dog have merely seven types of receptor cells where a human has six, and only three times the human acuity along each of its seven olfactory axes, then it would be able to discriminate 30^7 or 20 billion distinct odors" (*The Engine of Reason*, 26). It is no wonder then than dogs can distinguish between any two people on the planet by smell alone.

77. First published in *Cognition* 28 (1988): 3–71, it is reprinted in *Mind Design II*, ed. John Haugeland (Cambridge, Mass.: MIT Press, 1997). Page numbers refer to the latter.

78. Paul Smolensky, "Connectionism, Constituency, and the Language of Thought," in *Meaning in Mind: Fodor and His Critics* (Oxford: Blackwell Publishers, 1991); reprinted in *Minds, Brains, and Computers: The Foundations of Cognitive Science*, ed. Robert Cummins and Denise Cummins (Oxford: Blackwell Publishers, 2000), 286–306. Page numbers refer to the reprinted version.

79. Jeffrey Elman, "Language as a Dynamical System," in *Mind as Motion: Explorations in the Dynamics of Cognition*, ed. Robert F. Port and Timothy van Gelder (Cambridge, Mass.: MIT Press, 1995), 195–225.

80. See Churchland, *The Engine of Reason*, 97–114, for an introductory description of recurrent networks.

81. In *The Engine of Reason*, for example, Paul Churchland shows very concisely how the full range of our perceptible tastes, smells, and colors can be represented in the brain through vector coding of neural net activation patterns.

82. The account that follows is based on Petitot's article, "Morphodynamics and Attractor Syntax: Constituency in Visual Perception and Cognitive Grammar," also published in Port and van Gelder's *Mind as Motion*, 227–281.

83. Daniel C. Dennett, *Consciousness Explained* (Boston: Little, Brown, 1991). Churchland's critique is found *The Engine of Reason, the Seat of the Soul*, 208–226.

84. Dawkins defines the meme as a cultural replicator in *The Selfish Gene* (Oxford: Oxford University Press, 1977), 189–201. Examples are "tunes, ideas, catch-phrases, clothes, fashions, ways of making pots or of building arches." Dawkins continues: "Just as genes propagate themselves in the gene pool by leaping from body to body via sperms or eggs, so memes propagate themselves in the meme pool by leaping from brain to brain via a process which, in the broad sense, can be called imitation.... If the idea catches on, it can be said to propagate itself, spreading from brain to brain" (192).

85. See *The Engine of Reason, the Seat of the Soul*, 208–226, and Patricia S. Churchland and Terrence J. Sejnowski, *The Computational Brain* (Cambridge, Mass.: MIT Press, 1992), for further discussion.

86. Cf. Dennett: "We're just very, very complicated, evolved machines made of organic molecules instead of metal and silicon, and we are conscious, so there can be conscious machines" (*Consciousness Explained*, 431–432).

87. Available at http://cogsci.soton.ac.uk/~harnad/Papers/Py104/dennett.rob.html.

88. Rodney Brooks, *Flesh and Machines: How Robots Will Change Us* (New York: Pantheon, 2002).

89. Cynthia Breazeal, in an interview in Peter Menzel and Faith D'Aluisio, *Robo sapiens: Evolution of a New Species* (Cambridge, Mass.: MIT Press, 2000), 68. The book contains sections, including color photographs, devoted to both Cog and Kismet.

90. Brooks explains this system in *Flesh and Machines*, 94–95.

91. See, for example, Cynthia Breazeal and Brian Scassellati, "Infant-like Social Interactions between a Robot and a Human Caretaker," *Adaptive Behavior* 8, no. 1 (2000): 49–74.

92. John R. Searle, "Minds, Brains, and Programs," *Behavioral and Brain Sciences* 3 (1980): 417–424.

93. See Hofstadter in *The Mind's I*, ed. Douglas R. Hofstadter and Daniel C. Dennett (New York: Basic Books, 1981), 373–382; and Dennett in *Consciousness Explained*, 435–440.

94. David J. Chalmers, *The Conscious Mind* (New York: Oxford University Press, 1996), 322–328. Like Hofstadter, Chalmers emphasizes how seriously Searle underestimates the power of a Turing machine.

95. I take up this "turn" in chapter 8.

96. On this issue see Andy Clark, "Happy Couplings: Emergence and Explanatory Interlock," in *The Philosophy of Artificial Life*, ed. Margaret A. Boden (Oxford: Oxford University Press, 1996), 262–281.

Chapter 7

1. See, however, Gary F. Marcus's *The Algebraic Mind* (Cambridge: Mass.: MIT Press, 2001), which makes a compelling case for how to integrate connectionism and symbolic, or computational, cognitive science.

2. For example, in *Emergence: From Chaos to Order* (Reading, Mass.: Addison-Wesley, 1998), John Holland uses the term expressly to refer to "generators of emergent behavior" (4). Referring to Douglas Hofstadter's metaphor of the ant colony, Holland adds, "Somehow the simple laws of the agents generate an emergent behavior far beyond their individual capacities."

3. André Leroi-Gourhan discusses the reciprocal relationship between the first use of tools and the spreading of the cortex in *Gesture and Speech* (Cambridge, Mass.: MIT Press, 1993). First published in French in 1964, this valuable work on the relationship of technics and hominization has largely been ignored by AI and cognitive science.

4. It is published in Hofstadter's *Metamagical Themas* (New York: Basic Books, 1985), 631–665.

5. See Jacques Ferber, *Multi-Agent Systems: An Introduction to Distributed Artificial Intelligence* (Harlow, England: Addison-Wesley, 1999). Unfortunately, Ferber does not consider swarm intelligence.

6. John H. Holland, *Emergence: From Chaos to Order* (Reading, Mass.: Addison-Wesley), 5.

7. Eric Bonabeau, Marco Dorigo, and Guy Théraulaz, *Swarm Intelligence: From Natural to Artificial Systems* (Oxford: Oxford University Press, 1999).

8. For more on Swarm, see the official Web site at http://www.swarm.org, where it can be freely downloaded.

9. Rodney A. Brooks and Anita Flynn, "Fast, Cheap and Out of Control: A Robot Invasion of the Solar System," *Journal of the British Interplanetary System* 42 (1989). NASA eventually took up the suggestion, deploying a planetary mobot named Sojourner on the Pathfinder mission to Mars in 1996–1997.

10. Rodney Brooks, *Cambrian Intelligence: The Early History of the New AI* (Cambridge, Mass.: MIT Press, 1999), viii. The quotations, illustration, and diagram that follow are taken from this collection of Brooks's essays. The collection includes two particularly important early essays, "Intelligence without Reason" and "Intelligence without Representation," which are discussed below.

11. J. von Uexküll, "A Stroll through the Worlds of Animals and Men," in *Instinctive Behavior*, ed. Claire H. Schiller (New York: International Universities Press, 1957), 5–82.

12. Peter Menzel and Faith D'Aluisio, *Robo sapiens* (Cambridge, Mass.: MIT Press, 2000).

13. This is a central theme in Brooks's *Flesh and Machines: How Robots Will Change Us* (New York: Pantheon, 2002), where he argues that human beings are also machines—exceedingly complex machines—but machines nonetheless.

14. *The Artificial Life Route to Artificial Intelligence: Building Embodied, Situated Agents*, ed. Luc Steels and Rodney Brooks (Hillsdale, NJ: Lawrence Erlbaum Associates, 1995).

15. First published in the journal *Artificial Life*, the essay has been republished in *Artificial Life: An Overview*, ed. Christopher G. Langton (Cambridge, Mass.: MIT Press, 1995), 75–110.

16. While this statement applies to only a single agent, in Steels's typology of emergent functionality both of the second two types involve multiple agents.

17. Presented at the Burda Symposium on brain-computer interfaces at Munich, Germany, February 1995, and published in German in *Die Technik auf dem Weg zur Seele: Proceedings of the Burda Symposium on Brain-Computer Interfaces*, ed. Ch. Maar, E. Poppel, and T. Christaller (Hamburg, Germany: Rowohlt Taschenbuch Verlag, 1995), 327–344.

18. *Designing Autonomous Agents: Theory and Practice from Biology to Engineering and Back*, ed. Pattie Maes (Cambridge, Mass.: MIT Press, 1990).

19. *From Animals to Animats*, ed. Jean-Arcady Meyer and Stewart W. Wilson (Cambridge, Mass.: MIT Press, 1991).

20. The essay was first published in Langton's journal *Artificial Life* and republished in *Artificial Life: An Overview*, 135–162. Inserted page numbers refer to the latter.

21. Recently Brooks himself has discussed what he calls "missing stuff." I return to this problem in chapter 8.

22. A brief account is given in their essay "What Are Plans For?" published in Maes's collection *Designing Autonomous Agents*, 17–34. I have also drawn from the more extended account in Agre's *Computation and Human Experience* (Cambridge: Cambridge University Press, 1997), 263–301.

23. Herbert A. Simon, *The Sciences of the Artificial* (Cambridge, Mass.: MIT Press, 1969), 63–64.

24. In "New Approaches to Robotics," *Cambrian Intelligence* (64–66), Brooks cites Agre and Chapman's Pengi as an example of an intelligence that arises in and through an agent's interactions with its world.

25. It was published in *From Animals to Animats 2*, ed. Jean-Arcady Meyer, Herbert Roitblat, and Stewart W. Wilson (Cambridge, Mass.: MIT Press, 1993).

26. Maja Mataric, "Designing and Understanding Adaptive Group Behavior," *Adaptive Behavior* 4, no. 1 (1995): 51–80.

27. See Mataric, "Learning to Behave Socially," in *Animals to Animats 3*, ed. Dave Cliff, Philip Husbands, Jean-Arcady Meyer, and Stewart W. Wilson (Cambridge, Mass.: MIT Press, 1994), 453–462.

28. See Mataric, "Designing and Understanding Adaptive Group Behavior," 76.

29. The essay appeared in *Artificial Intelligence* 72 (January 1995): 173–215.

30. For a summary and examples of this new orientation, see *Mind as Motion: Explorations in the Dynamics of Cognition*, ed. Robert F. Port and Timothy van Gelder (Cambridge, Mass.: MIT Press, 1995). The volume includes Beer's "Computational and Dynamical Languages for Autonomous Agents," which republishes some of the material from the essay discussed here. See also Beer's "Dynamical Approaches to Cognitive Science" in *Trends in Cognitive Science* 4, no. 3 (2000): 91–99.

31. Published in *Toward a Practice of Autonomous Systems: Proceedings of the First European Conference on Artificial Life*, ed. Francisco Varela and Paul Bourgine (Cambridge, Mass.: MIT Press, 1992), 3–10.

32. See Varela's introduction (with Paul Bourgine) to *Toward a Practice of Autonomous Systems*, xi, discussed in chapter 4.

33. See Karl Sims, "Evolving 3D Morphology and Behavior by Competition," in *Artificial Life IV*, ed. Rodney Brooks and Pattie Maes (Cambridge, Mass.: MIT Press, 1994), 28–39. There are other examples as well, such as D. Terzopoulos's artificial fish and Jeffrey Ventrella's animated characters. A more recent important contribution is Maciej Komosinski's *Framsticks*, a software platform that enables the user to simulate and evolve 3D creatures. It can be freely downloaded from the Internet at http://www.frams.alife.pl/.

34. See Dave Cliff, Inman Harvey, and Phil Husbands, "Explorations in Evolutionary Robotics," *Adaptive Behavior* 2 (1993): 73–110.

35. Inman Harvey, "Artificial Evolution and Real Robots," *Artificial Life and Robotics* 1 (1997): 38.

36. Stefano Nolfi and Dario Floreano, *Evolutionary Robotics: The Biology, Intelligence and Technology of Self-Organizing Machines* (Cambridge, Mass.: MIT Press, 2000), 14.

37. Hod Lipson and Jordan B. Pollack, "Evolving Physical Creatures," in *Artificial Life VII*, ed. Mark A. Bedau et al. (Cambridge, Mass.: MIT Press, 2000), 282.

38. This research began at Brandeis University with Jordan Pollack and has continued with the Computational Synthesis Lab at Cornell University.

39. Jordon B. Pollack, Hod Lipson, Gregory Hornby, and Pablo Funes, "Three Generations of Automatically Designed Robots," *Artificial Life* 7, no. 3 (2001): 216. In this article the Golem Project is described as the second of three generations of automatically designed robots this team has built.

40. Victor Zykov, Efstathios Mytilinaios, Bryant Adams, and Hod Lipson, "Self-Reproducing Machines," *Nature*, May 12, 2005, 163–164.

41. See also Hod Lipson, "Evolutionary Robotics and Open-Ended Design Automation," in *Biomimetics*, ed. Yoseph Bar-Cohen (Boca Raton: CRC Press, 2005).

42. See Lipson's article, "Homemade," *IEEE Spectrum* (May 2005): 24–31, as well as Neil A. Gershenfeld, *Fab: The Coming Revolution on Your Desktop—From Personal Computers to Personal Fabrication* (New York: Basic Books, 2005).

43. Rolf Pfeifer and Fumiya Iida, "Morphological Computation: Connecting Body, Brain and Environment," available at http://www.ifi.unizh.ch/ailab/people/iida/research/pfeifer_iida_JSM05.pdf.

44. Michael Crichton, *Prey* (New York: HarperCollins, 2002). Crichton supplies an extensive bibliography of scientific sources for the novel.

45. Howard Rheingold, *Smart Mobs: The Next Social Revolution* (Cambridge, Mass.: Perseus, 2003).

46. Quoted by Kevin Kelly, *Out of Control: The Rise of Neo-biological Civilization* (Reading, Mass.: Addison-Wesley, 1994), 7.

47. See *Swarm Intelligence: From Natural to Artificial Systems* (Oxford and New York: Oxford University Press, 1999), from which the quotations that follow are taken.

48. Eric Bonabeau and Guy Théraulaz, "Swarm Smarts," *Scientific American*, March 2000, 72–79.

49. Derrick Story, "Swarm Intelligence: An Interview with Eric Bonabeau," *The O'Reilly Network* (02/21/2003), 3, available at http://www.oreillynet.com/pub/a/p2p/2003/02/21/bonabeau.html.

50. For Kelly, the terms *swarm model* and *swarm systems* simply denote self-organizing and emergent systems. Unfortunately there is no further discrimination, and the evocative power of the image of the swarm remains untapped.

51. James Kennedy and Russell C. Eberhart, *Swarm Intelligence* (San Francisco, Calif.: Morgan Kauffman, 2001).

52. Another important source is Robert Axelrod's Adaptive Culture Model, a computational model of the dissemination of culture presented in his book *The Complexity of Cooperation* (Princeton: Princeton University Press, 1997).

53. Mark M. Millonas, "Swarms, Phase Transitions and Collective Intelligence," in *Artificial Life III*, ed. Christopher G. Langton (Reading, Mass.: Addison-Wesley, 1994), 417–445.

54. G. Beni and J. Wang, "Swarm Intelligence," *Proceedings of the Seventh Annual Meeting of the Robotics Society of Japan* (Tokyo: RSJ press, 1989), 425–428.

55. See the Swarm-bot Web site at http://www.swarm-bots.org.

56. The images appear in Francesco Mondada et al., "SWARM-BOT: A Swarm of Autonomous Mobile Robots with Self-Assembling Capabilities," in *Proceedings of the International Workshop on Self-Organization of Social Behavior*, ed. C. K. Hemelrijk and E. Bonabeau (Ascona, Switzerland: Monte Verità, 2002), 11–22.

57. Gianluca Baldassarre et al., "Evolution of Collective Behavior in a Team of Physically Linked Robots," in *Applications of Evolutionary Computing*, ed. R. Gunther et al. (Heidelberg, Germany: Springer Verlag, 2003).

58. "UW Professors Develop Terrorist Defense Robots," *USA Today*, June 3, 2003.

59. Will Knight, "Military Robots to Get Swarm Intelligence," *New Scientist* (April 25, 2003).

60. David Hearst, "Sci-fi War Put under the Microscope," *Guardian*, May 20, 2003.

61. I draw here on Ray Kurzweil's account in *The Age of Spiritual Machines* (New York: Viking, 1999), 145.

62. See, respectively, von Neumann's *Theory of Self-Reproducing Automata* (Urbana: University of Illinois Press, 1966), and Ross Ashby's *Design for a Brain* (London: Chapman and Hall, 1952). Both are discussed in chapter 1.

Chapter 8

1. Rodney A. Brooks, "The Relationship between Matter and Life," *Nature*, January 18, 2001, and *Flesh and Machines* (New York: Pantheon Books, 2002).

2. Nevertheless, his research group at MIT is pursuing a number of ideas, which he describes in "Beyond Computation: A Talk with Rodney Brooks" at http://www.edge.org/3rd_culture/brooks_beyond/beyond_print.html. These include exploring mechanisms for

robotic self-repair, the transition in cellular reproduction from single cell to multicell organisms, self-organization in simple neural systems like that of the polyclad flatworm, as well as the use of soft materials, growth with tensegrity structures, biomaterials, and amorphous computing.

3. This is true, for example, of Stuart J. Russell and Peter Norvig's *Artificial Intelligence: A Modern Approach* (New York: Prentice Hall, 2002) and Nils J. Nilsson's *Artificial Intelligence: A New Synthesis* (San Francisco: Morgan Kauffmann, 1998).

4. As noted in chapter 6, in the early days of AI, Alan Turing, Claude Shannon, and Herbert Simon all emphasized the importance of chess programs for the development of machine intelligence. In fact, as Simon himself often remarked (in *Models of My Life*, 327, for example) chess machines and programs became AI's equivalent of evolutionary biology's Drosophila (fruit fly) experiments.

5. In the chess match Kasparov "sensed a new kind of intelligence across the table," he states in "IBM Owes Me a Rematch," *Time*, May 26, 1997.

6. Feng-Hsiung Hsu, *Behind Deep Blue* (Princeton: Princeton University Press, 2002).

7. See, for example, John David Funge, *Artificial Intelligence for Computer Games: An Introduction* (Wellesley, Mass.: AK Peters, 2004); Mat Buckland, *AI Techniques for Game Programming* (Cincinnati: Premier Press, 2002); and Brian Schwab, *AI Game Engine Programming* (Hingham, Mass.: Charles River Media, 2004).

8. Ray Kurzweil is perhaps best known for making such optimistic forecasts, in his books *The Age of Spiritual Machines* (New York: Viking Press, 1999) and *The Singularity Is Near* (New York: Viking Press, 2005).

9. Steve Grand, *Growing Up with Lucy* (London: Orion Books, 2003).

10. Eric B. Baum, *What Is Thought?* (Cambridge, Mass.: MIT Press, 2004).

11. In the first chapter Baum invites readers uncomfortable with his argument that "mind is a computer program" to think of the argument as an *ansatz* (or hypothesis) that serves as a starting point and will only be verified later.

12. For a specific focus on the emergence of the brain from instructions coded in DNA, see Gary Marcus, *The Birth of the Mind* (New York: Basic Books, 2004), which is something of a companion volume to Baum's book.

13. For more details see the foundational text, *The Adapted Mind: Evolutionary Psychology and the Generation of Culture*, ed. Jerome Barkow, Leda Cosmides, and John Tooby (Oxford: Oxford University Press, 1992).

14. Marvin Minsky, *The Society of Mind* (New York: Simon and Schuster, 1985); Daniel C. Dennett, *Consciousness Explained* (Boston: Little, Brown, 1991).

15. In *The Robot's Rebellion* (Chicago: University of Chicago Press, 2004), Keith E. Stanovich argues (without reference to Baum) that the goals of survival and reproduction represent only the human robot's interests (as programmed by its genes), not its singularly human interests, that is, those of the "vehicle" considered for itself. But the problem is complicated by the fact that the goals serving the vehicle's interests both conflict with and overlap with the goals serving only the genes' interests. Only a careful rational analysis can disentangle the two and lead to a reprioritizing of the distinctly human.

16. Steven Pinker, *How the Mind Works* (New York: W. W. Norton, 1997), 26.

17. Jeff Hawkins, *On Intelligence* (New York: Times Books, 2004).

18. Rodney Brooks, in the research referred to in note 2 above, offers a striking example of this plasticity, as revealed in experiments with the polyclad flatworm, whose very adaptable brain possesses a couple thousand neurons: "If you take a polyclad flatworm and cut out its brain, it doesn't carry out all its behaviors but it can still survive. If you then get a brain from another one and you put it into this brainless flatworm, after a few days it can carry out all of its behaviors pretty well. If you take a brain from another one and you turn it about 180 degrees and put it in backwards, the flatworm will walk backwards a little bit for the first few days, but after a few days it will be back to normal with this brain helping it out.

Or you can take a brain and flip it over 180 degrees, and it adapts, and regrows. How is that regrowth and self-organization happening in this fairly simple system?"

19. Many neuroscientists, of course, would understand the adjustments necessary to catching a particular ball as a computation. For a short but rich discussion of the application of the concept of computation to neural events, see Patricia S. Churchland, Christof Koch, and Terrence J. Sejnowski, "What Is Computational Neuroscience?" in *Computational Neuroscience*, ed. Eric L. Schwartz (Cambridge, Mass.: MIT Press, 1990), 46–55.

20. Grand describes Lucy in his book *Growing Up with Lucy* (London: Orion Books, 2003) and Lucy II on his Web site http://www.cyberlife-research.com. Page numbers to citations from the book will be inserted into the text, but for the Web site only Grand's section titles will be given.

21. Grand describes Creatures in his book *Creation: Life and How to Make It* (Cambridge, Mass.: Harvard University Press, 2001) as well as on his Web site. The game appeared in 1996 and is still commercially available.

22. However, as noted in previous chapters, many insect species like ants and termites can accomplish very complex tasks as a collectivity or in aggregate groups through self-organized and emergent activities.

23. For example, in our mind's eye we can easily rotate the letter *p* and compare it with the mental image of the letter *d*. It is highly unlikely, however, that any special mechanism has evolved expressly to perform this task.

24. See Josh Bongard, Victor Zykov, and Hod Lipson, "Resilient Machines through Continuous Self-Modeling," *Science*, November 17, 2006, 1118–1121.

25. Christoph Adami, "What Do Robots Dream Of?" *Science*, November 17, 2006, 1093–1094.

26. Luc Steels, "Evolving Grounded Communication for Robots," *Trends in Cognitive Science* 7, no. 7 (July 2003): 308–312.

27. See, for example, Steels's "The Synthetic Modeling of Language Origins," *Evolution of Communication* 1, no. 1 (1997): 1–34. Much of Steels's published research since the late 1990s addresses these issues. For a full listing, see his Web site at http://arti.vub.ac.be/~steels/.

28. Steels draws on Ludwig Wittgenstein's concept of language games, developed in his *Philosophical Investigations* (1953). A key notion is that language is an activity, and thus language and meaning are not based on context-independent abstractions but arise in and through interactive situations. Thus for Steels a language game is "both an integrating glue and a vehicle for supporting the grounding of symbolic representations." See Steels, "Language Games for Autonomous Robots," *IEEE Intelligent Systems* (September/October 2001): 16–22.

29. For a full description, see Yoshihiro Kuroki et al., "A Small Biped Entertainment Robot," *Journal of Robotics and Mechatronics* 14, no. 1 (2002): 6–12.

Index

Printed in the United States
By Bookmasters